Electricity Production from Renewables

Rui Castro

Electricity Production from Renewables

 Springer

Rui Castro 🆔
INESC-ID/IST
University of Lisbon
Lisboa, Portugal

ISBN 978-3-030-82418-1   ISBN 978-3-030-82416-7   (eBook)
https://doi.org/10.1007/978-3-030-82416-7

This Springer imprint is published by the registered company Springer Nature Switzerland AG
The registered company address is: Gewerbestrasse 11, 6330 Cham, Switzerland

to my wife Lina,
and to my two sons João Pedro
and Guilherme

Foreword by Filipe Faria da Silva

One of the main challenges of zero-emission electricity generation is the lack of a qualified workforce. The lack of engineers and other experts is already a serious bottleneck that is not expected to improve in the near future, and it threatens society's ability to reach many of the existing goals. This textbook can be a strong contribution for educating new generations or helping older generations to understand the uniqueness of producing electricity from renewable sources.

I can state from personal experience that Prof. Rui Castro has a long and successful experience educating new generations in different areas of power systems, with a special focus on renewables. Rui was already teaching and researching different renewables when many thought that, with the exception of hydro, these energy sources would never be more than a novelty incapable of a meaningful impact in electric power grids. Time has shown that he was right, and we are now on the way for wind, solar, and biomass, as well as the classic hydro, to become dominant energy sources in Europe. Such radical change requires us to understand new concepts and principles if we are to have a smooth and economic transition from the classic to the modern power grid.

Rui has prepared and used educating materials presented in this textbook for decades. He knows the most efficient way to expose them, so that one can understand the multiple facets of renewables and their interconnection. This can be immediately observed in the book structure that starts from basic concepts that many take as a given truth without knowing that there are important background reasons. As an example, many students ask why do we use AC, if then we need to conver into DC to charge our devices; with this book, they would know the reason from the start, which is excellent to keep motivation.

Another excellent way that the book reaches different audiences is by introducing both basic electric and basic economic concepts, without neglecting the development of competences for the future. By doing it, it levels the field and allows readers with different backgrounds to understand why different technologies present diverse advantages and disadvantages. Additionally, by covering the main renewable generation sources, one obtains a full overview of different solutions, their suitability, and not less, their complementarity.

Personally, as someone that is also responsible for educating new generations in this area, I can see this textbook being an important tool in many scenarios. Even in

Denmark, a country that leads the green transition and has in renewables generation and integration a trademark, a book capable of accelerating the acquisition of so many different concepts and technologies is of high value.

Filipe Faria da Silva
Associated Professor
Department of Energy Technology
Aalborg University
Aalborg, Denmark

Foreword by João Abel Peças Lopes

The need to reduce largely the amount of CO_2 emissions in the coming years all over the world, to tackle the threats of climatic changes, requires a large effort in decarbonizing the economy and society. The electric power sector has a great responsibility for the overall volume of CO_2 emissions since a large share of electricity is still produced using fossil fuels. It is then necessary to replace this fossil fuel generation with renewable power technologies. This means developing large renewable power plants, to be connected to the transmission network, but also to massively exploit distributed generation resources like solar PV, wind power, mini-hydro, and biomass or biofuels. This shift towards the use of renewable power sources will also support massive electrification of the economy, contributing in this way to an effective decarbonizing of the society.

For doing so, it is necessary to start by educating future electrical engineers to get the right skills to deal with the development of the new electricity generation paradigm, where renewable power sources will have paramount importance. Textbooks with the relevant material for students to learn about the basic and advanced concepts related to the main renewable power sources conversion technologies are then one of the most important tools for the success of a new epoch in power systems engineering. This book responds therefore to the challenges and needs of the education and training of future electrical engineers.

The book encompasses nine chapters with very interesting material, allowing the reader to start from the very basic electrotechnical concepts and after drive through the most relevant aspects related to the different renewable sources characteristics and study the corresponding power converters and their control solutions. Very important is the fact that the book also presents a list of exercises or problems for students to solve in each chapter, in order to help consolidating the learning results.

The first two chapters are related to the fundamentals of power systems concepts, dealing with the chain of the electrical energy system, AC systems and electrical circuits, three-phase systems, per-unit system, and the induction machine operating either as a motor or as a generator.

Chapter 3 deals with the economics of renewable energy projects, dealing with concepts like the LCOE and economic assessment indicators like the Net Present Value (NPV), Internal Rate of Return (IRR), payback period, and return on investment.

Chapter 4 describes the solar PV conversion technologies and the PV cell operating principles, the equivalent electrical circuits, the concept of Maximum Power Point Tracker (MPPT), layouts of utility-scale PV power plants, and inverter models. Floating PV and Concentrated solar power technologies are also presented and discussed.

Chapter 5 is devoted to wind power. The basic concepts related to wind power aerodynamics are described, together with wind data and resource assessment methods. The main components of wind turbines and the need and interest in using variable-speed machines are described also.

Chapter 6 addresses the different types of variable-speed wind energy conversion systems, namely double-fed induction generators and variable-speed synchronous generators linked to the grid via full electronic converters, including the permanent magnet synchronous generator. Operating principles and models are described together with the corresponding control solutions.

Chapter 7 provides a tutorial on offshore wind power plants, addressing the internal wind park layout options and the dimensioning of the interconnection cables. The two main options related to HVAC or HVDC to connect the wind park to onshore are presented and discussed.

Chapter 8 describes small hydro plants, providing a thorough description of the main issues related to this technology, ranging from environmental impacts to the project of small hydro plants, and addressing the characterization of the yearly water inflows, type of hydro turbines, and its selection and computation of the installed capacity.

Chapter 9 illustrates the combined heat and power technologies, starting by addressing the fossil-fuelled solutions, like the natural gas-fired combustion engines, steam turbines, micro-turbines, and fuel cells, and also includes biomass and biofuels. Finally, a model for the economic assessment about the use of these solutions is also provided.

João Abel Peças Lopes
Full Professor of Electrical
Engineering of the
Faculty of Engineering
University of Porto
Porto, Portugal

Acknowledgments

The writing of a book is a task that cannot be executed by a single person working alone. So, I was helped by several people: my former professors, current colleagues, ex- and present students, institutions, my family, and so on. I owe a special thanks to all of them, and I am sorry if I forgot to mention anyone that deserves an acknowledgment.

In what regards the institutions, I wish to thank Técnico Lisboa, the science and engineering school of my home University of Lisbon, and INESC-ID, my research institution, for making available all the necessary material resources, including computers and dedicated software, indispensable for the writing of the book, and for all the facilities and support granted to the project. This project was funded by national funds through Fundação para a Ciência e a Tecnologia (FCT) with reference UIDB/50021/2020.

When it comes to individuals, I should begin by paying a heartfelt tribute to the memory of the late Técnico full Prof. Domingos Moura, an emblematic figure in the field of Energy in Portugal. I owe him much of what I know today. I would also like to thank him for allowing me to use his old texts as a source of inspiration to write some parts of this book. His input was most valuable in Chap. 1—Introduction: The Power System, Chap. 2—AC Electrical Circuits for Non-electrical Engineers, and Chap. 3—Economic Assessment of Renewable Energy Projects.

University of Lisbon Emeritus Professor Sucena Paiva is to be recognized for his immense knowledge of power systems. His reference masterpiece "Redes de Energia Elétrica" (in Portuguese) was also a source of inspiration that I followed, namely in Chap. 1—Introduction: The Power System—and Chap. 2—AC Electrical Circuits for Non-electrical Engineers.

I thank Fernando Camilo, a former Ph.D. student of mine, whose contribution was of utmost importance. He has drawn all the figures in the book, and I really appreciated his effort in helping me. Without his contribution, the book would not have been possible.

A special thanks is owed to a retired colleague of mine and my Ph.D. supervisor, my friend Ferreira de Jesus. In my academic life, I learned a lot from him, and I am in debt to him for that. The source of inspiration to write Chap. 6—Wind Energy Conversion Equipment—and Chap. 7—Offshore Wind Electrical Systems—was his primary drafts on the said subjects. As a recognized expert in the area, my colleague at Técnico Lisboa, Eduarda Pedro, revised Chap. 2—AC Electrical Circuits for Non-electrical Engineers—of the book and was always a supportive friend with whom I discussed the main topics of the book.

A part of Chap. 7—Offshore Wind Electrical Systems—was based on the following Master Thesis I have supervised:

- Filipe Faria da Silva, Offshore wind parks electrical connection, 2008.
- Miguel Marques, Steady State Analysis of the Interconnection of Offshore Energy Parks, 2010.
- João Jesus, Grid Architectures for Offshore Energy Parks: High Voltage Direct Current Voltage Source Converter (HVDC-VSC), 2010.
- João Limpo, Assessment of Offshore Wind Energy in Portuguese Shallow Waters: Site Selection, Technical Aspects and Financial Evaluation, 2011.
- Tiago Pereira, A comparison of internal grid topologies of offshore wind farms regarding reliability analysis, 2017.

The wind forecast part of Chap. 5—Wind Power—received the input of the following Master Thesis, I have also supervised:

- Gonçalo Nazaré, Wind power forecast using Neural Networks tuned with advanced optimization techniques, 2016.
- Miguel Godinho, Wind Generation Forecast from the perspective of a DSO, 2018.

I am grateful to my former MSc students for their excellent work and for providing the background material I further developed. Also, the contribution of Gonçalo Calado, who is currently writing his Master Thesis on "Techno-economic assessment of interconnection of offshore wind farms using Hydrogen-based solutions", was most welcome in the final part of Chap. 7—Offshore Wind Electrical Systems.

To conclude, I want to leave a note of profound gratitude to my family: Lina, my wife, and my sons, João Pedro and Guilherme, who, with all their support, understanding, and affection, were able to create the psychological and material conditions to the accomplishment of this project.

Lisbon, Portugal   Rui Castro
June 2021

About This Book

The path towards a decarbonized power system is nowadays viewed as irreversible. The current power system is aged, and it has been conceived in the last century: designed in the '50s and installed in the '60s and '70s, before de microprocessor era. It is a centralized power system, some operations still need to be performed manually, and it is fragile, mainly because it is exposed to the anger of the weather. Therefore, the goal is to upgrade the current power system because it makes no sense to dismiss it and build a new one from scratch. Moreover, the idea is to upgrade the power system in a smart way, giving birth to the concept of the smart grid.

The smart grid is viewed as the energy Internet. A smart grid puts Information and Communication Technologies (ICT) on top of the power system to make it cleaner, safer, and more reliable and efficient.

To put the Intelligence infrastructure inside the Electrical infrastructure requires an advanced metering infrastructure to establish two-way communications between advanced meters and utility business systems while ensuring the confidentiality, integrity, and availability of the electronic information. This is to be achieved in a context where the consumers want to play an active role in the management of the power system, large-scale integration of plug-in electric vehicles is targeted, as well as a massification of storage systems. Last, but not least, the smart grid must create adequate conditions for the decarbonization of the power system which is fully dependent on the introduction of larger and larger amounts of variable power generation from renewable sources.

This book is precisely about electricity production from renewables. The main renewable power generation technologies are approached in this book, by presenting the respective operating principles and introducing the appropriate models to compute the electricity produced by each source. Solar, including PV and Concentrating Solar Power (CSP); Wind, including onshore and offshore; and Small-Hydro and Combined Heat and Power (CHP) are the renewable technologies dealt with in the book. As most of the renewable generators use power electronics interfaces to perform grid connection, this feature was taken into consideration and a dedicated chapter is included.

The scope of the book is the utility-scale renewable power plants connected to the existing AC system. So, it was found necessary to introduce the basics of the power system operation and management and the fundamentals of AC circuits, which are required to understand the running of the buffer where the renewable energy is injected. Also, investment decisions are based on both technical and economic criteria, thus the necessity and justification of the topic on the economic assessment of renewable energy projects need to be addressed in the book.

This book is mainly directed to University Master's students attending Renewable Energy courses. It may serve as a textbook for these courses as all the Renewable Energy fundamentals are addressed in the book. The electricity-related aspects are indeed dealt with in more detail, but a thorough overview of each renewable technology engineering is offered at a university course level.

The book is also adequate for Ph.D. students that wish to deepen their knowledge on Renewable Energy in general and electricity production in particular. The book is helpful because some of the presented models are reported with high-level depth and the specifics of electricity generation are detailed.

Finally, engineers of all specialities may find the book interesting because it provides answers to problems found in their professional life, namely the worked examples and solved problems that are based on real-world engineering issues, with real data taken from existing installations. From a broader perspective, the book will certainly be useful to all those, and there are more and more of them, who are interested, by duty or self-education, in the topic of Renewable Energy.

Contents

About the Author

Rui Castro received a Ph.D. degree from Instituto Superior Técnico (Técnico Lisboa—IST), University of Lisbon, in 1994, and the Habilitation title from the same University in 2017. Currently, he is an Associate Professor at the Power Systems Section, Electrical and Computer Engineering Department at Técnico Lisboa, and a researcher at INESC-ID. He lectures the IST Master's Courses on "Renewable Energy and Dispersed Power Generation" and "Economics and Energy Markets" and the Ph.D. Course on "Renewable Energy Resources". He published two books, one on Renewable Energy and the other on Power Systems (in Portuguese). He has participated in several projects with the industry, namely with the EDP group, REN (Portuguese Transmission System Operator), and ERSE (Portuguese Energy Regulator). He published more than 100 papers in top international journals, covering topics on renewable energy, PV systems modelling and analysis, the impact of PV systems on the distribution grid, demand-side management, offshore wind farms, wind power forecasting, energy resource scheduling on smart grids, water pumped storage systems, battery energy storage systems, electrical vehicles, hydrogen systems, energy transition, etc. Rui Castro was the recipient of a special mention of the University of Lisbon Scientific Awards in 2018 and another one in 2020.

Personnel: https://sites.google.com/site/ruigameirocastro

Google Scholar: https://scholar.google.pt/citations?user=oIRnXSkAAAAJ&hl=pt-PT

ORCID: https://orcid.org/0000-0002-3108-8880

SCOPUS: https://www.scopus.com/authid/detail.uri?authorId=55937371000

IST: https://sotis.tecnico.ulisboa.pt/researcher/ist12375

INESC-ID: https://www.inesc-id.pt/member/13084/

Abbreviations

AC	Alternating Current
AEL	Alkaline Electrolyser
AFC	Alkaline Fuel Cell
AI	Artificial Intelligence
AMTD	Arithmetic Mean Temperature Difference
ANFIS	Adaptive Neuro-Fuzzy Inference System
ANN	Artificial Neural Network
AR	Auto-Regressive
ARIMA	Auto-Regressive Integrated Moving Average
ARMA	Auto-Regressive Moving Average
ASAI	Average Service Availability Index
a-Si	Amorphous Silicon
BIPV	Building Integrated Photovoltaics
CAIDI	Customer Average Interruption Duration Index
CCGT	Combined Cycle Gas Turbine
CDF	Cumulative Distribution Function
CdTe	Cadmium Telluride
CHCP	Combined Heat, Cooling, and Power
CHG	Green House Gas
CHP	Combined Heat and Power
CIGS	Copper–Indium–Gallium Selenide
CPV	Concentration Photovoltaics
CSI	Current Source Inverter
CSP	Concentrating Solar Power
DC	Direct Current
DDSG	Direct-Driven DC-Link Synchronous Generator
DFIG	Double-Fed Induction Generator
DOIG	Double Output Induction Generator
DNI	Direct Normal Irradiance
DPP	Discounted Payback Period
DSO	Distribution System Operator

EESG	Electrically Excited Synchronous Generator
EHV	Extremely High Voltage
ELM	Elman Recurrent Network
EMF	Electromotive Force
ENS	Energy Not Supplied
EPR	Ethylene Propylene Rubber
EUF	Energy Utilization Factor
EWEA	European Wind Energy Association
FACTS	Flexible AC Transmission System
FDC	Flow Duration Curve
FPV	Floating Photovoltaics
HT	High-Temperature
HV	High Voltage
HV&AC	Heating, Ventilation, and Air Conditioning
HVAC	High-Voltage Alternate Current
HVDC	High-Voltage Direct Current
HVDC-LCC	High-Voltage Direct Current–Line-Commutated Converter
HVDC-VSC	High-Voltage Direct Current–Voltage Source Converter
ICE	Internal Combustion Engine
IGBT	Insulated Gate Bipolar Transistor
IRR	Internal Rate of Return
IST	Instituto Superior Técnico (Técnico Lisboa)
KCL	Kirchoff Current Law
KVL	Kirchoff Voltage Law
LCOE	Levelized Cost Of Energy
LHV	Lower Heating Value
LMTD	Log Mean Temperature Difference
LPFF	Low-Pressure Fluid-Filled
LPOF	Low-Pressure Oil-Filled
LT	Low-Temperature
LV	Low Voltage
LVRT	Low-Voltage Ride-Through
MA	Moving Average
MCFC	Molten Carbonate Fuel Cell
MI	Mass Impregnated
MIBEL	Iberian Electricity Market
MOSFET	Metal Oxide Semiconductor Field Effect Transistor
MPPT	Maximum Power Point Tracker
MTBF	Mean Time Between Failure
MV	Medium Voltage
NLN	Neural Logic Network
NOC	Normal Operating Condition

NOCT	Normal Operating Conditions Temperature
NPV	Net Present Value
NWP	Numerical Weather Prediction
NZEB	Near Zero Energy Building
O&M	Operation and Maintenance
OF	Oil Filled
OPV	Organic Photovoltaics
p.u.l.	Per unit of length
PAFC	Phosphoric Acid Fuel Cell
PB	Payback period
PCC	Point of Common Coupling
pdf	Probability density function
PEMEL	Proton Exchange Membrane Electrolyser
PEMFC	Proton Exchange Membrane Fuel Cell
PHS	Pumped Hydroelectricity Storage
PMSG	Permanent Magnet Synchronous Generator
pu	Per unit
PV	Photovoltaic
PWM	Pulse Width Modulation
RBFN	Radial Basis Function Network
RES	Renewable Energy Source
RMS	Root Mean Square
ROI	Return On Investment
SAIDI	System Average Interruption Duration Index
SAIFI	System Average Interruption Frequency Index
SCFF	Self-Contained Fluid-Filled
SCIG	Squirrel-Cage Induction Generator
SHP	Small Hydro Plant
SOE	Solid Oxide Electrolyser
SOFC	Solid Oxide Fuel Cell
STATCOM	STATic Synchronous COMpensator
STC	Standard Test Condition
SVC	Static Var Compensator
TCR	Thyristor Controlled Reactor
THD	Total Harmonic Distortion
TSC	Thyristor Switched Capacitor
TSO	Transmission System Operator
TSR	Tip Speed Ratio
VSC	Voltage Source Converter
VSI	Voltage Source Inverter
WTG	Wind Turbine Generator
XLPE	Cross-Linked Polyethylene

List of Figures

Chapter 6

Chapter 7

Chapter 8

Chapter 9

List of Tables

Introduction: The Power System

1

Abstract

The power system is one of the most complex systems the human being has created. It aims at supplying electricity to the consumers while preserving a fundamental restriction: in each time instant, the generated power must exactly match the demand. This power balance is not easy to achieve, requiring a great number of complex procedures to be undertaken at every time instant. In this chapter, we address the power system. The power system comprises the generators, the consumers, and the grid, composed of transmission lines and transformers, that condition and carry the electricity from the former to the latter. We begin by introducing the main electrical quantities of the power system—voltage, current, and power. The essential options of the power system are discussed: why Alternating Current? Why 50 or 60 Hz? Why three phases? Also, the typical organization of the power system in generation, transmission, distribution, retail, and consumption is presented. An overview of the different types of generation facilities—thermal, hydro, wind, and solar—is offered. Finally, a brief explanation of the power balancing mechanism is proposed.

1 Introduction

When we talk about renewable energy, we are talking about a large number of technologies that can provide energy services, in the form of electricity, heating and cooling, and transport solutions, in a sustainable way. The questions facing the renewable energy sector should not only focus, for example, on whether the energy system should have a centralized or decentralized structure, or which renewable technology will be dominant in the future. All solutions and all types of renewable energy should be seen as interdependent with a view to the diversification of energy supply, climate change mitigation, and sustainable development.

In this text, we address electricity production using Renewable Energy Sources (RESs). In most countries in the world, the RES are connected to the power system injecting all the electricity they can produce given the available resource. The power system is perhaps the most complex system operating in the world. To fulfil its mission, which is to supply electricity to the consumers whenever they request, a vast number of technical restrictions must be obeyed. The most important one is that the electricity generated at any given time instant must exactly match the electricity consumed.

As the electricity produced by the RES is mainly to be injected into the power system, it is justified that we begin this text by introducing the power system. We begin by highlighting the physical structure of the power system, with its main components: generators (both conventional and RES powered), overhead lines, underground cables, and transformers, which compose the grid, and the final consumers of the electricity.

Then, we move on by discussing the essential options of the power system. Why does it use Alternating Current (AC) instead of Direct Current (DC)? Why using 50 or 60 Hz and not higher or lower frequencies? Why is the power system three phase and not single phase instead?

It is usual to divide the power system into five parts: generation, transmission, distribution, retail, and consumption. We discuss the function of each of these power system's parts, particularly the generation facilities: coal-fired, combined cycle, nuclear hydro, wind, and solar.

The chapter ends with a brief outline of how the said crucial power balancing task is performed in the context of the power system.

2 General Concept

The power system is a very complex system, which is designed with the main objective of delivering electricity to the consumers. The electricity, or electrical energy, is produced[1] in power plants, which are usually located far from the places where the consumers are concentrated. As so, it is necessary to transport the energy from the places where it is produced to the locations where it is consumed. Electricity transportation requires a physical infrastructure,[2] composed namely by overhead power lines and underground cables. This infrastructure further includes the electrical devices that allow the power transmission to be performed with

---

[1]Actually, the electricity is not produced, nor generated, but instead it is converted (transformed) from other forms of energy. For simplicity, we will use the words "production" (or "generation") and "consumption", but we should keep in mind that we are referring to the word "conversion" or "transformation".

[2]It is technically possible to have wireless electricity transmission, using advanced technologies, as direct induction or resonant magnetic induction. However, at the present time, there are still no commercial high-power applications, mainly due to the low efficiency of the process: only a small parcel of the transmitted power is received by the electrical recipient.

reduced losses; these devices are the transformers. Finally, the electricity is delivered to the customers with the appropriate power quality.

2.1 Fundamental Quantities

To observe, study, and understand the operation of the power system, two fundamental quantities are used: the current and the voltage. The electric current is an organized flow of electrons in a material, usually, a metal. In metals, some electrons can get out of the electron cloud structure and move across the material, therefore, giving rise to an electric current. To keep the electrons' flow, it is necessary to continuously supply energy, because the electrons lose energy in the collisions with the material structure. Therefore, the wires get hot when there is a current flowing across them. The voltage is somehow (*lato* sensu) a measure of the energy that should be supplied to keep the current, i.e., to keep the electrons moving on. There is a third quantity, named power, that is related to the product of voltage and current. Power is the availability to produce electricity. When we refer to the intended full-load sustained output of a facility, such as a power plant, we use the term capacity.

To represent the voltage, the letters V or U are used. Here, we will use the letter V. Volt (V) is the International System of Units (SI) for voltage. Current is represented by I and is measured in Ampère (A). As for power, the representative letter is S and the SI unit is Volt-Ampère (VA).[3]

2.2 Requirements

As mentioned before, the power system operation is a very complex task, for several reasons, one of the main ones being the necessity of assuring that a fundamental condition is met: power production must be equal, at each time instant, to power consumption plus transmission losses, at that very same instant. The accomplishment of this condition is essential because electricity characteristics do not make it appropriate to be stored, or, at least, currently, electricity storage in large amounts is not economical.

Actually, electricity can be stored, using for this purpose the well-known batteries. However, with the current economical technology, the energy that can be stored in batteries is minimal, as compared with the electricity movements associated with the consumption at a country-level. As it is not possible to store electricity in significant amounts, the mentioned condition must be assured at every time instant.

---

[3]The multiples are kilo (k) $= 10^3$; Mega (M) $= 10^6$; Giga (G) $= 10^9$; Tera (T) $= 10^{12}$ and the submultiples are mili (m) $= 10^{-3}$; micro (μ) $= 10^{-6}$; nano (n) $= 10^{-9}$; pico (p) $= 10^{-12}$.

Despite advanced load[4] forecasting techniques being currently used, it is virtually impossible to avoid a mismatch between production and consumption. Therefore, sophisticated control systems are required to regulate the power production in the plants holding that feature, as well as to manage the power flow in the interconnections.

Besides this fundamental requirement, which is related to the overall operation of the power system, there are other second-order requirements, but nevertheless important:

- Electricity should be supplied at every place where it is required.
- Electricity should meet quality criteria: constant frequency, voltage under control, sinusoidal waveform,[5] and high reliability.
- Production costs must be minimal.
- Environmental impact must be low.

2.3   Structure and Components

Figure 1 shows a schematic of the power system structure.

Power is produced in the three following types of infrastructures, ranked by capacity level (from the highest to the lowest).

- Large power stations with a rated capacity of hundreds or one thousand MVA. These power stations can be: (i) thermal power plants when the thermal energy associated with a fossil fuel (coal or natural gas) combustion is converted into electricity; (ii) hydro plants, when the kinetic and potential energy associated with a river waterfall is transformed into electrical energy. The location of thermal and hydropower plants is concentrated in specific sites, which must hold certain characteristics: for instance, easy fuel supply and availability of refrigeration water, in the case of thermal, and a waterfall and a water flow, in the hydro case.
- Medium- or Low-power stations, with a rated capacity of tens of MVA. These power stations can be small-hydropower plants; wind power, which uses wind energy; and Photovoltaic (PV), which transform the sun irradiation. These generating units are called dispersed power production because they are not concentrated in specific sites, instead, they are distributed by several places, selected due to the prime mover abundance.
- Micropower stations, with rated capacity around several kVA. These power plants (also known as micro-generation) are located at the consumption point, which is usually a household or a small industry. Normally, these power plants are of PV type, installed on the rooftops of the buildings.

---

[4]Load or demand are words used to refer to electricity consumption.
[5]We will see later what these concepts mean.

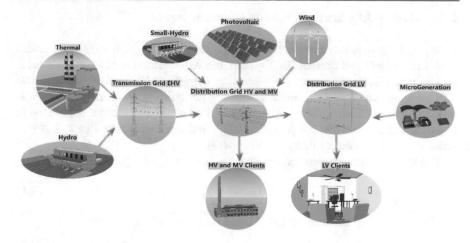

Fig. 1 Structure of a power system

The power produced in the large thermal and hydropower plants is delivered to the transmission network, composed of Extremely High Voltage (EHV) overhead power lines. EHV is the voltage level appropriate to the transportation of large electricity amounts over long distances because the transmission losses decrease when the voltage increases.[6] The transformers are the devices that allow changing the voltage level so that the optimal losses are attained from both the technical and economic points of view.

Extremely High Voltages are not economically adequate to transport electricity at a regional or local level, where both the distances and the amounts of electricity involved are smaller. Therefore, the electricity is transferred to the distribution grid that operates in High Voltage (HV), Medium Voltage (MV), and Low Voltage (LV). Wind and PV parks are usually connected in HV or MV. As for the micro-generation, it is connected to the LV level.

3 Basic Options

The power system was developed based on three basic options that have been decided a long time ago: the power system operates in Alternating Current (AC), at the frequency of 50 Hz[7], and is three phase. Let us look at the meaning of these concepts and the basic options that were taken to converge in the current situation.

---

[6]We will prove this later in this text.

[7]In some parts of the world, as the USA and part of Asia, 60 Hz are used, as we will see later.

3.1   Alternating Current Versus Direct Current

In an AC system, the fundamental quantities change in time, following a sinusoidal waveform; in a Direct Current (DC) system, these quantities are constant over time.

The power system operates in Alternating Current (AC). In fact, the electrical quantities, as voltages and currents, change periodically in time as sinusoidal functions (sine or cosine waves). An example of an alternating current is depicted in Fig. 2. It can be seen that the period is 0.02 s and the maximum value is 1 A (we recall that current is measured in A—Ampère).

The current instantaneous values repeat periodically with period T:

$$i(t) = i(t + T) \tag{3.1}$$

The relationship between period (T) and frequency (f) is well known:

$$f = \frac{1}{T} \tag{3.2}$$

The frequency is measured in Hertz (Hz—Hertz). As the period is 0.02 s, the frequency is 50 Hz. 1 Hz is 1 cycle per second, meaning that in one second there are 50 full cycles or 50 periods. 50 Hz is the frequency used in Europe, but, for instance, in the USA, the frequency is 60 Hz.

The angle (rad—radians), i.e., the argument of the sinusoidal function, is given by

$$\theta = \omega t \tag{3.3}$$

in which, ω is the angular frequency (rad/s) and t is the time (s—second). During one full cycle ($t = T$), an angle of 2π is described. Therefore,

$$\omega = 2\pi f \tag{3.4}$$

In the example of Fig. 2, the current as a function of time is given by (note that f = 50 Hz):

$$i(t) = I_{MAX} sin\theta = I_{MAX} sin(\omega t) = 1 sin(100\pi t)\text{A} \tag{3.5}$$

$I_{MAX} = 1$A is the maximum value or peak value or amplitude of current $i(t)$. The angular frequency is $\omega = 100\pi\,\text{rad/s}$, because $T = 0.02s$ and, consequently, $f = 50$Hz.

We use small letters to represent the time evolution of the quantities. For instance, $i(t)$ is the current time evolution.

We denote by I(capital letter) the Root Mean Square (RMS) value of current $i(t)$, i.e., the square root of the mean square (the arithmetic mean of the squares of the current instantaneous values), in a period T:

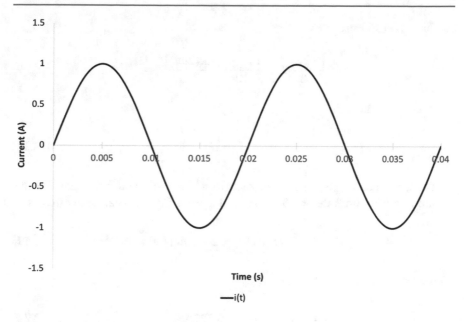

Fig. 2 Example of an alternating current

$$I = \left[\frac{1}{T} \int_0^T [i(t)]^2 dt \right]^{\frac{1}{2}} \tag{3.6}$$

For an alternating electric current, RMS is equal to the value of the Direct Current (DC) that would produce the same average heat power dissipation in a resistive load, in a period T.

RMS values are of utmost importance. Usually, common measuring instruments, voltmeters, and ammeters (from Ampère meter) measure the RMS values of the voltage and current, respectively.[8] Moreover, the RMS values are the ones that effectively contribute to the transfer of useful power between two systems. The useful power is normally represented by the letter P.

From Eq. (3.6), one can write

$$I^2 = \frac{1}{T} \int_0^T [I_{MAX} sin(\omega t)]^2 dt = \frac{I_{MAX}^2}{T} \int_0^T [sin^2(\omega t)] dt \tag{3.7}$$

As

$$sin^2(\omega t) = \frac{1 - cos(2\omega t)}{2} \tag{3.8}$$

---

[8]Note that it makes no sense to measure the average value because it would be zero.

Equation (3.7) becomes

$$I^2 = \frac{1}{T}\frac{I_{MAX}^2}{2}T + \frac{1}{T}\frac{I_{MAX}^2}{2}\left[\frac{sin(2\omega t)}{2\omega}\right]_0^T = \frac{I_{MAX}^2}{2} \tag{3.9}$$

and we conclude that

$$I = \frac{I_{MAX}}{\sqrt{2}} \tag{3.10}$$

The amplitude of a sinusoidal current (or a sinusoidal voltage) is equal to the square root of 2 times the RMS value. In this way, Eq. (3.5) can be written as

$$i(t) = \sqrt{2}I sin(\omega t) = \sqrt{2}\frac{1}{\sqrt{2}}sin(100\pi t)\text{A} \tag{3.11}$$

the current RMS value being

$$I = \frac{1}{\sqrt{2}} = 0.707\text{A} \tag{3.12}$$

We recall that a capital letter denotes the RMS value of a quantity. For instance, I is the current RMS. A graphic showing the time evolution of the alternating current that we have been using as an example with the RMS value marked is depicted in Fig. 3.

We will see later other applications where the RMS values play a very important role in AC electrical circuits.

The main advantage to justify the AC option is that it facilitates electricity transmission. Actually, transmission losses are inversely dependent on the square of the voltage as we will now demonstrate in a simplified way.

It is known that the transmitted power P depends on the product of RMS voltage and current:

$$P = k_1 VI \tag{3.13}$$

Moreover, we know that power losses P_L depend on the square of the RMS current:

$$P_L = k_2 I^2 \tag{3.14}$$

As so, we can write

$$P_L = \frac{k_2 P^2}{k_1^2}\frac{1}{V^2} = k\frac{1}{V^2} \tag{3.15}$$

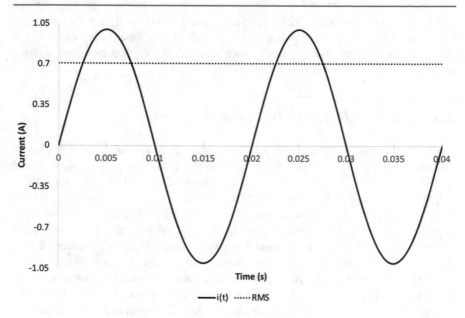

Fig. 3 Example of an alternating current with RMS value marked

This means that the voltage should be as high as necessary to keep the losses under an acceptable value. Transformers are the electrical equipment that allows for changing the voltage level, whether it is to increase (to transport in the transmission grid) or to decrease (to distribute to the customers in the distribution grid). The issue is that conventional transformers only operate in AC; they do not perform the required function in DC.[9]

Another reason to justify the use of AC is that it facilitates the current interruption. Sometimes, namely when a fault in the grid occurs, it is necessary to interrupt the current to clear the fault. High currents are hard to interrupt due to the high magnetic energy that is stored in the circuits. As an alternating current is periodically equal to zero, circuit breakers take advantage of this AC characteristic to perform current interruption.

3.2 50 Hz Versus 60 Hz

In the countries that followed the European trends, the power system frequency of 50 Hz is used. In the USA and the countries it technologically influences, 60 Hz is the frequency. This is an unfortunate circumstance, which is nowadays irreversible. One can ask why frequencies lower than 50 Hz or higher than 60 Hz are not used.

---

[9]Actually, there are currently DC transformers. They perform the DC voltage level change using power electronics devices. However, they are still not an economic solution for EHV and HV.

Frequencies lower than 50 Hz produce an inconvenient flicker in traditional lights, used in the past. Frequencies higher than 60 Hz would give rise to high losses in the magnetic circuits (losses increase with the frequency as we will see later in this text).

3.3 Single Phase Versus Three Phase

The power system is three phase, i.e., to transmit power between a source and a recipient, three wires are used; in certain conditions, a return wire—called neutral wire—is required. The advantages of using a three-phase system as compared to a single-phase one, in which one active wire and one return wire are used, can be found at the generation, transmission, and use of energy levels. However, the most important advantage is at the transmission level, as we will explain below.

Let us suppose we want to transmit power P at the distance d. We need a wire with a length $2d$ because a return path is required to close the loop. If instead, we use three active wires, the return can be made by a single common return wire, and we can transmit $3P$ with a $4d$ length conductor. This means that we multiply the transmitted power by 3, but the wire length is only multiplied by 2. This is a significant advantage of three-phase systems. Moreover, in some conditions, if the system is said to be balanced, the return wire can be suppressed, because it would carry a current equal to zero (we will return to this issue later). In these circumstances, the economy is even more apparent: the conductors' length is increased by 50% ($3d$) and the transmitted power is multiplied by 3. This is the main reason why the power system is three phase.

The power system being three phase, two voltages can be defined: a phase-to-neutral voltage (V_{p-n}) and a phase-to-phase voltage (V_{p-p}). We will prove later that the relationship between these two voltages is

$$V_{p-p} = \sqrt{3}V_{p-n} \tag{3.16}$$

4 Organization

As mentioned before, the power system is a highly complex system, governed by a fundamental constraint: at each instant, production must match the consumption added of the losses. In the past, the generating system was composed of centralized and controlled power plants. However, over the last years, the structure of the generating system has undergone significant changes: nowadays, distributed generation is playing an increasingly important role in the modern power system. Distributed generation is composed of many low-power units, dispersed along with the network, generally dependent on non-controllable renewable resources, like the sun and the wind. This structural change is motivating a paradigm shift from a consumption-driven system to a generation-driven system. Finding the best strategies to deal with the new paradigm, which seems irreversible, is a challenging task for electrical engineers.

Unless otherwise stated, the power system organization, whose grounds follow, applies to the so-called industrialized countries—Europe, the USA, and certain parts of Asia. For outside countries, the power system organization might be diverse.

The current power system is usually divided into five blocks: (1) Generation, (2) Transmission, (3) Distribution, (4) Retail, and (5) Consumption.

4.1 Generation

Power production is a market-driven activity, open to the private initiative, but the participants must be certified. The role of the State is to create adequate conditions for market development and to monitor the system to ensure the security of supply.

Power is generated in power plants. Several technologies can be used. Some of them are economical and are currently being used; others are not economical and therefore are currently being researched to make them economical in the future. Hereafter, we present the main technologies currently used for power production.

Thermal power plants
   Coal-fired.
   Combined-cycle gas turbine.
   Combined heat and power.
   Nuclear.
Hydropower plants
   Run-of-river.
   Reservoir.
   Pumped hydroelectricity storage.
Wind power plants
   Onshore.
   Offshore.
Solar power plants
   Photovoltaic.
   Concentrating solar power.

These are the technologies that, to a less or great extent, are currently being used in a fully commercial phase or a pre-commercial phase. Examples of currently non-commercial technologies are waves, tides, fuel cells, etc.

4.1.1 Thermal Power Plants

Thermal power plants are usually divided into three main groups: coal-fired, combined cycle, and nuclear.

In conventional coal-fired power plants, a fossil fuel, the coal (which has been ground into a fine powder by a pulverizer), is burned to obtain thermal energy (heat). The heat, which is released from the coal combustion, heats water that runs through a series of pipes, located inside the boiler, therefore making the water change from the liquid phase to steam. The high-pressure steam is carried to a steam turbine, turning the turbine's blades, so as to obtain mechanical power (power associated with the rotation of the turbine's blades). An electrical generator[10] is mounted in the same shaft where the turbine rotates, therefore converting mechanical power into electrical power.[11] Meanwhile, the low-pressure vapour is cooled down into a condenser,[12] returning to the liquid phase. This water is pumped back into the boiler and the process is reinitiated. Figure 4 shows a schematic of a typical coal-fired power plant.

The efficiency of coal-fired power plants, i.e., the ratio between the output electricity and the input heat, is low, around 35%. Furthermore, coal is an extremely pollutant fossil fuel, significantly contributing to Greenhouse Gas (GHG) emissions. From the fossil fuels used to produce electricity, coal is the most aggressive from an environmental point of view. For this reason, the construction of new coal-fired power plants has decreased significantly, for several years. In many countries, coal-fired power plants are being decommissioned before the end of their useful lifetime, due to environmental reasons.

CCGT (Combined Cycle Gas Turbine) is the technology that replaced coal technology. In a CCGT power plant, atmospheric air is compressed and passed into a combustion chamber where it is burned with natural gas. The mixture of air and combustion gases is carried to a gas turbine, where it turns the turbine's blades and produces mechanical power. An alternator then converts this movement energy into electricity. The efficiency of this process is low, about 30%. This means that a significant part of the input heat is not converted into electrical energy and is available to be used elsewhere. This heat available in the hot exhaust gases is transferred to a heat recovery boiler, where steam is obtained by heating water that runs in a pipeline inside the boiler. The blades of a second turbine, this time a steam turbine, are turned by the steam, therefore producing rotational movement, which drives a second alternator to produce more electricity. We highlight that this technology uses two thermodynamic cycles: the gas cycle and the steam cycle, hence the designation of combined cycle. In Fig. 5, we can see a diagram of a CCGT power plant.

The efficiency, i.e., the ratio between the total output electricity (natural gas plus steam-based) and the input natural gas heat (the only fossil fuel used) is now far higher, around 55 to 60%. Moreover, natural gas is less pollutant than coal, and typically natural gas GHG emissions (namely CO_2 emissions) are one-half of the

---

[10]Electrical generators in thermal and hydropower plants are called alternators. This corresponds to a type of electrical generator that rotates at constant speed, related to the constant 50 Hz frequency, the so-called synchronous speed.

[11]We will see later how this conversion is actually performed.

[12]Do not confuse condenser with capacitor. The former is a heat exchanger, in which the vapour is condensed; the latter is an electrical device capable of storing electrical energy.

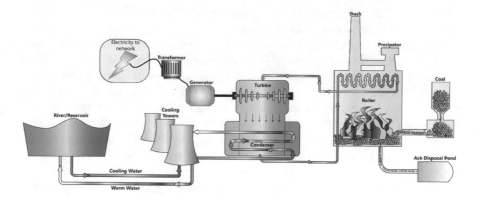

Fig. 4 Diagram of a coal-fired power plant

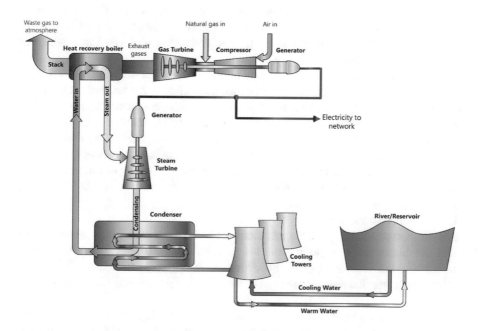

Fig. 5 Diagram of a CCGT power plant

coal ones.[13] This is the reason why the latest thermal power plants installed in the industrialized countries use CCGT technology.

An extension of the combined cycle concept is the Combined Heat and Power (CHP).[14] In these low-power plants (typically some MW sized), the heat that is not converted into electricity is used in a useful application. Useful applications are, for instance, hot water for Heating, Ventilation, and Air Conditioning (HV&AC) in buildings or process heat in the industry. Instead of a turbine, an internal combustion engine, usually fed by natural gas, is mostly used as the thermal machine, whose input is heat and output is movement. An alternator then converts the mechanical power into electricity. Heat exchangers are used to recovering the heat available, for instance, in the exhaust gases rejected to the stack of the engine.

A nuclear power plant is just like a coal-fired power plant with one fundamental difference. Instead of burning fossil fuel to produce heat, the heat source in the nuclear power plant is a nuclear reactor. A nuclear reactor produces and controls the release of heat energy from splitting the atoms of certain elements, in a process called nuclear fission. Uranium is the dominant choice of nuclear fuel in the world today. As for the rest of the process, it is exactly the same: the heat is used to generate steam, which drives a steam turbine connected to a generator, which in turn produces electricity.

Figure 6 shows a diagram of a typical nuclear power plant. In the industrialized countries, there are many countries with installed nuclear power. Following some tragic accidents, with devasting consequences, in some countries, we are assisting in an early closure of nuclear power plants, due to safety reasons.

Nuclear power plants do not release carbon or pollutants like nitrogen and sulphur oxides into the air. This evident benefit of using nuclear power is tempered by a few issues that need to be considered, including the safety of nuclear reactors and the disposal of radioactive waste.

4.1.2   Hydropower Plants

In hydropower plants, one takes advantage of a river waterfall and a water flow to turn the blades of a hydro turbine and obtain electricity through an alternator mounted in the same shaft of the hydro turbine (see Fig. 7). The water flow is the volume of water that crosses a section of the river in the time unit; it is measured in m^3/s.

The efficiency of the power generation process is far higher than in the case of the thermal power plants, reaching, in this case, values around 80%. Furthermore, a Renewable Energy Source (RES), the water, is used, which turns these power plants into valuable assets.

The main issue is that they cannot be located everywhere, instead, they require proper geographic conditions to be met, namely a water flow and a waterfall must

---

[13]Typically, a coal-fired power plant emits 0.9 ton of CO_2 for each MWh of electricity produced. This figure reduces to 0.4 ton/MWh in a CCGT power plant. MWh (Megawatt-hour) is the unit in which electrical energy is usually measured: it corresponds to an average power of 1 MW used for 1 h, as we will show later in this text.

[14]More details about CHP power plants will be given in a separate chapter of this book.

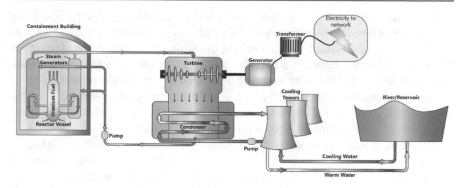

Fig. 6 Diagram of a nuclear power plant

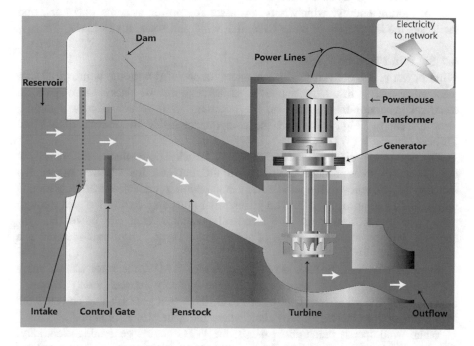

Fig. 7 Diagram of a hydropower plant

exist. Also, if further geographic conditions are present, large hydropower plants require the construction of big reservoirs of water, which do interfere with aquatic life, namely by blocking the migration of animals. These are environmental impacts that should not be neglected. In this sense, small hydro plants are less aggressive to the environment.

There are two types of hydropower plants: run-of-river and reservoir. In run-of-river plants, the resource is used as it comes, and little or no water storage is provided. When the resource is not enough, the power plant stops; when it exceeds the power plant rated capacity, the surplus energy is wasted. Reservoir power plants are much more valuable. The reservoir is a natural lake, made possible due to adequate geographic conditions, in which incoming water can be stored. In this way, optimal management of the power plant is achieved, as it can perform power regulation or even operate in dry periods, using the stored water.

Reservoir power plants allow for Pumped Hydroelectricity Storage (PHS), in which water is pumped from a lower reservoir to the higher one, consuming excess power during off-peak periods. Then, at peak hours, when electricity is more valuable, the stored water is released through the turbines to produce power, as is explained in Fig. 8. Roundtrip efficiencies around 70–75% are to be expected.

Even though the overall process has an efficiency of less than 100%, this is a way of storing energy, which is a key asset for power systems operators, and facilitates the integration of other RES, like wind and solar, for instance.

4.1.3 Wind Power Plants

Wind power plants convert the kinetic energy associated with the wind speed into mechanical energy, through the spinning of the wind turbine blades. Once again, an electrical generator is mounted on the common shaft, which allows obtaining electrical energy. Figure 9 shows the components inside the nacelle of a Wind Turbine Generator (WTG).

The type of electrical generator that equips the wind converters is different from the alternators that produce electricity in thermal and hydropower plants. To increase the efficiency of the wind energy conversion, a variable-speed operation is required. This imposes the electrical generators to be connected to the grid through power electronics converters. The electrical generator itself can be of the synchronous type or, more frequently, of asynchronous type.[15] Both are operated at a variable speed.

A modern WTG may have a rated capacity of 3 MW, the rotor blades being 45 m long. The rotor blade's speed may change typically between 9 and 18 rpm,[16] depending on the wind speed. In most WTG, there is a gearbox that converts the low-speed shaft to the high-speed shaft where the generator is mounted.

When wind speeds reach about 4 m/s, the WTG begins generating electricity. Rated power is obtained at about 15 m/s. The WTG control system then regulates the output power to the rated power till the cut-off wind speed (typically around 25 m/s) is attained. For higher wind speeds, the WTG shuts down and turns out of the wind to protect from over speed failures.

The efficiency depends on the wind speed. For the most frequent wind speeds, the efficiency lays in the interval 40 to 50%. Of course, the main advantage of wind power is that it is carbon emissions-free. On the drawbacks side, they are unable to

---

[15]Double-Fed Induction Generator.
[16]Revolutions per minute.

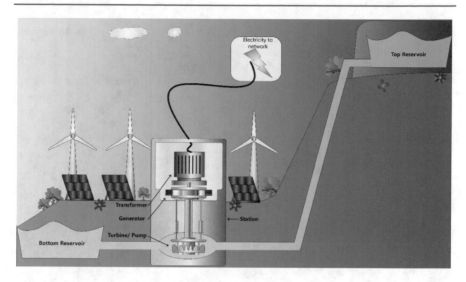

Fig. 8 Diagram of a pumped hydroelectricity storage power plant

regulate the output power following the system load needs. As a matter of fact, the WTG output power is linked to the wind speed resource, which is uncontrollable. Thermal and reservoir hydro can perform this important task that allows to balance generation and consumption. As we have seen before, this is an essential condition to the power system operation.

With the progressive shortness of ideal windy sites onshore, the next step of wind energy development is offshore wind, i.e., the deployment of WTG in the open sea. There are several countries, like, for instance, the UK, that are betting strong in offshore to take advantage of higher and more uniform (unperturbed) winds. Of course, installation and operation and maintenance costs are also higher, due to harsh conditions at the open sea. For the time being, the cost–benefit analysis is still doubtful, but the situation is expected to reverse in the coming years.

To take advantage of higher and steadier offshore winds, the offshore WTGs are larger. Currently, manufacturers are offering WTG of about 10 MW unit power, with blades of 85 m in length. For installation in deep waters, innovative floating semi-submersible designs are being proposed, as is the case of the *WindFloat* project. A pre-commercial installation is being tested in Portugal.

4.1.4 Solar Power Plants
In all technologies that we have approached so far—thermal, hydro, and wind—the conversion principle is always the same: a turbine generator group is used to perform the conversion. Photovoltaic (PV) power plants configure a totally different operating principle because there is no turbine nor generator.

In PV power plants, the sunlight is directly converted into electricity. A material —the most used is silicon—after proper treatment, shows special characteristics

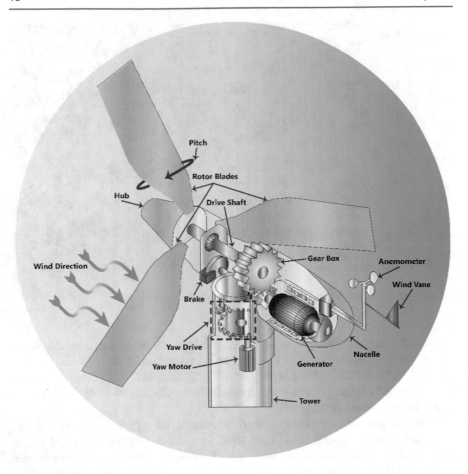

Fig. 9 Wind Turbine Generator components

when exposed to the sun. Actually, the photons, which are the particles that compose the sun irradiation, can displace electrons that acquire the capacity to move, therefore giving rise to an electrical current (Fig. 10). This current is DC; therefore, to connect the PV power station to the AC grid, an interface device is required. This power electronics device is called an inverter and performs DC to AC conversion.

In commercial applications, other materials may be used, instead of silicon. These materials are known as thin films. However, despite great expectations, the real-world operation was disappointed due to the low efficiencies achieved; therefore, the market share is currently low.

PV cells are assembled in modules. The capacity of a PV module is a few hundred Watts (200–300 W). To obtain more power, PV modules are assembled in

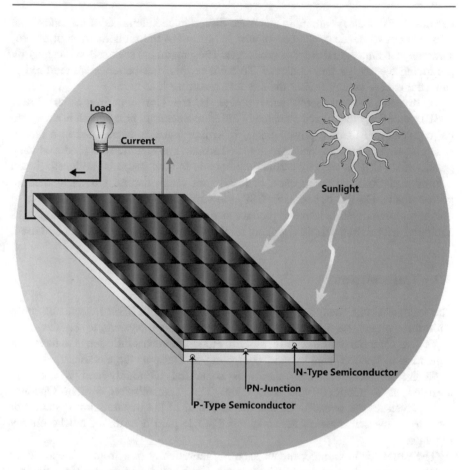

Fig. 10 PV cell operating principle

series and parallel composing PV arrays. Lots of PV arrays compose a utility-scale PV park with an installed capacity that can reach tens or even hundreds of MW.

Historically, PV power presented two significant weaknesses: high cost and low efficiency. Nowadays, the first drawback is completely overcome: PV cost is lower than other forms of electricity production, namely wind power. As for the efficiency, the advances were not so impressive, the maximum efficiency of monocrystalline silicon modules still being close to 16–18%. Despite the low efficiency and the fact that there is no sun at night, the production cost of each MWh of PV electricity is nowadays considered to be the lowest of all electricity production sources. This is due to the impressive drop in PV investment costs in recent years.

PV power is the preferred technology for micro-generation applications in buildings, namely in households, commercial buildings, and industrial factories. PV facades or PV rooftops are more and more disseminated elements of the landscape. This is a consequence of the consumers' evolution to the modern concept of pro-sumers—consumers that are also producers. The prosumers seek self-sufficiency by producing the energy they consume. To achieve this aim, batteries are required to store the excess energy during the day and restoring it at night.

A diverse way of using solar power is the Concentrating Solar Power (CSP) technology that uses a completely different operating principle. It is similar to the conventional thermal power plants, in which water is overheated in a boiler; steam is produced and expanded in a steam turbine; and electricity is produced via a generator. In CSP power plants, the difference is in how steam is produced. Steam is produced from the combustion of fossil fuel (coal, natural gas…) in thermal power plants. On the opposite, in CSP, the sunlight is focussed on a receiver to obtain high-temperature heat and produce steam. For this purpose, mirrors or lenses equipped with a solar position tracking system to focus the solar radiation are used.

4.2 Transmission

The transmission system is the core of the power system. Grid investments raise with the voltage level, thus making the transmission system to be the most expensive component of the power system. Hence, it makes little sense to have two transmission systems, side by side, competing to transport the electricity.

Therefore, electricity transmission is a natural monopoly and is normally awarded as a public service concession. The Transmission System Operator (TSO) oversees the overall technical management of the transmission system. For the use of the transmission network, the TSO is paid a tariff set by the energy regulator.

The regulator is an independent agency theoretically free from influence from external sources (government, private sector, the public) in its decision making. Some of the regulator's key functions are as follows:

- Issue licenses related to regulatory functions.
- Set performance standards.
- Monitor the performance of regulated firms.
- Establish the level and structure of regulated tariffs.
- Arbitrate disputes among stakeholders.

The transmission grid is the EHV network. In Fig. 11, the example of the Portuguese transmission grid is depicted. Typical EHV levels are 400, 220, and 150 kV.[17] The transmission grid is mostly composed of overhead lines; some underground cables are used near the main cities.

---

[17]These numbers are the RMS value of the phase-to-phase voltage. In power systems, when we refer a voltage by a number, it is always the RMS value of the phase-to-phase voltage.

Fig. 11 Portuguese transmission grid. *Source* REN (Portuguese TSO), www.ren.pt

In the transmission grid, there are transformers to perform voltage transformation between EHV levels, as well as transformers to interface the transmission grid with the distribution grid (EHV/HV). The transformers are located in the so-called substations. Figure 12 shows a schematic of an EHV/HV substation with the identification of its main components.

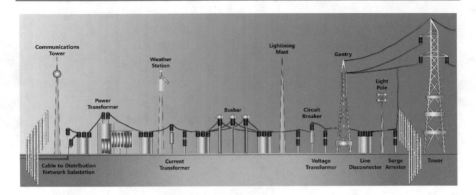

Fig. 12 Schematic of a transmission substation

4.3   Distribution

For the same reasons as the transmission system, the electricity distribution is also a natural monopoly and is normally awarded as a public service concession. The Distribution System Operator (DSO) oversees the operation and maintenance of the distribution grid and the respective power flow management. The DSO is remunerated for the use of the distribution network by a tariff set by the energy regulator.

The HV, MV, and LV distribution networks are composed of overhead lines and underground cables, the latter in a percentage greater than in the transmission case. Typically, 60 kV is used in HV, and 30, 15, and 10 kV are three examples of MV voltage levels. 400 V[18] is typically the LV level. Beyond the lines and cables, the distribution grid includes the substations (where the HV/MV transformers are installed) and the transformation stations (where the MV/LV transformers are located). A piece of the HV distribution grid in the Lisbon area (Portugal) is depicted in Fig. 13.

4.4   Retail

The retailing activity is a separate activity from distribution, being the last electricity-related activity in the supply chain to the consumers. As the generation, retailing is usually a market-driven activity, open to the private initiative, but the participants must be certified.

In an electricity market, as in the case of the Iberian Electricity Market (MIBEL —Mercado Ibérico da Eletricidade), at every hour, the power producers offer to sell electricity at a certain price; the retailers offer to purchase electricity at another price. The selling offers are placed in an increasing price order; the purchase offers

---

[18]400 V is the RMS phase-to-phase voltage, which correspond to 230 V, RMS phase-to-neutral voltage.

Fig. 13 A piece of the HV distribution grid in the Lisbon, Portugal, area. *Source* E-Redes, Portuguese DSO, www.edp.pt

are ordered price-decreasingly. The intersection point between the two curves (supply and demand) defines the market-clearing price. All the producers that have been selected to operate (because their offers match the demand) will receive at the market price, and the retailers will also pay the market price. Hence, the market price changes on an hourly basis, despite the consumer tariffs still do not reflect this hourly price changing.

The retailer is the clients' representative in the market. It buys electricity at the market price, pays the tariffs for the use of the transmission and distribution grids, pays other power system costs, and reflects all these parcels in the tariffs it charges to the clients.

4.5   Consumption

The domestic consumers are connected to the LV distribution network. The big consumers are usually connected in the HV distribution network; the medium industrial consumers are normally connected in the MV network.

The load diagram is the representation of the load power[19] as a function of time. The load diagram supplies very important information for the power system operation; they allow to find the peak power (maximum power) and the consumed electricity (area below the load diagram).[20]

As an example, the load diagram on the day of annual peak demand for the Portuguese power system (2016 and 2017) is presented in Fig. 14. We can see that Portugal was exporting energy to Spain during the whole day. Also, note the typical behaviour of the daily load diagram of the Portuguese power system: during the night, the load power decreases to about half of the peak power; then, a first peak occurs at about noon; and finally, the peak power is attained by dinner time.

5   Balancing Generation and Consumption

The energy in its electrical form is not economically appropriate for large-scale storage, at least with current technology. As so, as already mentioned, the electricity that is obtained in the power stations must match, in each time instant, the consumption plus the losses.

The power demanded by the consumers is always changing, because of a random switch on and switch off of the electrical devices. Moreover, some electricity production is time variable in an uncontrolled way. This is, namely, the case of wind power, which depends on the available wind resource, and to a somehow less extent, PV power, which is, nevertheless, more predictable than wind.

To compensate for all these changes, there are several tools available using the market features: to trade electricity at the interconnections and to use the water storage devices are examples of actions undertaken to compensate for the mismatch between generation and consumption. Also, the conventional power station regulators are constantly controlling the valves that supply water to the hydro turbines or fossil fuels to the thermal power plants, so that a perfect match between generation and consumption is attained at each instant.

The prime movers (water in the hydropower plants; water steam in the coal-fired power plants; gases mix in the gas power plants) turn the turbine blades and originate the driving torque that moves the alternator. The driving torque is opposed by the resistance torques. When the driving torque is equal to the summation of the

---

[19]In power systems engineering, the power demanded by the consumers is usually known as load power.

[20]We will see later that the electrical energy is the integral of the electrical power.

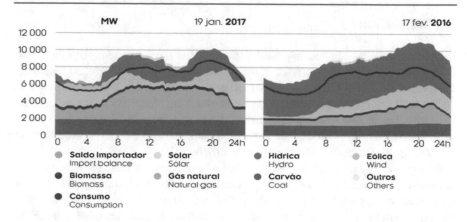

Fig. 14 Load diagram on the day of annual peak demand (Portuguese power system; 2016 and 2017). *Source* REN, Portuguese TSO, www.ren.pt

resistance torques, the turbine and the generator rotate at constant speed[21]: the electricity produced from the resources is equal to the electricity consumed by the receivers and the grid frequency is constant.

However, the regulators are not fast enough to keep a constant speed. During the time of opening or closing the valves, the driving and the resistance torques are not equal. As so, if the consumption is reduced, the resistance torque is reduced, and the machines experiment a transient speed increase (acceleration). The opposite happens when consumption increases.

When the angular speed changes, it changes the grid frequency. This is most undesirable because it disturbs the electricity supply to the consumers. As so, the frequency must be kept constant. This is accomplished by designing the rotating masses with a large moment of inertia. It is the kinetic energy of the rotating masses that compensates for the unbalance between production and consumption, while the regulators are performing the required valve opening or closing actions. Let us look at the so-called swing equation:

$$T_d - T_r = J \frac{d\omega}{dt} \tag{3.17}$$

where T_d and T_r are the driving torque and the resistance torque, respectively, J is the moment of inertia, and ω is the angular speed ($d\omega/dt$ is the angular acceleration).

It is possible to conclude that even for a high unbalance in the torques, a moderate angular acceleration can be achieved providing that the moment of inertia is high enough. The design of conventional power plants includes choosing the appropriate values for the moment of inertia of the rotating masses so that the frequency changes are kept within a very narrow range.

---

[21]The so-called synchronous speed.

6 Conclusions

In this introductory chapter, we have presented the power system. This was necessary because, in many countries of the world, namely in the developed countries, the Renewable Energy Sources (RESs) are grid connected, i.e., they are explored in such a way that they inject in the power system all the electricity they can produce. It is true that in less developed countries there is a huge market for RES being explored supplying electricity to isolated places that do not have access to an electrical grid. But this is not the topic of this book.

The power system is a complex and sophisticated system that allows consumers to consume electricity whenever they want. To perform this task, power system controllers must comply with a fundamental and difficult-to-achieve constraint: in each time instant, the power generated by the generating facilities must balance the power consumed by the power system clients.

When the power system was designed, some decisions were taken. The power system operates in Alternating Current (AC), with a frequency of 50 or 60 Hz, depending on the regions of the planet, and is three phased. The reasons behind these options were presented and discussed in this chapter.

Modern power systems are usually divided into generation, transmission, distribution, retail, and consumption. There are several types of generation plants: thermal includes coal-fired, combined cycle, and nuclear, and are known as conventional power plants; the RES with current economic interest are hydro, wind, and solar. Hydro is an asset for every system operator due to its valuable fast power regulating features. Wind power plants are currently mostly placed onshore, but the offshore development boost is simply waiting for the economic conditions to become available. Solar resource is abundant in many regions of the planet, and Photovoltaics (PV) is currently viewed as the cheapest way to produce electricity.

The transmission and distribution system compose the grid, which is necessary to deliver the electricity from the places where it is produced to the places where it is consumed. The division of the grid into transmission and distribution systems is related to the voltage level. It was shown that the higher the voltage level, the lower the power losses. So, the transmission system, which operates with very high voltages, is required to transport large quantities of energy over long distances. The distribution system, which operates with lower voltages, is in charge of the electricity distribution to the consumers.

In the modern organization of the electricity sector, with energy being traded in organized markets, the retailers are the consumers' representatives that buy electricity in the market on their behalf. The other participants of the market are the generators that offer to sell energy. The transmission and distribution systems are operated under public service concessions and receive regulated tariffs for the transmission and distribution services they provide. Finally, the consumers end this value chain by consuming the electricity they want at the time they require. And the power system is the system that makes this happen.

AC Electrical Circuits for Non-electrical Engineers

2

Abstract

Electrical circuits are the basis of electrical engineering. The topic is approached in every electrical engineering course. In this chapter, we provide an overview of the basics of electrical circuits theory to readers that do not hold a background in electrical engineering. Therefore, the subject is softly approached, and the contents are not delivered with the scientific depth a detailed course on electrical circuits would require. Alternate Current (AC) circuits topic is introduced, pointing out the need for complex numbers to ease the analysis, and presenting the consequent complex amplitude (phasor) concept. The power in AC circuits (active, reactive, and complex) is also a matter of concern in this chapter. We proceed with balanced three-phase circuits, where, for instance, phase and line voltages are introduced. The per unit system, which facilitates the power system analysis, is a topic that is also addressed in this chapter. Finally, the induction motor is presented as an application case of AC electrical circuits. To illustrate the methods, there are several numeric examples, spread all over the text.

1 Introduction

This text is intended to provide basic Alternate Current (AC) electrical circuits knowledge to readers that do not hold a background in electrical engineering. Therefore, the main target of this chapter is the reader with a non-electrical engineering background. Nevertheless, electrical engineers may find this chapter useful to recall the theory of electrical circuits hereby presented in a simplified way.

The inspiring source is the text "Noções de Eletrotecnia" published in 1992, by Técnico Lisboa (IST), late full professor Domingos Moura. In some aspects, the reference book "Redes de Energia Elétrica" from IST's emeritus professor José Sucena Paiva has been consulted too.

Certainly, the topics are not approached with the same theoretical groundings as when they are presented in an electrical engineering course. Actually, we are concerned about just introducing the basic operating principles, so that a non-electrical engineer can understand them and providing the main tools to deal with the typical electrical-related problems that an engineer may have to face in his/her job. All in all, the text softly approaches the main electrical circuits topics and by no means introduces them with the accurate required scientific deepness.

The AC electrical circuits are introduced in Sect. 2, beginning with the production of a sinusoidal voltage from the rotation of a winding inside a magnetic field. The need for complex numbers to solve AC electrical circuits is justified and the basic types of AC electrical circuits are solved using the phasor concept.

As far as the study of power systems is concerned, the concept of power and the concept of a three-phase system are among the most used. We introduce the real, reactive, and complex power and discuss the theory of three-phase systems, introducing the concepts of phase and line voltages and the star and delta connection of loads. Also, the per unit (pu) system, which was created to facilitate the computations in power system engineering, is presented.

A known application of the AC electrical circuits theory is the induction motor. The steady-state model of this electrical machine, which is precisely based on an equivalent electrical circuit, is introduced in this chapter, together with the main quantities that characterize its operation.

2   AC Electrical Circuits

Time variation of the electrical quantities makes it more difficult to analyze the steady-state behaviour of AC electrical circuits. However, Electrical Engineering developed some methods to smooth this difficulty. We will present in the sequence the most used concepts and methods to analyze the steady-state behaviour of AC electrical circuits.

2.1   The Sinusoidal Voltage

Figure 1 shows a uniform (constant) magnetic field created by permanent magnets (represented by magnetic poles N and S). Inside this magnetic field, a rectangular winding is rotating at constant angular speed ω (rad/s) imposed by an external agent.

Now look at Fig. 2: when the angle θ is 0 (lighted winding), the flux crossing the winding is maximum; when the angle θ is $\pi/2$ (dimmed winding), the flux is null.

As so, the magnetic flux, φ (Wb—Webber), can be expressed as:

$$\varphi(t) = \varphi_{MAX}\cos\theta = \varphi_{MAX}\cos\omega t \tag{2.1}$$

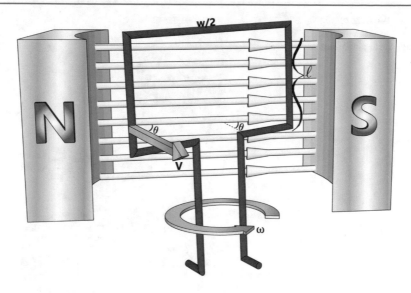

Fig. 1 Rectangular winding rotating inside a uniform magnetic field

where, φ_{MAX} is the maximum flux and θ is the angle between the magnetic field and a perpendicular to the winding V (nothing to do with voltage) in Fig. 1.

Following a known electrical law (Faraday's Law), an electromotive force (EMF), with the dimensions of a voltage (V–Volt) is induced in the winding, which is given by

$$e(t) = -\frac{d\varphi}{dt} = \omega\varphi_{\text{MAX}}\sin(\omega t) = E_{\text{MAX}}\sin(\omega t) = \sqrt{2}E\sin(\omega t) \qquad (2.2)$$

where E_{MAX} is the maximum EMF, linearly dependent on the maximum flux and on the angular speed. The EMF RMS value is E. It should be noted that no EMF is produced if the winding is static ($\omega = 0$).

We conclude that the rotation of a winding inside a uniform magnetic field produces a sinusoidal EMF. This is the operating principle of an AC generator. To transmit power from the rotating winding to a stationary circuit, slip rings and carbon brushes are used: slip rings are rotating metallic devices connected to the winding, and carbon brushes are fixed contacts connected to an external static circuit (see Fig. 3).

At the terminals of the external circuit, a sinusoidal voltage, that slightly differs from the EMF due to circuit losses, is obtained

$$v(t) = \sqrt{2}V\sin(\omega t) \qquad (2.3)$$

Fig. 2 Winding rotation
makes the magnetic flux
change from maximum to
zero

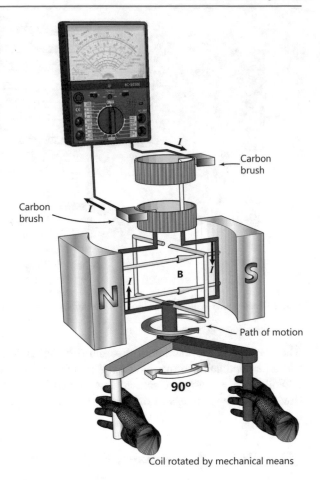

Carbon brush

Carbon brush

Path of motion

90°

Coil rotated by mechanical means

2.2 Pure Resistive Circuit

At this stage, no current exists. To obtain a current, a load must be connected to the external circuit. If the simplest load, a constant resistance, R (Ω – Ohm), is connected, current will flow through it. Following another electrical law (Ohm's Law), this current, i_R, will be given by

$$i_R(t) = \frac{v_R(t)}{R} = \sqrt{2}\frac{V_R}{R}\sin(\omega t) = \sqrt{2}I_R\sin(\omega t) \qquad (2.4)$$

Let us look at a numerical example. Let us suppose that a $V_R = 230$ V RMS voltage is produced in a larger but similar system to the one in Fig. 2. Then, a resistance $R = 23\ \Omega$, is connected to the system through an external circuit, as seen in Fig. 4. The current RMS value is $I_R = V_R/R = 10$ A. The time evolution of the voltage and current is presented in Fig. 5.

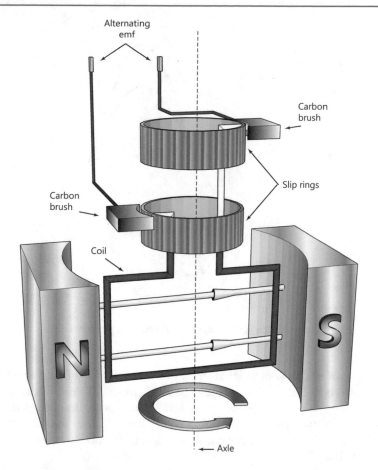

Fig. 3 Slip rings and carbon brushes

Fig. 4 Resistive circuit

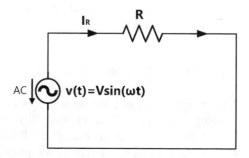

From this figure, we can see that voltage and current are zero at the same time instant; and that they are maximum also at the same time instant. We say that voltage and current are "in phase".

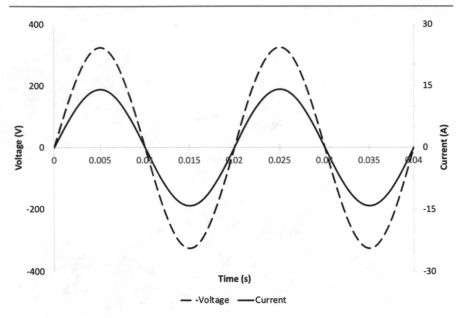

Fig. 5 Voltage and current changing as a function of time in a resistive circuit

We can now use the knowledge we have just learned to investigate what is going on with regards to EMF and flux. The situation is quite different: when the flux is maximum, the EMF is zero and when the EMF is maximum, the flux is zero. This means that these two quantities are not in phase, actually, they are "out of phase", which is due to the derivative relationship between the cause (flux) and the effect (EMF) (see Eq. 2.2).

In Fig. 6, it is depicted the time evolution of the flux (Eq. 2.1) and EMF (Eq. 2.2), where the out of phase of these two quantities is apparent. One can say that the flux leads the EMF (and the EMF lags the flux). For instance, the maximum flux occurs at $t = 0$, and the maximum EMF occurs later, at $t = 0.005$ s, therefore, the flux leads the EMF.

2.3 Pure Inductive Circuit

It is known that when a current $i_L(t)$ flows through a rolled-up conductor, called an inductor, (AB in Fig. 7), a magnetic flux, $\varphi(t)$, is created.

If in the space where the flux is established, no ferromagnetic materials exist, the flux is proportional to the current, the proportionally constant being L, as follows:

$$\varphi(t) = Li_L(t) \tag{2.5}$$

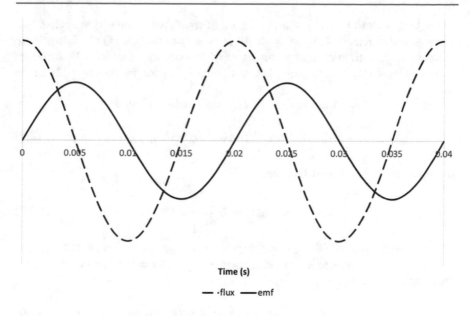

Time (s)

— ·flux ——emf

Fig. 6 Flux and EMF changing as a function of time

Fig. 7 Pure inductive circuit

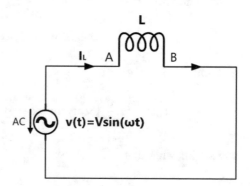

L is the inductance; it is a measure of the induced magnetic flux due to the current that flows across the circuit itself. The unit of the coefficient of inductance is H—Henry.[1]

From Faraday's Law, we know that providing the flux changes in time, an EMF is produced between AB terminals, which is given by

$$e(t) = -\frac{d\varphi}{dt} = -L\frac{di_L}{dt} \tag{2.6}$$

---

[1]For the majority of the applications, an inductance equal to 1 H is very high. This is why normally we use the milliHenry (1 mH = 10^{-3} H).

Looking back to Fig. 7, it shows a circuit in which a voltage $v(t)$ is applied to a coil with an inductance L; we suppose the resistance of the coil is null. Under these circumstances, a time-changing current $i_L(t)$ flows across the coil. This current creates a time-changing magnetic flux, $\varphi(t)$, which, in turn, induces an EMF in the coil itself.

Applying Kirchoff Voltage Law (KVL), we obtain

$$v_L(t) + e(t) = 0 \tag{2.7}$$

and then taking Eq. 2.6 into account

$$v_L(t) = L\frac{di_L}{dt} \tag{2.8}$$

It is possible to demonstrate that a sinusoidal voltage causes a sinusoidal current. It is a usual procedure to assume the voltage angle at $t = 0$ to be equal to zero. Therefore, the voltage is given by

$$v_L(t) = \sqrt{2}V_L sin(\omega t + 0) = \sqrt{2}V_L sin(\omega t) \tag{2.9}$$

We will prove that, for the circuit of Fig. 7, the current is

$$i_L(t) = \sqrt{2}I_L sin(\omega t - \frac{\pi}{2}) \tag{2.10}$$

From Eq. 2.8, we write

$$v_L(t) = L\frac{d}{dt}\left(\sqrt{2}I_L sin(\omega t - \frac{\pi}{2})\right) = \sqrt{2}\omega L I_L cos(\omega t - \frac{\pi}{2}) \tag{2.11}$$

and so, knowing that

$$cos(\omega t - \frac{\pi}{2}) = sin(\omega t) \tag{2.12}$$

we obtain

$$v_L(t) = \sqrt{2}\omega L I_L sin(\omega t) \tag{2.13}$$

which proves our hypothesis.

Comparing Eqs. 2.9 and 2.13, we conclude that

$$V_L = \omega L I_L = X_L I_L \tag{2.14}$$

where X_L is known as reactance, in this case, an inductive reactance.

The units of the reactance are the same as for the resistance, i.e., Ω–Ohm, because, as for the case of the resistance, the reactance is a ratio between the RMS values of the voltage and current.

In a resistive circuit, we have

$$R = \frac{V_R}{I_R} \tag{2.15}$$

and in a purely inductive circuit, it is

$$\omega L = X_L = \frac{V_L}{I_L} \tag{2.16}$$

Furthermore, looking at Eqs. 2.10 and 2.13, we conclude that, in a purely inductive circuit, the current is lagging the voltage, or the voltage is leading the current, by an angle of $\pi/2$.

Let us look again at a numerical example. Consider that the RMS voltage is $V_L = 230$ V, $f = 50$ Hz, and a coil with a reactance $X_L = 23\ \Omega$ ($L = 73.21$ mH) is connected to the voltage source. The current RMS value is $I_L = V_L/X_L = 10$ A. The time evolution of the voltage and current are presented in Fig. 8.

It can be seen that the current is lagging the voltage because the peak current occurs after the peak voltage.

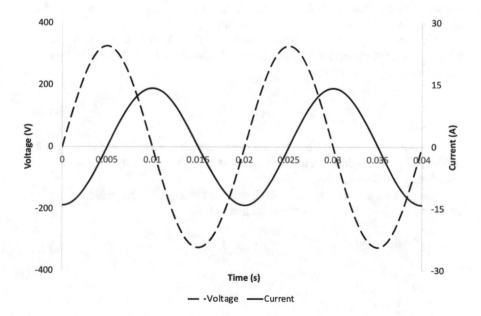

Fig. 8 Voltage and current changing as a function of time in a purely inductive circuit

Fig. 9 Pure capacitive circuit

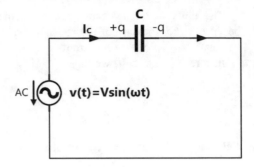

2.4 Pure Capacitive Circuit

A capacitor consists of two conductive plates which are separated by a dielectric (insulator) medium. The electrical charge of a capacitor, $q(t)$, is directly proportional to the voltage, $v_C(t)$, at the capacitor's terminals. Hence, we can write

$$q(t) = Cv_C(t) \qquad (2.17)$$

where C is the proportionality coefficient, known as capacitance. The capacitance is measured in F–Farad [2] and the electrical charge in C–Coulomb.

The electrical charge is the charge in one of the two conductors that compose a capacitor (Fig. 9). In the other conductor, there is a symmetrical (with a minus sign) but equal charge. However, the relevant charge is q, and not $2q$, which is the charge that is moving during the charge and discharge processes.

The current $i_C(t)$ in each cross-section of a conductor is the quantity of electricity (electrical charge) flow that crosses that section in each time instant. This definition allows us to write

$$i_C(t) = \frac{dq}{dt} \qquad (2.18)$$

As a physical quantity, a current is a rate at which a charge flows past a point on a circuit. For the circuit in Fig. 9, the current $i_C(t)$ measures the rate at which the capacitor's electrical charge is changing.

If one admits that the capacitance, C, does not change in time, which is true most of the times, from Eq. 2.17, we can conclude that

$$i_C(t) = C\frac{dv_C}{dt} \qquad (2.19)$$

---

[2]Usually, the capacitance is given in microFarad ($1\mu F = 10^{-6}$ F).

As for the case of the purely inductive circuit, let us now prove for the pure capacitive circuit, that if the source voltage is $v(t) = \sqrt{2}V\sin(\omega t)$, the current will be

$$i_C(t) = \sqrt{2}I_C\sin(\omega t + \frac{\pi}{2}) \qquad (2.20)$$

From Eq. 2.19, we obtain

$$i_C(t) = C\frac{d}{dt}\left(\sqrt{2}V_C\sin(\omega t)\right) = \sqrt{2}\omega CV_C\cos(\omega t) \qquad (2.21)$$

and so, as

$$\cos(\omega t) = \sin(\omega t + \frac{\pi}{2}) \qquad (2.22)$$

It follows that

$$i_C(t) = \sqrt{2}\omega CV_C\sin(\omega t + \frac{\pi}{2}) \qquad (2.23)$$

which proves our hypothesis.

From Eqs. 2.20 and 2.23, we conclude that

$$I_C = \omega CV_C = B_CV_C; \; V_C = \frac{1}{\omega C}I_C = X_CI_C \qquad (2.24)$$

where B_C is known as susceptance and X_C is our already known reactance, in this case, a capacitive reactance.

The units of the susceptance are S—Siemens. In a purely capacitive circuit, a susceptance is a ratio between the RMS values of the current and voltage, and the reactance is the ratio between the RMS values of the voltage and current.

If we look at Eq. 2.9 and Eq. 2.23, we notice that the current is leading the voltage, or the voltage is lagging the current, by an angle of $\pi/2$. This holds true for a purely capacitive circuit.

Let us focus again on a numerical example. The RMS voltage is $V_C = 230$ V, $f = 50$ Hz, and a capacitor with a reactance $X_C = 23\,\Omega$ ($B_C = 43.48$ mS; $C = 138.40\,\mu F$) is connected to the voltage source. The current RMS value is $I_C = V_C/X_C = 10$ A. The time evolution of the voltage and current are presented in Fig. 10.

It is apparent that the current is leading the voltage because the peak current occurs before the peak voltage.

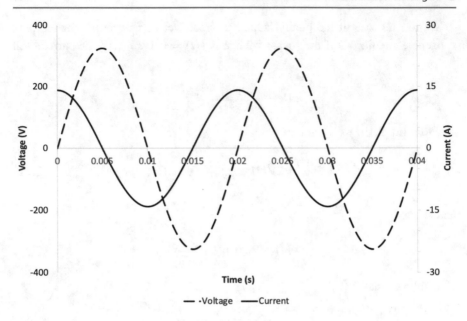

Fig. 10 Voltage and current changing as a function of time in a purely capacitive circuit

2.5   The Need for Complex Numbers

We have seen that knowing the resistance (in a purely resistive circuit) or the reactance (in a purely inductive or capacitive circuit) allows us to know the RMS value of the current from the knowledge of the RMS value of the voltage. However, this is not enough to completely describe the sinusoidal current.

In general, a sinusoidal current can be written as (let us recall that for a purely inductive circuit, $\phi = \pi/2$; for a purely capacitive circuit, $\phi = -\pi/2$; for a purely resistive circuit, $\phi = 0$):

$$i(t) = \sqrt{2}I sin(\omega t - \phi) \tag{2.25}$$

To describe $i(t)$, we need:

- the RMS value of the current, I;
- the frequency, f, that allows us to compute the angular frequency, $\omega = 2\pi f$;
- the initial phase, ϕ, i.e., the angle at the origin of times, $t = 0$.

We conclude that we need three kinds of information to completely describe a sinusoidal current (for a voltage, the same applies). However, in many situations, we are interested in assessing the power system in a steady state. In steady-state and linear circuits (constant parameters), the frequency is always equal to $f = 50$ Hz.

Consequently, $\omega = 100\pi$, and we need only to know two kinds of information to completely describe a sinusoidal current (or voltage): RMS and initial phase.

As so, we need numbers able to carry two kinds of information. These numbers are the complex numbers.

2.6 Review on Complex Numbers

We start from that there is an (imaginary) number, j, which, multiplied by itself, equals -1

$$j = \sqrt{-1} \tag{2.26}$$

We use j for the imaginary unit. Many books use the letter i, but this letter is normally assigned in electrical engineering to denote the current.

A so-called complex number

$$z = a + jb \tag{2.27}$$

has both a real part ($Re(z) = a$) and an imaginary part ($Im(z) = b$). The format in Eq. 2.27 is called the rectangular format of complex numbers. The representation of this complex number in the complex plan is shown in Fig. 11.

The complex number is represented by the dot in Fig. 11. We can assign to that complex number a vector with a certain length (amplitude) and angle with respect to the x-axis (phase).

Leonhard Euler (1707–1783) discovered the relation, which relates complex numbers to the (periodic) trigonometric functions, known as Euler's formula

$$e^{j\phi} = cos\phi + jsin\phi \tag{2.28}$$

With the aid of Euler's formula, it is possible to transform any complex number in the rectangular format into the polar format

Fig. 11 Representation of a complex number in the complex plane

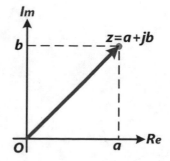

Fig. 12 Representation of a
complex number in the polar
format in the complex plane

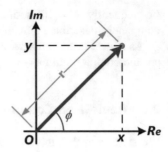

$$z = re^{j\phi} \tag{2.29}$$

where r is the amplitude and ϕ is the phase, as in Fig. 12.

A complex number with unitary modulus ($r=1$) and angle ϕ is represented in Fig. 13.

Plugging Eqs. 2.28 into 2.29, we obtain

$$z = rcos\phi + jrsin\phi = a + jb \tag{2.30}$$

where

$$Re\{z\} = a = rcos\phi; Im\{z\} = b = rsin\phi \tag{2.31}$$

Fig. 13 Representation of a
complex number with unitary
modulus and angle ϕ in the
complex plane

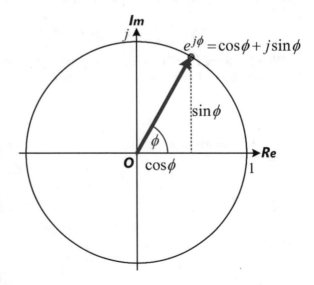

Fig. 14 A complex number can be represented either in the rectangular or polar formats

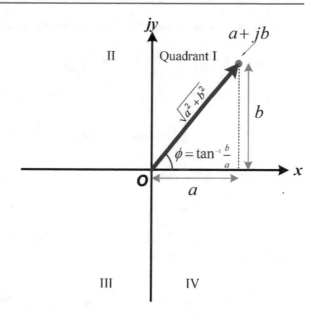

Solving the system of two equations in Eq. 2.31 (dividing the second equation by the first and then squaring and summing the two equations), we are led to

$$r = \sqrt{a^2 + b^2}; \phi = tan^{-1}\left(\frac{b}{a}\right) \qquad (2.32)$$

The same result is reached by visual inspection of Fig. 13, taking trigonometric relationships and Pythagoras theorem into account (Fig. 14).

Let us now recall the basic operations with complex numbers. Consider two complex numbers

$$z_1 = a_1 + jb_1 = r_1 e^{j\phi_1}; z_2 = a_2 + jb_2 = r_2 e^{j\phi_2} \qquad (2.33)$$

Summations and subtractions are easier performed in the rectangular format

$$z_1 + z_2 = a_1 + jb_1 + a_2 + jb_2 = (a_1 + a_2) + j(b_1 + b_2) \qquad (2.34)$$

$$z_1 - z_2 = a_1 + jb_1 - (a_2 + jb_2) = (a_1 - a_2) + j(b_1 - b_2) \qquad (2.35)$$

Products and divisions are easier performed in the polar format

$$z_1 z_2 = r_1 e^{j\phi_1} r_2 e^{j\phi_2} = r_1 r_2 e^{j(\phi_1 + \phi_2)} \qquad (2.36)$$

$$\frac{z_1}{z_2} = \frac{r_1 e^{j\phi_1}}{r_2 e^{j\phi_2}} = \frac{r_1}{r_2} e^{j(\phi_1 - \phi_2)} \qquad (2.37)$$

We define complex conjugate of a complex number $z = a + jb = re^{j\phi}$ as

$$z^* = (a - jb) = re^{-j\phi} \tag{2.38}$$

and the reciprocal as

$$\frac{1}{z} = \frac{1}{a + jb} = \frac{1e^{j0}}{re^{j\phi}} = \frac{1}{r}e^{-j\phi} \tag{2.39}$$

Now some singular cases that will be important in the sequence ($a>0$, $b>0$)

$$a = ae^{j0}; \ jb = be^{j\frac{\pi}{2}}; -a = ae^{j\pi}; -jb = be^{-j\frac{\pi}{2}} \tag{2.40}$$

Multiplying a vector by j, moves the vector forward by 90°; multiplying by $-j$, holds the vector back by −90°.

2.7   Complex Amplitudes to Solve AC Circuits

We have seen that complex numbers can be associated with vectors. When complex numbers are used to solve AC circuits, the quantities that are to be represented by complex numbers are, among others, voltages and currents. It is quite apparent that a sinusoidal voltage cannot be represented by a vector, because it has no associated direction in space as, for instance, a force, a velocity, or an acceleration. However, as for the rest, the vector representation is applicable. Because of this, the nomenclature "phasors" will be used instead of "vectors".

One way of representing an AC quantity is in terms of a rotating phasor, (we use bold capital letters with an upper bar to denote rotating phasors, as shown in Fig. 15). A rotating phasor is rather like the hand on a clock, though the phasors we will consider will all rotate in the anticlockwise direction. A rotating phasor $\overline{\mathbf{A}}$ has a magnitude A and rotates at a fixed angular speed ω, so that the angle $\theta = \omega t$ from the x-axis to the phasor increases with time. The projection of the rotating phasor onto the y–axis (or in the x-axis) is a sinusoidal function of the angle θ. The projection in the y-axis results in a sine function, whereas we obtain a cosine from the projection in the x-axis. Therefore, a rotating phasor is a complex number that can represent the amplitude and phase of a sinusoid.

In this example, in $t = 0$, the angle $\theta = \omega$ t is zero. As so, the initial phase is $\phi = 0$. However, we could easily generalize for $\phi = \theta(0) \neq 0$, as seen in Fig. 16.

Based on what you have presented whatsoever, let us consider a sinusoidal time-dependent current.

$$i(t) = \sqrt{2}I sin(\omega t - \phi) \tag{2.41}$$

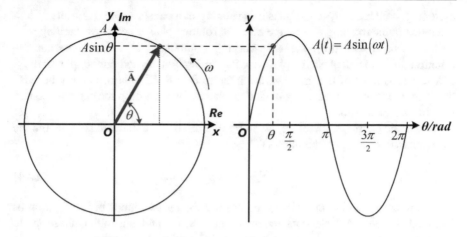

Fig. 15 A phasor as a representation of a sinusoidal wave with $\phi = (0) = 0$

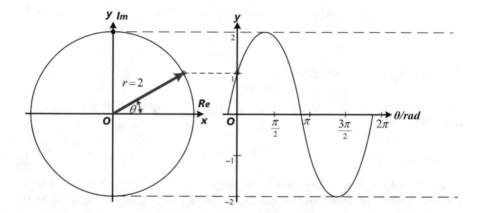

Fig. 16 A phasor as a representation of a sinusoidal wave with $\phi = \theta(0) \neq 0$

We have shown that this is equivalent to having the projection on the y-axis of the rotating phasor

$$\mathbf{\bar{I}} = \sqrt{2}I \cos(\omega t - \phi) + j\sqrt{2}I \sin(\omega t - \phi) \qquad (2.42)$$

According to Euler's formula, this is equal to

$$\mathbf{\bar{I}} = \sqrt{2}I\, e^{-j\phi} e^{j\omega t} \qquad (2.43)$$

When multiplied by $e^{j\omega t}$, the phasor starts rotating anticlockwise in the complex plane. As the sinusoidal function has a constant frequency (for steady-state power systems $f=50$ Hz always), the phasor is rotating at a constant rate. If observed in a reference frame rotating at the same rate, the rotating phasor becomes standing still, i.e., we can simply drop the time-varying component $e^{j\omega t}$, and represent the function in terms of a static phasor with an amplitude and initial phase only. Moreover, as most of the time, we are interested in RMS values (namely because these are the values a measuring device measures), we can drop the constant square root of two factors too.

This static phasor is, therefore, given by (we use bold capital letters to denote the static phasor, or simply, the phasor)

$$I = Ie^{-j\phi} = Icos\phi - jIsin\phi \tag{2.44}$$

It is important to highlight that the phasor I carries two kinds of information as intended to solve AC electrical circuits: the RMS I and the initial phase ϕ. In Electrical Engineering, the phasor is named "**Complex Amplitude**".

To recover the original sinusoidal current (Eq. 2.41), we simply compute

$$i(t) = Im\{\bar{I}\} = Im\left\{\sqrt{2}Ie^{j\omega t}\right\} = \sqrt{2}I \, sin(\omega t - \phi) \tag{2.45}$$

2.8 Complex Impedance

Let us consider an AC electrical circuit, whose voltage and current complex amplitudes are

$$V = Ve^{j\alpha}; I = Ie^{j\beta} \tag{2.46}$$

It is common practice to consider the angle's reference located in the voltage complex amplitude, therefore, the voltage initial phase is usually assumed to be 0. Here, for generality, we have considered it to be equal to α.

As seen before, Eq. 2.46 contains all the necessary information to define the phasors I and V, as well as the time evolution $i(t)$ and $v(t)$, providing the angular frequency is known, which is common.

The ratio between the complex amplitude of the voltage and the complex amplitude of the current is a complex number known as complex impedance and denoted as Z

$$Z = \frac{V}{I} = \frac{Ve^{j\alpha}}{Ie^{j\beta}} = \frac{V}{I}e^{j(\alpha - \beta)} = Ze^{j\phi} \tag{2.47}$$

$$Ze^{j\phi} = Zcos\phi + jZsin\phi = R + jX \tag{2.48}$$

The impedance modulus is $Z = V/I$ (ratio between the respective RMS values) and the impedance angle, ϕ, is the difference between the voltage angle (α) and the current angle (β). On the other hand, the real part of an impedance is the resistance, and the imaginary part is the reactance.

The complex impedance may be regarded as an operator that allows converting one quantity into another one.

The phi angle (the impedance angle) is a very important angle. Its cosine is called power factor

$$pf = cos\phi \tag{2.49}$$

Phi ranges from $-\pi/2$ (pure capacitive circuit) to $\pi/2$ (pure inductive circuit); if $\phi = 0$, the circuit is resistive. We will retake this subject later.

The reciprocal (inverse) of the complex impedance is the complex admittance, denoted by Y

$$Y = \frac{1}{Z} = \frac{I}{V} = \frac{Ie^{j\beta}}{Ve^{j\alpha}} = \frac{I}{V}e^{j(\beta-\alpha)} = \frac{1}{Z}e^{-j\phi} = Ye^{-j\phi} = G + JB \tag{2.50}$$

The real part of the admittance is the conductance, and the imaginary part is the susceptance. The relationship between impedance and admittance can be further developed as

$$Y = \frac{1}{Z} = \frac{1}{R+jX} = \frac{R-jX}{(R+jX)(R-jX)} = \frac{R}{R^2+X^2} - j\frac{X}{R^2+X^2} = G + jB \tag{2.51}$$

We highlight that neither the impedance nor the admittance represent sinusoidal waves. They are just complex numbers.

2.9   Application to Basic AC Circuits

We will now apply the complex amplitude method to the pure resistive, pure inductive, and pure capacitive circuits that we have studied before. As usual, we consider that the voltage holds the angles' reference, therefore, the initial phase of the voltage is zero.

We recall that, in a purely resistive circuit, the voltage and current are in phase. As so, we can write the respective complex amplitudes as

$$V_R = V_R e^{j0} = V_R; I_R = I_R e^{j0} = I_R \tag{2.52}$$

The complex impedance is

$$Z_R = \frac{V_R}{I_R} = \frac{V_R}{I_R} = R \tag{2.53}$$

In a purely resistive circuit, the complex impedance is the so-called resistance, which is a real number (see Fig. 17). In this case, the phi (ϕ) angle is zero.

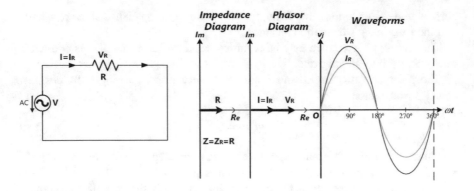

Fig. 17 Voltage, current, and impedance in a purely resistive circuit

As far as a purely inductive circuit is concerned, the current lags the voltage by 90°. Therefore, we can write

$$V_L = V_L e^{j0} = V_L; I_L = I_L e^{-j\frac{\pi}{2}} \tag{2.54}$$

The complex impedance is (take note that $j\omega L = \omega L e^{j\frac{\pi}{2}}$)

$$Z_L = \frac{V_L}{I_L} = \frac{V_L}{I_L} e^{j\frac{\pi}{2}} = \omega L e^{j\frac{\pi}{2}} = j\omega L \tag{2.55}$$

Please take note that in a purely inductive circuit, the phi angle (the impedance's angle) is 90°.

Let us recall here for convenience the fundamental equation of the purely inductive circuit (Eq. 2.8)

$$v_L(t) = L\frac{di_L}{dt} \tag{2.56}$$

In terms of complex amplitudes, from Eq. 2.55, one can write

$$V_L = Z_L I_L = j\omega L I_L \tag{2.57}$$

Comparing Eqs. 2.56 and 2.57, we conclude that in the complex amplitude framework the derivative d/dt is translated by the operator $j\omega$. As a matter of fact, when we multiply the phasor I_L by $j\omega L$, this is equivalent to rotate phasor I_L by 90° in the anticlockwise direction, therefore, obtaining phasor V_L (see Fig. 18, but please disregard the power curve, we will get back to this concept later).

In what concerns the pure capacitive circuit, it was previously demonstrated that the current leads the voltage by 90°. Therefore, the complex amplitudes of the voltage and current are given by

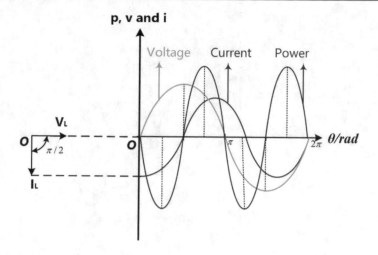

Fig. 18 Voltage, current, and impedance in a purely inductive circuit

$$\boldsymbol{V_C} = V_C e^{j0} = V_C; \boldsymbol{I_C} = I_C e^{j\frac{\pi}{2}}$$

(2.58)

The complex impedance is (take note that $-j(1/\omega C) = (1/j\omega C) = \left(\frac{1}{\omega C}\right)e^{-\frac{j\pi}{2}}$):

$$\boldsymbol{Z_C} = \frac{\boldsymbol{V_C}}{\boldsymbol{I_C}} = \frac{V_C}{I_C}e^{-j\frac{\pi}{2}} = \frac{1}{\omega C}e^{-j\frac{\pi}{2}} = -j\frac{1}{\omega C}$$

(2.59)

It is apparent that the phi angle is –90° in a purely capacitive circuit.

In a purely capacitive circuit, it is common to express the relationship between the complex amplitudes of the voltage and current in the opposite way, leading to the complex admittance

$$\boldsymbol{Y_C} = \frac{\boldsymbol{I_C}}{\boldsymbol{V_C}} = \frac{I_C}{V_C}e^{j\frac{\pi}{2}} = \omega C e^{j\frac{\pi}{2}} = j\omega C$$

(2.60)

At this point, we recall the fundamental equation of the pure capacitive circuit for convenience (Eq. 2.19)

$$i_C(t) = C\frac{dv_C}{dt}$$

(2.61)

We can write Eq. 2.61 in the form

$$\boldsymbol{I_C} = j\omega C \boldsymbol{V_C}$$

(2.62)

Comparing Eqs. 2.61 and 2.62, we conclude again that the multiplication by the operator $j\omega$ performs a 90° rotation in the anticlockwise direction.

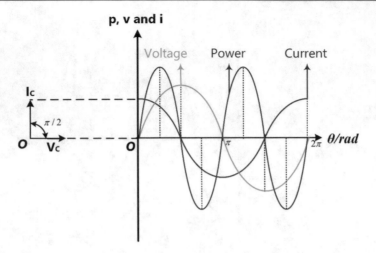

Fig. 19 Voltage, current, and impedance in a purely capacitive circuit

Figure 19 shows the representation of the voltage and current (static) phasors, as well as the time evolution of these two quantities in a purely capacitive circuit (again disregard the power curve, at this stage).

2.10 RLC Circuit

With what we have learned so far, we are now able to solve the RLC series circuit, presented in Fig. 20.

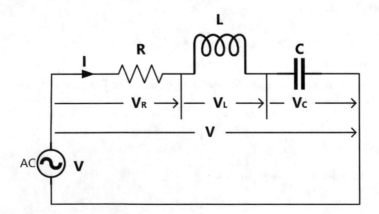

Fig. 20 RLC series circuit

At this stage, it should be apparent that it is equivalent to write the time evolution equations of the voltage and current or the respective complex amplitudes. Of course, we will use complex amplitudes because it is much easier.

KVL[3] allows us to write

$$V = V_R + V_L + V_C \tag{2.63}$$

And now we can write the complex amplitudes of the three voltages as (we recall that $-j = 1/j$)

$$V = RI + j\omega LI + \frac{1}{j\omega C}I = \left(R + j\omega L - j\frac{1}{\omega C}\right)I = ZI \tag{2.64}$$

The equivalent impedance of the RLC series circuit is

$$Z = R + j\left(\omega L - \frac{1}{\omega C}\right) = R + jX \tag{2.65}$$

The same conclusion could be immediately reached if we recall a basic electrical knowledge stating that the equivalent impedance of a set of series-connected impedances is equal to the sum of the individual impedances

$$Z_{eq} = Z_1 + Z_2 + Z_3 = Z_R + Z_L + Z_C \tag{2.66}$$

We will use Eq. 2.65 to state that an impedance is a complex number whose real part is called resistance and the imaginary part is the reactance. In this case, the equivalent circuit's resistance and reactance are

$$R = R; X = \omega L - \frac{1}{\omega C} \tag{2.67}$$

The impedance modulus and phase are, respectively (recall Eqs. 2.32, 2.47 and 2.48)

$$Z = \sqrt{R^2 + \left(\omega L - \frac{1}{\omega C}\right)^2} ; \phi = tan^{-1}\left(\frac{\omega L - \frac{1}{\omega C}}{R}\right) \tag{2.68}$$

As seen, reactance X depends on L and C; therefore, X can be negative, zero, or positive, meaning that the behaviour of the RLC circuit can be capacitive, resistive, or inductive, respectively.

The impedance, resistance, and reactance are all measured in Ω—ohm. We highlight that the complex impedance is not a time-varying quantity. It is given by

---

[3]Kirchoff Laws—KVL and KCL—also apply to complex amplitudes: in a closed loop $\sum V_i = 0$ (KVL) and in a node $\sum I_i = 0$ (KCL).

the ratio of two complex amplitudes and does not represent a quantity that is changing with time.

The complex amplitude of the current is given by

$$I = \frac{V}{Z} = \frac{Ve^{j0}}{\sqrt{R^2 + \left(\omega L - \frac{1}{\omega C}\right)^2}e^{j\phi}} = \frac{V}{\sqrt{R^2 + \left(\omega L - \frac{1}{\omega C}\right)^2}}e^{-j\phi} \qquad (2.69)$$

The correspondent current sine wave is

$$i(t) = \sqrt{2}\frac{V}{\sqrt{R^2 + \left(\omega L - \frac{1}{\omega C}\right)^2}}\sin(\omega t - \phi) \qquad (2.70)$$

Let us put some numbers on this. Consider that the RMS voltage is $V=230$ V, $f=50$ Hz, $R=10$ Ω; $L=63.67$ mH, $C=636.5$ µF. Under these circumstances, $L = 2\pi fL = X_L=20$ Ω, $1/\omega C = X_C=5$ Ω. The impedance will be

$$Z = 10 + j15 = 18.03e^{j56.31°} \, \Omega$$

The current (in fact it is the complex amplitude of the current, but we will drop the "complex amplitude" mention when there is no misunderstanding risk) is

$$I = \frac{V}{Z} = 12.76e^{-j56.31°} \, A$$

considering, as usually, that the angles' reference is the zero-voltage angle

$$V = 230e^{j0} \, V$$

As the current's phase is negative (–56.31°), it follows that the current lags the voltage. Therefore, the circuit is inductive.

The phi angle (remember that this is the impedance's angle) is positive (56.31°). The power factor is

$$pf = \cos(56.31°) = 0.55 \text{ ind.}$$

You may have noticed that we added "ind." (inductive) to the numerical value of the power factor. Why? Let us open a parenthesis to approach this.

Let us now assume that $X_L=5$ Ω and $X_C=20$ Ω, while keeping $R=10$ Ω. In this case

$$Z_2 = 10 - j15 = 18.03e^{-j56.31°}\Omega$$

and the current would be

$$I_2 = \frac{V}{Z_2} = 12.76e^{j56.31°}\,A$$

The current's initial angle is now $56.31°$ meaning that the current is leading the voltage (remember that the voltage's angle is $0°$). The circuit is now said to be capacitive. However, notice that the numerical value of the power factor is the same. We, therefore, write ("cap." meaning capacitive)

$$pf_2 = cos(-56.31°) = 0.55 \text{ cap.}$$

Closing parenthesis and retaking the RLC series circuit case, we can compute the three voltages

$$V_R = RI = 10 \times 12.76e^{-j56.31°}\,V = 127.6e^{-j56.31°}\,V$$
$$V_L = jX_LI = 20e^{j90°}12.76e^{-j56.31°} = 255.16e^{j33.69°}\,V$$
$$V_C = -jX_CI = 5e^{-j90°}12.76e^{-j56.31°} = 63.79e^{-j146.31°}\,V$$

Of course, V_R is in phase with I, and V_L is advanced by $90°$, and V_C is delayed by $90°$, both with respect to the current's complex amplitude I.

If we want to recover the time evolution of the current, we will write ($56.31° = 0.98$ rad)

$$i(t) = \sqrt{2}12.76sin(100\pi t - 0.98)\,A$$

We will now show that the alternative to solve the very same problem is much harder. Let us assume that we do not want to use the concepts of complex amplitude and want to solve the RLC circuit the hard way. Looking at Fig. 20 and recalling what we have learned so far, we can write, without complex amplitudes (the unknown is the current $i(t)$)

$$v(t) = Ri(t) + L\frac{di(t)}{dt} + \frac{1}{C}\int i(t)dt \qquad (2.71)$$

We expect the current to oscillate with the same frequency, but at a different phase, so we admit the solution (trial solution) would be of the form

$$i(t) = Asin\omega t - Bcos\omega t \qquad (2.72)$$

Plugging in Eq. 2.71, we get

$$v(t) = R(A \sin \omega t - B \cos \omega t) + L\frac{d(A \sin \omega t - B \cos \omega t)}{dt}$$
$$+ \frac{1}{C}\int (A \sin \omega t - B \cos \omega t)dt \tag{2.73}$$

or:

$$\sqrt{2}V\sin\omega t = R(A \sin \omega t - B \cos \omega t) + \omega L(A \cos \omega t + B \sin \omega t)$$
$$+ \frac{1}{\omega C}(-A \cos \omega t - B \sin \omega t) \tag{2.74}$$

The sine and cosine coefficients must be equal in both terms. Therefore

$$\sqrt{2}V = AR + B\omega L - B\frac{1}{\omega C} ; 0 = -BR + A\omega L - A\frac{1}{\omega C} \tag{2.75}$$

Now, from the second equation, we write

$$B = \frac{A}{R}\left(\omega L - \frac{1}{\omega C}\right) \tag{2.76}$$

and replacing on the first, and back substituting, yields to

$$A = \frac{\sqrt{2}VR}{R^2 + \left(\omega L - \frac{1}{\omega C}\right)^2} \tag{2.77}$$

$$B = \frac{\sqrt{2}V\left(\omega L - \frac{1}{\omega C}\right)}{R^2 + \left(\omega L - \frac{1}{\omega C}\right)^2} \tag{2.78}$$

We will now be factoring out our trial solution in Eq. 2.72 as

$$i(t) = \sqrt{A^2 + B^2}\left(\frac{A}{\sqrt{A^2 + B^2}}\sin\omega t - \frac{B}{\sqrt{A^2 + B^2}}\cos\omega t\right) \tag{2.79}$$

$\frac{A}{\sqrt{A^2 + B^2}}$ and $\frac{B}{\sqrt{A^2 + B^2}}$ are the cosine and the sine of a ϕ angle defined on a right triangle with legs A and B. This angle is given as

$$\phi = tan^{-1}\left(\frac{B}{A}\right) = tan^{-1}\left(\frac{\omega L - \frac{1}{\omega C}}{R}\right) \tag{2.80}$$

and Eq. 2.79 can be written as

$$i(t) = \sqrt{A^2 + B^2}(cos\phi sin\omega t - sin\phi cos\omega t) \tag{2.81}$$

$$i(t) = \sqrt{A^2 + B^2} sin(\omega t - \phi) \tag{2.82}$$

On the other way, note that

$$\sqrt{A^2 + B^2} = \frac{\sqrt{2}V}{\sqrt{R^2 + \left(\omega L - \frac{1}{\omega C}\right)^2}} \tag{2.83}$$

which allow us to finally conclude that

$$i(t) = \frac{\sqrt{2}V}{\sqrt{R^2 + \left(\omega L - \frac{1}{\omega C}\right)^2}} sin(\omega t - \phi) \tag{2.84}$$

The reader is invited to compare this equation with Eq. 2.70. They are equal, as expected, but the calculations burden involved in the hard way is much higher than when we have used the complex amplitude approach. We hope that, at this time, the reader is convinced that the complex amplitude method is the smart way of solving AC circuits.

Now let us look at the RLC parallel circuit depicted in Fig. 21. Of course, we will use, from now on, only the complex amplitude method.

We will keep working with complex amplitudes because the computations are easier using this AC circuits solving method. Using Kirchoff Current Law (KCL), we can write about the total current that

$$I = I_R + I_L + I_C \tag{2.85}$$

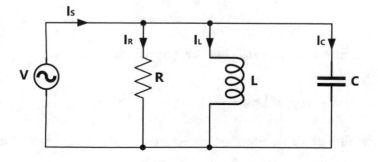

Fig. 21 RLC parallel circuit

We have learned to relate the complex amplitudes of current and voltage in resistive, inductive, and capacitive circuits. As so, Eq. 2.85 becomes (again we recall that $-j = 1/j$)

$$I = \frac{1}{R}V + \frac{1}{j\omega L}V + j\omega CV = \left(\frac{1}{R} - j\frac{1}{\omega L} + j\omega C\right)V = YV \qquad (2.86)$$

The equivalent admittance of the circuit is

$$Y = \left[\frac{1}{R} + j\left(\omega C - \frac{1}{\omega L}\right)\right] = G + jB \qquad (2.87)$$

When the elements of the circuit are parallel connected, it is more comfortable to work with admittances instead of impedances. The equivalent admittance of a set of admittances connected in parallel is the sum of all the individual admittances.

$$Y_{eq} = Y_1 + Y_2 \qquad (2.88)$$

This is equivalent to

$$\frac{1}{Z_{eq}} = \frac{1}{Z_1} + \frac{1}{Z_2}$$

$$\frac{1}{Z_{eq}} = \frac{Z_2 + Z_1}{Z_1 Z_2}$$

$$Z_{eq} = \frac{Z_1 Z_2}{Z_1 + Z_2} \qquad (2.89)$$

Admittance is a complex number whose real part is called conductance (G) and the imaginary part is the susceptance (B). In this case, the conductance and susceptance are, respectively

$$G = \frac{1}{R}; B = \omega C - \frac{1}{\omega L} \qquad (2.90)$$

The admittance modulus and phase are, respectively

$$Y = \sqrt{\frac{1}{R^2} + \left(\omega C - \frac{1}{\omega L}\right)^2}; \theta = tan^{-1}\left[R\left(\omega L - \frac{1}{\omega C}\right)\right] \qquad (2.91)$$

The admittance, conductance, and susceptance are all measured in S – Siemens.

Let us retake the same numbers we have considered before: the RMS voltage is $V = 230$ V, $f = 50$ Hz, $R = 10$ Ω; $L = 63.7$ mH, $C = 634$ μF. Under these circumstances $1/\omega L = B_L = 0.05$S, $\omega C = B_C = 0.2$ S. The admittance will be

$$Y = 0.1 + j0.15 = 0.18e^{j56.31°}\,\text{S}$$

The current is (we recall that $V = 230e^{j0}$ V)

$$I = YV = 41.46e^{j56.31°}\,\text{A}$$

In this case, the current's phase is positive (56.31°), meaning that the current leads the voltage. Therefore, the circuit is capacitive.

The power factor is (recall that the phi angle is negative)

$$pf = cos(-56.31°) = 0.55\,\text{cap.}$$

We can now compute the three currents

$$I_R = GV = 23e^{j0}\,\text{A}$$

;

$$I_L = -jB_LV = 11.50e^{-j90°}\,\text{A};$$

$$I_C = jX_CV = 45.40e^{j90°}\,\text{A}$$

Of course, I_R is in phase with V, and I_L is delayed by 90°, and I_C is advanced by 90°, both with respect to the voltage's complex amplitude V.

If we want to recover the time evolution of the current, we will write (56.31° = 0.98 rad)

$$i(t) = \sqrt{2}41.46sin(100\pi t + 0.98)\,\text{A}$$

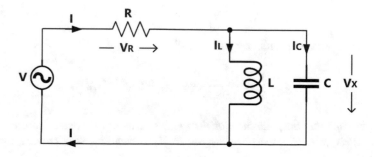

Fig. 22 Circuit of Example 2—1: LC parallel with losses circuit

Example 2—1

A sinusoidal voltage V=48 V, f=50 Hz, feeds the circuit depicted in Fig. 22. The given parameters of the circuit are R=2 Ω; L=0.05 H, C=900 μF. Compute:

1. *The equivalent complex impedance of the circuit.*
2. *Current I and voltages V_R and V_X.*
3. *Currents I_L and I_C.*

1. *Let us begin by computing the impedances of the elements, which are*

$$R = 2\,\Omega;\ X_L = \omega L = 15.71\,\Omega;\ X_C = \frac{1}{\omega C} = 3.54\,\Omega$$

Therefore, the equivalent impedance is (we recall that $j^2=-1$ and Eq. 2.89):

$$\mathbf{Z} = R + jX_{LC} = R + \frac{-jX_L jX_C}{jX_L - jX_C} = R - j\left(\frac{X_L X_C}{X_L - X_C}\right) \tag{2.92}$$

Taking the imaginary part of and replacing the expressions of the reactances, we obtain

$$\frac{X_L X_C}{X_L - X_C} = \frac{\frac{\omega L}{\omega C}}{\omega L - \frac{1}{\omega C}} = \frac{\frac{L}{C}}{\frac{\omega^2 LC - 1}{\omega C}} = \frac{\omega L}{\omega^2 LC - 1} \tag{2.93}$$

C and L could be such that the denominator is zero ($\omega^2 LC=1$). In this case, the reactance would be infinity, meaning an open circuit. This is not the case of the proposed problem.

Replacing values, we obtain

$$\mathbf{Z} = 2 - j4.56 = 4.98e^{-j66.34°}\,\Omega$$

2. *Current I can be easily computed by*

$$I = \frac{V}{Z} = \frac{48e^{j0}}{4.98e^{-j66.34°}} = 9.63e^{j66.34°}\,A$$

We assumed that the reference for the angles is, as usual, located in the voltage. The circuit is capacitive because the phi angle is negative (the current leads the voltage).

The voltage at the resistance terminals, V_R, is

$$V_R = RI = 19.26e^{j66.34°} = 7.73 + j17.64\,V$$

This voltage is in phase with the current.
As for voltage V_X, we may write, using KVL

$$V - V_R - V_X = 0$$

$$V_X = V - V_R = 40.27 - j17.64 = 43.96e^{-j23.66°} \text{ V}$$

Another way of obtaining the same result is

$$V_X = jX_{LC}I = 4.56e^{-j90°}9.63e^{j66.34°} = 43.96e^{-j23.66°} \text{ V}$$

3. *Currents I_L and I_C are respectively (we recall that $jX_L = X_Le^{j90°}$)*

$$I_L = \frac{V_X}{jX_L} = 2.80e^{-113.66°} \text{ A};$$

$$I_C = \frac{V_X}{jX_C} = 12.43e^{j66.34°} \text{ A}$$

If required, we are now able to compute the time-dependent equations of voltages and currents as:

$$v(t) = \sqrt{2}48 \sin(100\pi t + 0) \text{ V}$$
$$v_R(t) = \sqrt{2}19.26 \sin(100\pi t + 1.16) \text{ V}$$
$$v_X(t) = \sqrt{2}43.96 \sin(100\pi t - 0.41) \text{ V}$$
$$i(t) = \sqrt{2}9.63 \sin(100\pi t + 1.16) \text{ A}$$
$$i_L(t) = \sqrt{2}2.80 \sin(100\pi t - 1.98) \text{ A}$$
$$i_C(t) = \sqrt{2}12.43 \sin(100\pi t + 1.16) \text{ A}$$

Figure 23 depicts the corresponding graphics.

3   Power in AC Circuits

When a capacitor of capacitance C is applied a voltage v, energy W_e is supplied to the capacitor

$$W_e = \frac{1}{2}Cv^2 \tag{3.1}$$

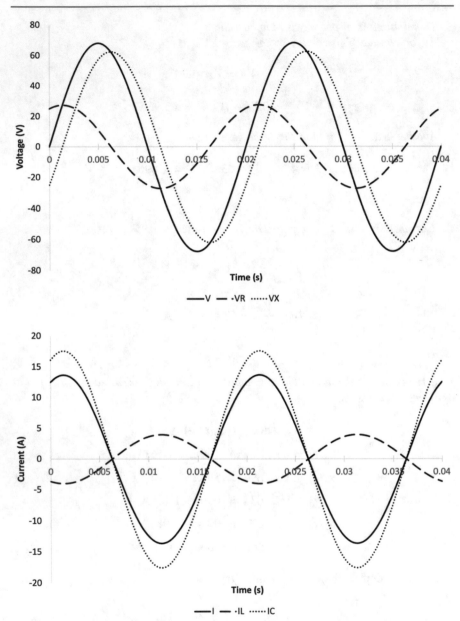

Fig. 23 Time evolution of voltages (on top) and currents (on bottom) on the circuit of Example 2—1

This energy is stored in the electrical field created by the charges in the conductive plates of the capacitor.

Moreover, when a current i flows into an inductor, a magnetic field is created in the space outside the conductor. In this magnetic field, the magnetic energy W_m is stored

$$W_m = \frac{1}{2}Li^2 \tag{3.2}$$

where L is the inductance of the inductor.

In AC circuits, the voltages and currents are periodically changing with time. This implies that the energies stored in the magnetic and electric fields are periodically changing, giving rise to energy exchanges that do not correspond to irreversible energy delivery to the loads. Loads are the usual designation for electricity consumption at the end-user level. The source is the electricity feeding system, normally the public electricity grid.

Therefore, in the AC circuits we can distinguish three types of power, in each time instant:

- The active power, which is related to the energy consumption in the load, or, better saying, to the conversion of electricity in other forms of energy that take place in the load.
- A power given by

$$P_c = \frac{dW_e}{dt} = Cv\frac{dv}{dt} \tag{3.3}$$

which corresponds to the variation of the energy stored in the electrical fields between the load and the source.

- A power given by

$$P_m = \frac{dW_m}{dt} = Li\frac{di}{dt} \tag{3.4}$$

which corresponds to the variation of the energy stored in the magnetic fields between the load and the source.

3.1   Active Power and Reactive Power

Let us suppose that we apply a sinusoidal voltage to a given circuit

$$v(t) = \sqrt{2}V sin(\omega t) \tag{3.5}$$

and that the corresponding current is

$$i(t) = \sqrt{2}I sin(\omega t - \phi) \tag{3.6}$$

The power delivered to the circuit is

$$p(t) = v(t)i(t) = 2VI sin(\omega t) sin(\omega t - \phi) \tag{3.7}$$

Applying some trigonometric rules after some manipulation, we obtain

$$p(t) = VI cos\phi (1 - cos(2\omega t)) - VI sin\phi sin(2\omega t) \tag{3.8}$$

$$p(t) = p_a(t) + p_r(t) \tag{3.9}$$

A graphic representation is given in Fig. 24 of $v(t)$, $i(t)$, and $p(t)$ (V=230 V, I=100 A, f=50 Hz, $\phi = \pi/3$).

Figure 25 depicts the graphical representation of $p(t)$, $p_a(t)$, and $p_r(t)$.

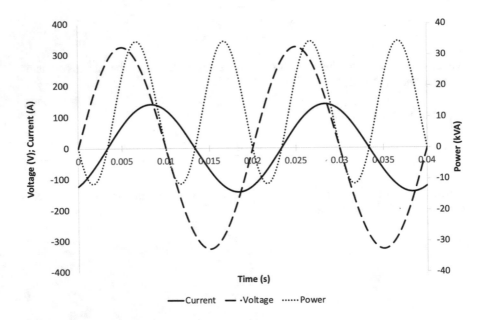

Fig. 24 Voltage, current, and power in an AC circuit

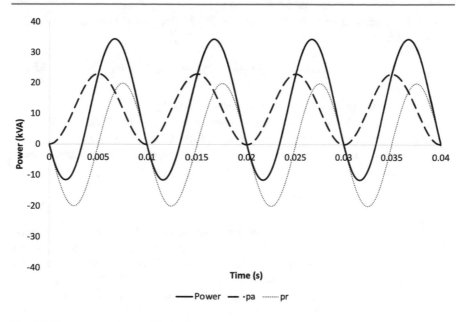

Fig. 25 Power, p_a, p_r in an AC circuit

Power $p(t)$ changes periodically in time and can be divided into two parcels as shown in Eq. 3.8. The angular speed of the first parcel, $p_a(t)$, is 2ω and the mean value is

$$P = VI cos\phi = \{p_a(t)\}_{avg} \tag{3.10}$$

We denote the mean value of $p(t)$ by P. It is called active power and is measured in W—Watt, its multiples used in power systems being kW (1 kW = 10^3 W), and MW (1 MW = 10^6 W). The active power may be also called real power. The integration in time of the active power (the active power is constant in a particular time instant, but changes over time) is the energy that is irreversibly delivered to the load circuit

$$E = \int_{t_1}^{t_2} P(t)dt \tag{3.11}$$

This energy will produce work or heat; therefore, it is called active energy.

Let us highlight the practical use of Eq. 3.11, when the active power can be considered constant over a time interval. In this case

$$E = P(t_2 - t_1) = P\Delta t \tag{3.12}$$

If P is constant over two-time intervals (P_1 in the first interval and P_2 in the second), it is

$$E = P_1(t_2 - t_1) + P_2(t_3 - t_2) = P_{avg}(t_3 - t_1) \qquad (3.13)$$

in which P_{avg} is the average power between P_1 and P_2.

This clearly shows why electricity is usually measured in MWh. 1 MWh corresponds to an average power of 1 MW used for 1 h.

The angular speed of the second parcel, $p_r(t)$, is also 2ω, but its mean value is zero. The maximum value of the second parcel is

$$Q = VI\sin\phi = \{p_r(t)\}_{MAX} \qquad (3.14)$$

Q is called the reactive power and is measured in var –Volt-Ampère reactive. The multiples used in power systems are kvar and Mvar. It corresponds to the energy that oscillates between the voltage source and the electric and magnetic fields of the load circuit. This is related to the two power parcels, P_c and P_m, that we have identified previously. The time integration of $p_r(t)$ over an integer number of cycles is zero, the reactive power is defined as the peak-value of $p_r(t)$. We call this power a reactive power in the sense that it is a non-active power.

From Eq. 3.10 and Eq. 3.14, an important relationship between active and reactive power may be derived

$$\frac{Q}{P} = \frac{VI\sin\phi}{VI\cos\phi} = tan\phi \qquad (3.15)$$

3.2   Active Power and Reactive Power in the Basic AC Circuits

In the purely resistive circuit, it is $\phi = 0$, therefore (take note that $V_R = RI_R$)

$$p(t) = p_a(t); p_r(t) = 0; P = VI = RI^2; Q = 0 \qquad (3.16)$$

The relevant graphics are shown in Fig. 26 (V=230 V, I=100 A, f=50 Hz, ϕ=0).

We highlight that the energy flows always from the voltage source to the resistive load; there are no energy oscillations.

As for the purely inductive circuit, it is ϕ=π/2, which originates (recall that $V_L = \omega L I_L$)

$$p(t) = p_r(t); p_a(t) = 0; P = 0; Q = VI = \omega L I^2 \qquad (3.17)$$

Figure 27 shows the quantities of interest regarding the purely inductive circuit (V=230 V, I=100 A, f=50 Hz, ϕ=π/2).

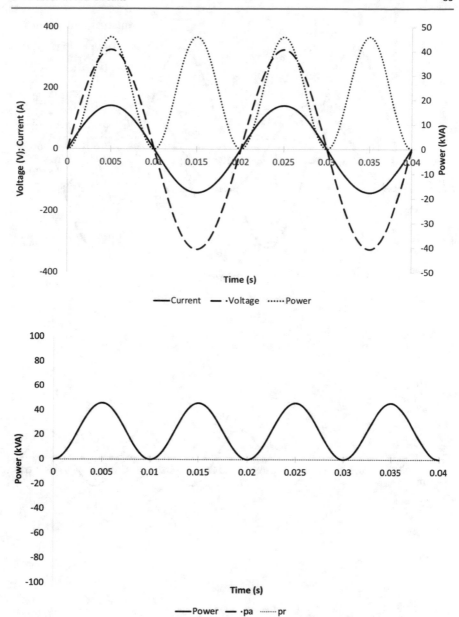

Fig. 26 Time evolution of voltage, current, and power (top) and power, p_a, p_r (bottom) in a purely resistive circuit

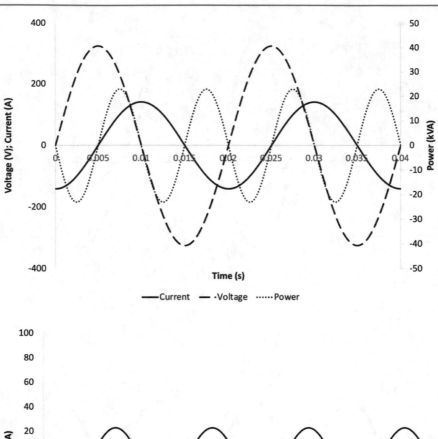

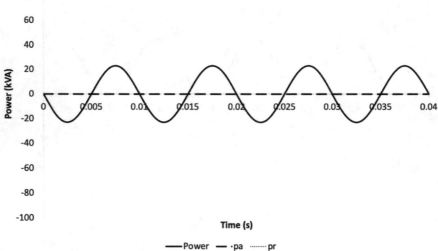

Fig. 27 Time evolution of voltage, current, and power (top) and power, p_a, p_r (bottom) in a purely inductive circuit

The coil, which composes the purely inductive circuit, sometimes supplies energy to the source, other times receives energy from it, with a null mean value. It follows that no energy is supplied or received irreversibly. Now you should be able to understand Fig. 18.

We note that the reactive power is positive. This is the usual convention: the reactive power is positive when the load exchanges magnetic energy with the source. By convention, we say that the reactive power is absorbed in an inductive circuit.

As far as the pure capacitive circuit is concerned, we recall that $\phi = -\pi/2$. Under these circumstances, we can write (do not forget that $I_C = \omega C V_C$)

$$p(t) = p_r(t); p_a(t) = 0; P = 0; Q = -VI = -\omega C V^2 \qquad (3.18)$$

The monitored quantities are displayed in Fig. 28 (V=230 V, I= 100 A, f=50 Hz, $\phi = -\pi/2$).

Sometimes the capacitor, which composes the pure capacitive circuit, behaves like a load (stores electrical energy), other times, it behaves as a generator (supplies electrical energy), with a null mean value. The conclusion is that no electrical energy is irreversibly stored in the capacitor. You should take a new look at Fig. 19.

Now the reactive power is negative. This is the usual convention when electrical energy is exchanged. In this case, by convention, we say that reactive power is supplied in a purely capacitive circuit.

In Table 1, we summarize the main aspects we have learned so far about active and reactive power in basic AC circuits.

Example 3—1

Retake the circuit of Example 2—1 depicted in Fig. 22 and compute:

1. *The active power, the reactive power, and the power factor of the circuit.*
2. *The reactive power "absorbed" by the coil and the reactive power "supplied" by the capacitor.*

Recall that V=48 V, f=50 Hz, R=2 Ω; L=0.05 H, C=900 μF.
We recall that in Example 2—1, we computed

$$I = \frac{V}{Z} = \frac{48e^{j0}}{4.98e^{-j66.34°}} = 9.63e^{j66.34°} \text{ A}$$

$$V_X = jX_{LC}I = 4.56e^{-jS90°} 9.63e^{j66.34°} = 43.96e^{-j23.66°} \text{ V}$$

$$I_L = \frac{V_X}{jX_L} = 2.80e^{-j113.66°} \text{ A}$$

1.

$$P = VI\cos\phi = 48 \times 9.63\cos(-66.34°) = 185.55 \text{ W}$$

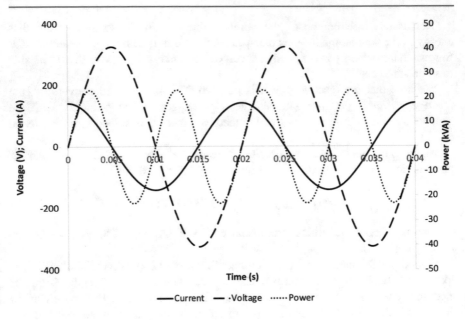

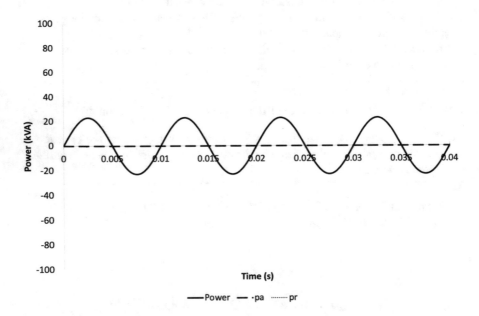

Fig. 28 Time evolution of voltage, current, and power (top) and power, p_a, p_r (bottom) in a purely capacitive circuit

Table 1 Phi angle, active power, and reactive power for several load types

Load type	Phi angle	Active power	Reactive power
R	$\phi = 0$	$P = VI > 0$	$Q = 0$
L	$\phi = \pi/2$	$P = 0$	$Q = VI > 0$
C	$\phi = -\pi/2$	$P = 0$	$Q = -VI < 0$
RL series	$0 < \phi < \pi/2$	$P = VI\cos\phi > 0$	$Q = VI\sin\phi > 0$
RL parallel	$0 < \phi < \pi/2$	$P = VI\cos\phi > 0$	$Q = VI\sin\phi > 0$
RC series	$-\pi/2 < \phi < 0$	$P = VI\cos\phi > 0$	$Q = VI\sin\phi < 0$
RC parallel	$-\pi/2 < \phi < 0$	$P = VI\cos\phi > 0$	$Q = VI\sin\phi < 0$

$$Q = VI\sin\phi = 48 \times 9.63\sin(-66.34°) = -423.46\,\text{var}$$

$$pf = \cos(-66.34°) = 0.40\,\text{cap.}$$

2.

$$Q_L = \omega L I_L^2 = 123.05\,\text{var}$$

$$Q_C = -\omega C V_X^2 = -546.52\,\text{var}$$

$$Q = Q_L + Q_C = 423.46\,\text{var.}$$

3.3 Complex Power

Let us consider a voltage and a current, whose complex amplitudes are given by

$$V = Ve^{j\alpha}; I = Ie^{j\beta} \tag{3.19}$$

We define the quantity "complex power" as

$$S = VI^* \tag{3.20}$$

where I^* is the complex conjugate of I

$$I^* = Ie^{-j\beta} \tag{3.21}$$

Developing Eq. 3.20, we obtain (we recall that $\alpha - \beta = \phi$)

$$S = VI^* = Ve^{j\alpha}Ie^{-j\beta} = VIe^{j(\alpha-\beta)} = VIe^{j\phi} \tag{3.22}$$

Recurring to Euler's formula leads to

$$S = VIe^{j\phi} = VIcos\phi + jVIsin\phi = P + jQ \tag{3.23}$$

The complex power is a complex number whose real part is the active power, and the imaginary part is the reactive power.

The modulus (absolute value) of the complex power is the apparent power

$$S = |S| = VI = \sqrt{P^2 + Q^2} \tag{3.24}$$

The apparent power is measured in VA—Volt-Ampère, or kVA or MVA, for power systems applications.

3.4 Power Factor Correction

In a transmission line, the transmission losses (Joule losses) are given by (R_L is the line resistance)

$$P_L = R_L I^2 \tag{3.25}$$

Taking Eq. 3.24 into account, we can write

$$P_L = R_L I^2 = R\frac{P^2 + Q^2}{V^2} \tag{3.26}$$

For instance, if R_L is the resistance of a power line that transmits power P, at voltage V, the transmission losses are strongly dependent on reactive power flow Q. Minimum losses are achieved if $Q=0$, meaning no reactive power flow in the line. As so, the reactive power needs should be locally generated, therefore, avoiding reactive power flowing in the lines, at all, or at least, minimizing reactive power flow to achieve reduced losses.

Consider the circuit represented in Fig. 29, but disregard the capacitor C, for the time being. The line resistance is R_L. The circuit is inductive (RL load), therefore, the current is lagging the voltage, which we admit, as usual, to hold the zero-reference angle. The phase shift is ϕ_1 and the RMS current is I_1, the circuit is consuming reactive power. The active power, reactive power, and transmission losses are given, respectively, by

$$P_1 = VI_1cos\phi_1; Q_1 = VI_1sin\phi_1; P_{L1} = R_L I_1^2 \tag{3.27}$$

Now consider that we install a capacitor C and we connect it in parallel with the RL load. As the capacitor is submitted to voltage V, a current I_C flows in the

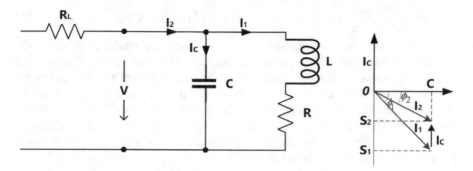

Fig. 29 Power factor correction

capacitor branch. Applying KCL, we obtain (note that these are the complex amplitudes of the currents)

$$I_2 = I_C + I_1 \tag{3.28}$$

Current I_C has an angle equal to 90° because a capacitive current leads the voltage by 90°. The new line current is I_2 which has an angle ϕ_2. The phasor diagram is shown in Fig. 29. Note that $I_2 < I_1$ and $\phi_2 < \phi_1$.

From the phasor diagram of Fig. 29, it is apparent that

$$I_2 cos\phi_2 = I_1 cos\phi_1 = \overline{OC} \tag{3.29}$$

$$I_2 sin\phi_2 = \overline{OS_2} < I_1 sin\phi_1 = \overline{OS_1} \tag{3.30}$$

$$I_2^2 < I_1^2 \tag{3.31}$$

The new active power, reactive power, and transmission losses are now given, respectively, by

$$P_2 = VI_2 cos\phi_2 = P_1 = VI_1 cos\phi_1 \tag{3.32}$$

$$Q_2 = VI_2 sin\phi_2 < Q_1 = VI_1 sin\phi_1 \tag{3.33}$$

$$P_{L2} = R_L I_2^2 < P_{L1} = R_L I_1^2 \tag{3.34}$$

The reduction of the line current (from I_1 to I_2) while keeping the transmitted active power unchanged ($P_1 = P_2$) is called the power factor correction. A full corrected power factor is a power factor equal to 1. In this case, we have not performed full power factor correction, actually, we have improved it (make it closer to 1)

$$\phi_2 < \phi_1; pf_2 = cos\phi_2 > pf_1 = cos\phi_1 \qquad (3.35)$$

We note that the transmitted active power remains unchanged, while the transmission losses decreased due to the decrease in the line current. The reactive power flow across the line also decreased because the capacitor partly supplied the reactive power consumed by the coil. In the event the capacitor supplies all the reactive power absorbed by the coil, the reactive power demanded to the source would be zero, the power factor would be 1, and the losses would be minimal.

To keep the losses at the minimum possible, the grid operators want to avoid reactive power flow in their networks. Therefore, they heavily charge the consumers with a poor power factor. That is why the consumers install capacitors to locally supply the reactive power needs. This holds true mainly for industries, factories, medium/big consumers connected in MV or HV.

Example 3—2
A device fed by an RMS voltage $V = 230$ V, $f = 50$ Hz, absorbs $I_1 = 16$ A with a power factor equal to $pf_1 = 0.6$ ind. Compute the capacitance of a capacitor to be connected in parallel with the device so that the power factor is corrected to $pf_2 = 0.9$ ind.

Please refer to the phasor diagram of Fig. 29 and recall some trigonometric relationships.

$$I_2 cos\phi_2 = I_1 cos\phi_1$$

$$I_2 = I_1 \frac{cos\phi_2}{cos\phi_1} = I_1 \frac{pf_2}{pf_1} = 10.67 \text{ A}$$

$$\phi_1 = cos^{-1}(0.6) = 53.13°$$

$$\phi_2 = cos^{-1}(0.9) = 25.84°$$

$$I_C = I_1 sin\phi_1 - I_2 sin\phi_2 = 8.15 \text{ A}$$

$$C = \frac{I_C}{\omega V} = 112.8 \,\mu F$$

The same result could be obtained if we deal with the reactive power concept.

$$Q_1 = VI_1 sin\phi_1 = 230 \times 16 sin(53.13°) = 2944.0 \text{ var}$$

$$Q_2 = VI_2 sin\phi_2 = 230 \times 10.67 sin(25.84°) = 1069.38 \text{ var}$$

$$Q_C = Q_2 - Q_1 = -1874.62 \text{ var}$$

$$C = \frac{|Q_C|}{\omega V^2} = 112.8 \,\mu F$$

4 Balanced Three-Phase AC Circuits

4.1 Three-Phase Systems

Consider three identical coils, a_1a_2, b_1b_2, and c_1c_2, as shown in Fig. 30. a_1, b_1, and c_1 are the starting terminals, whereas a_2, b_2, and c_2 are the end terminals of the three coils. The spatial phase difference of 120° ($2\pi/3$ rad) has to be maintained between the starting terminals a_1, b_1, and c_1.

Now, let the three coils, mounted on the same axis, rotate at ω rad/s, inside the uniform magnetic field created by the two poles NS. Previously, in this chapter, we have seen that, for one rotating coil, a sinusoidal EMF is induced in the coil. Similarly, three sinusoidal 120° phase-shifted EMFs are now induced in the coils, whose equations are given by

$$e_1(t) = \sqrt{2}E\sin(\omega t) \tag{4.1}$$

$$e_2(t) = \sqrt{2}E\sin(\omega t - \frac{2\pi}{3}) \tag{4.2}$$

$$e_3(t) = \sqrt{2}E\sin(\omega t - \frac{4\pi}{3}) \tag{4.3}$$

if the time origin for $e_1(t)$ is conveniently chosen so that at $t=0$, $e_1(t)=0$. The time evolution of the EMFs is shown in Fig. 31.

The EMF is the no-load voltage. If loads are connected, the load voltages will slightly differ from the EMFs, due to circuit losses, and will be given by

$$V_1 = Ve^{j0}; V_2 = Ve^{-j120°}; V_3 = Ve^{-j240°} \tag{4.4}$$

Fig. 30 Three coils in a three-phase system

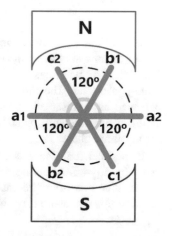

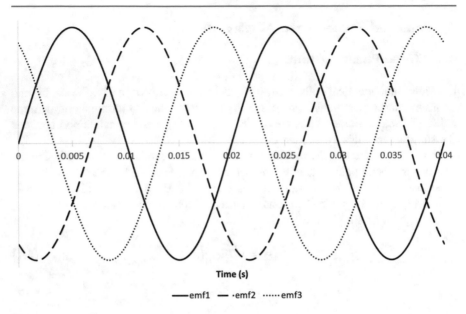

Time (s)

——emf1 — ·emf2 ······emf3

Fig. 31 EMF in a three-phase system

If each coil is connected to an impedance $\mathbf{Z} = Ze^{j\phi}$, the current (complex amplitude) in each coil will be

$$I_1 = \frac{V_1}{Z} = \frac{V}{Z}e^{-j\phi} = Ie^{-j\phi} \tag{4.5}$$

$$I_2 = \frac{V_2}{Z} = \frac{V}{Z}e^{j(-120-\phi)} = Ie^{-j(120+\phi)^\circ} \tag{4.6}$$

$$I_3 = \frac{V_3}{Z} = \frac{V}{Z}e^{j(-240-\phi)} = Ie^{-j(240+\phi)^\circ} \tag{4.7}$$

We say that I_1 is the complex amplitude of the current in phase 1.

Let us now suppose a three-phase system composed of a three-phase generator (the coils, or windings, are the three phases of the generator) connected to a three-phase load. The three loads are equal; therefore, we say that the system is balanced as represented in Fig. 32. For the time being, 4 wires are used, the 4th acting as a return path, called the neutral wire "N". This return path can be common for the three phases.

If the load impedance is equal in the three phases ($\mathbf{Z} = Ze^{j\phi}$), the currents are given by Eq. 4.5. Applying KCL, the current flowing in the neutral wire is the sum of the three phasors representing the complex amplitudes of the currents. By performing this calculation, we will obtain zero.

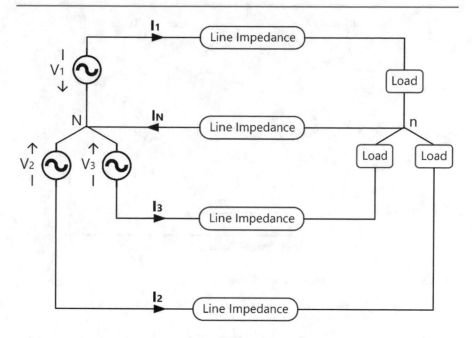

Fig. 32 Balanced three-phase system with 4 wires

$$I_N = I_1 + I_2 + I_3 = 0 \qquad (4.8)$$

Reporting to Fig. 32, the positive directions for the voltages and currents are indicated.

Note that Eq. 4.8 holds true if the three-phase system is symmetrical and balanced. Symmetrical means that the three currents (and voltages) are 120° phase shifted; balanced means that the three loads are equal and so the currents hold the same RMS value and are 120° shifted. As a consequence, the neutral wire can be suppressed, and nothing changes in the electric circuit. We say that we have a 3-wire three-phase system as in Fig. 33.

In the transmission network, the system is symmetrical and balanced, so 3 conductors are used.

Let us now suppose that we have three different loads (the system is unbalanced).

$$Z_1 = Z_1 e^{j\phi_1}; Z_2 = Z_2 e^{j\phi_2}; Z_3 = Z_3 e^{j\phi_3} \qquad (4.9)$$

The three currents would be now (the system is unsymmetrical)

$$I_1 = \frac{V_1}{Z_1} = I_1 e^{-j\phi_1} \qquad (4.10)$$

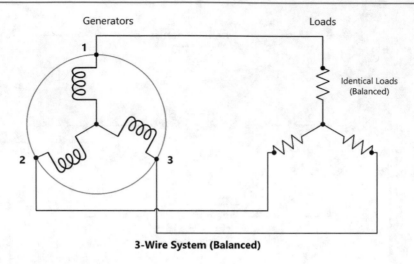

Fig. 33 Balanced three-phase system with 3 wires

$$I_2 = \frac{V_2}{Z_2} = I_2 e^{-j(120° + \phi_2)} \tag{4.11}$$

$$I_3 = \frac{V_3}{Z_3} = I_3 e^{-j(240° + \phi_3)} \tag{4.12}$$

The neutral current is now different from zero and the 4th wire cannot be suppressed, as shown in Fig. 34.

$$I_N = I_1 + I_2 + I_3 \neq 0 \tag{4.13}$$

In the distribution network, the system is unbalanced, and so 4 wires are used.

Example 4—1
A three-phase symmetrical, balanced, 4-wire system, with V=230 V, f=50 Hz, feeds a three-phase balanced load $Z_1 = Z_2 = Z_3 = 14.14 + j14.14$ Ω. Compute the neutral current.

The three voltages are:

$$V_1 = 230e^{j0} \text{ V}; V_2 = 230e^{-j120°} \text{ V}; V_3 = 230e^{-j240°} \text{ V}$$

Each impedance is

$$Z = 14.14 + j14.14 = \sqrt{14.14^2 + 14.14^2}\, e^{jtan^{-1}\left(\frac{14.14}{14.14}\right)} = 20e^{j45°} \text{ Ω}$$

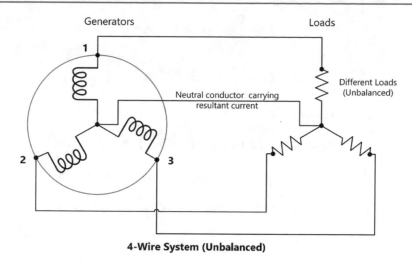

4-Wire System (Unbalanced)

Fig. 34 Unbalanced three-phase system with 4 wires

The three currents are

$$I_1 = \frac{V_1}{Z_1} = 11.5e^{-j45°} \text{ A}; I_2 = \frac{V_2}{Z_2} = 11.5e^{-j165°} \text{ A}; I_3 = \frac{V_3}{Z_3} = 11.5e^{-j285°} \text{ A}$$

or in the rectangular form

$$I_1 = 11.5cos(-45°) + j11.5sin(-45°) = 8.13 - j8.13 \text{ A}$$

$$I_2 = 11.5cos(-165°) + j11.5sin(-165°) = -11.11 - j2.98 \text{ A}$$

$$I_3 = 11.5cos(-285°) + j11.5sin(-285°) = 2.98 + j11.11 \text{ A}$$

The neutral current is

$$I_N = I_1 + I_2 + I_3 = 0$$

Example 4—2
Retake Example 4—1 but assume now that the impedance is unbalanced. The modulus of the impedance remains the same, while the angles change Z_1=20 Ω; Z_2=j20 Ω; Z_3= − j20 Ω. *Compute the neutral current.*

The three currents are

$$I_1 = \frac{V_1}{Z_1} = 11.5e^{-j0°} \text{ A}$$

$$I_2 = \frac{V_2}{Z_2} = 11.5e^{-j210°} \text{ A}$$

$$I_3 = \frac{V_3}{Z_3} = 11.5e^{-j150°} \text{ A}$$

or in the rectangular form

$$I_1 = 11.5cos(0°) + j11.5sin(0°) = 11.5 + j0 \text{ A}$$

$$I_2 = 11.5cos(-210°) + j11.5sin(-210°) = -9.96 + j5.75 \text{ A}$$

$$I_3 = 11.5cos(-150°) + j11.5sin(-150°) = -9.96 - j5.75 \text{ A}$$

The neutral current is

$$I_N = I_1 + I_2 + I_3 = -8.42 + j0 = 8.42e^{j180°} \text{ A}.$$

4.2   Phase-To-Neutral Voltage and Phase-To-Phase Voltage

Let us assume that the loads, instead of being connected as in Fig. 33, we now connect them as in Fig. 35. We say that the loads are connected in delta or in a triangle. In the previous case, we say that the loads are connected in star.

In this case (delta connection), the loads are connected between two phases.

Let us consider that the three symmetrical, balanced voltages are

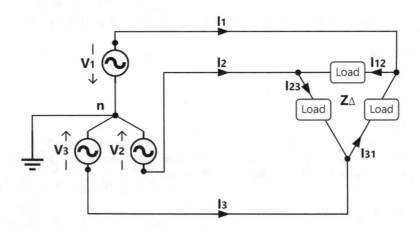

Fig. 35 Three-phase system with loads delta connection

$$V_1 = Ve^{j0}; V_2 = Ve^{-j120°}; V_3 = Ve^{-j240°} \tag{4.14}$$

Applying KVL, we obtain

$$V_1 - V_2 = V_{12} = Z_\Delta I_{12} \tag{4.15}$$

$$V_2 - V_3 = V_{23} = Z_\Delta I_{23} \tag{4.16}$$

$$V_3 - V_1 = V_{31} = Z_\Delta I_{31} \tag{4.17}$$

For instance, let us take as an example, voltages V_1 and V_2. Voltage V_1 is taken between a phase and the neutral point; as such, it is called phase-to-neutral voltage, or phase voltage. Voltage V_{12} is the difference between voltage V_1 and V_2 and is taken between two phases; as so, it is called phase-to-phase voltage or line voltage.

Let us keep going on with phase voltages V_1 and V_2. For these two voltages, we can write

$$V_1 = V\cos(0°) + jV\sin(0°) = V \tag{4.18}$$

$$V_2 = V\cos(-120°) + jV\sin(-120°) = -\frac{1}{2}V - j\frac{\sqrt{3}}{2}V \tag{4.19}$$

$$V_{12} = V_1 - V_2 = \frac{3}{2}V + j\frac{\sqrt{3}}{2}V \tag{4.20}$$

$$V_{12} = \sqrt{\left(\frac{3}{2}V\right)^2 + \left(\frac{\sqrt{3}}{2}V\right)^2}\, e^{j\tan^{-1}\left(\frac{\frac{\sqrt{3}}{2}V}{\frac{3}{2}V}\right)} = \sqrt{\frac{12}{4}V^2}e^{j\tan^{-1}\left(\frac{\sqrt{3}}{3}\right)} = \sqrt{3}Ve^{j30°} \tag{4.21}$$

If we make the same computations for the other line voltages, we will find that it is possible to always write

$$V_{p-p} = \sqrt{3}V_{p-n} \tag{4.22}$$

meaning that the RMS phase-to-phase voltage (V_{p-p}) is the square root of three times higher than the RMS phase-to-neutral voltage (V_{p-n}).

Here are all the three-line voltages:

$$V_1 - V_2 = V_{12} = \sqrt{3}Ve^{j30°} \tag{4.23}$$

$$V_2 - V_3 = V_{23} = \sqrt{3}Ve^{-j90°} \tag{4.24}$$

$$V_3 - V_1 = V_{31} = \sqrt{3}Ve^{j150°} \tag{4.25}$$

We highlight that the line voltages are also 120° phase shifted.

In Low Voltage, the phase-to-neutral RMS voltage is 230 V and the RMS phase-to-phase voltage is 400 V.

4.3   Star Connection and Delta Connection

From what we have learned, we conclude that if the loads are star connected, as in Fig. 34, for instance, the loads are subjected to the phase-to-neutral voltages. Also, the current that flows in the line (connecting one phase of the generator to the respective load) is equal to the current across the load. We say that, in this case, the line current is equal to the load (or phase) current (see current I_1 in Fig. 36).

As far as the delta-connected loads are concerned, the situation is different. At each load, the phase-to-phase voltage is applied, as we have seen. Regarding Fig. 36 and applying KCL, we obtain

$$I_1 + I_{31} = I_{12}; I_1 = I_{12} - I_{31} \tag{4.26}$$

In a similar way to what we have seen in the case of the relationship between phase-to-phase and phase-to-neutral voltages (Eq. 4.21), here the same relationship applies: the RMS line current (for instance I_1) is square root of 3 times higher than the RMS delta load (or phase) current

$$I_{line\Delta} = \sqrt{3} I_{phase\Delta} \tag{4.27}$$

This holds true for a balanced, symmetrical, and three-phase system with delta-connected loads.

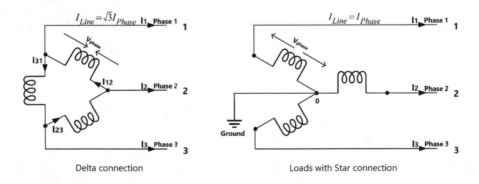

Delta connection   Loads with Star connection

Fig. 36 Loads with star connection (right) and delta connection (left)

4.4 Three-Phase Power in Balanced and Symmetrical Systems

Let us recall the unbalanced system, star connected, depicted in Fig. 34. The three-phase active power delivered to the loads is

$$P_{3ph} = V_1 I_1 cos\phi_1 + V_2 I_2 cos\phi_2 + V_3 I_3 cos\phi_3 \tag{4.28}$$

where V_1, V_2, V_3 are the RMS phase-to-neutral voltages, I_1, I_2, I_3 are the RMS line currents and ϕ_1, ϕ_2, ϕ_3 are the phase shifts between the phase-to-neutral voltage and the corresponding current.

If the system is symmetrical and balanced, the three RMS voltages are equal (V_{p-n}), the three RMS line currents are equal (I_{line}) and the three-phase shifts are equal (ϕ_Y). Therefore, for a star-connected system, we will write, recalling the relationship between phase-to-phase and phase-to-neutral voltages

$$P_{3phY} = 3V_{p-n} I_{line} cos\phi_Y = \sqrt{3} V_{p-p} I_{line} cos\phi_Y \tag{4.29}$$

Now, let us recall the balanced, delta connected, system displayed in Fig. 35. The three-phase active power delivered to the loads is given by

$$P_{3ph\Delta} = 3V_{p-p} I_{phase} cos\phi_\Delta \tag{4.30}$$

where V_{p-p} is the RMS phase-to-phase voltage, I_{phase} is the RMS phase (load) current and ϕ_Δ is the phase shift between the phase-to-phase voltage and the phase current.

We recall that

$$\phi_\Delta = \phi_Y = \phi \tag{4.31}$$

that is, the angle between the phase-to-neutral voltage and the line current is equal to the angle between the phase-to-phase voltage and the phase current.

Recalling Eq. 4.27 $(I_{line\Delta} = \sqrt{3} I_{phase\Delta})$ and that $3/\sqrt{3} = \sqrt{3}$, we conclude that whether the loads are star or delta connected, the three-phase active power delivered to the loads is always

$$P_{3ph} = \sqrt{3} V_{p-p} I_{line} cos\phi \tag{4.32}$$

The same logic can be applied to derive the three-phase apparent power and the three-phase reactive power

$$S_{3ph} = \sqrt{3} V_{p-p} I_{line}$$

$$Q_{3ph} = \sqrt{3}V_{p-p}I_{line}\sin\phi \tag{4.33}$$

Example 4—3

Suppose that the three balanced loads of Example 4—1, instead of being star connected, are now delta connected. Compute the phase currents, the line currents, and the three-phase active power delivered to the loads. Compare with the star-connected loads' case.

Let us begin by recalling the voltages and currents in the star-connected case (Example 4—1)

$$V_1 = 230e^{j0}\ \text{V}; V_2 = 230e^{-j120°}\ \text{V}; V_3 = 230e^{-j240°}\ \text{V}$$

$$I_{1Y} = 11.5e^{-j45°}\ \text{A}; I_{2Y} = 11.5e^{-j165°}\ \text{A}; I_{3Y} = 11.5e^{-j285°}\ \text{A}$$

And now the line voltages

$$V_{12} = \sqrt{3}230e^{j30°}\ \text{V}; V_{23} = \sqrt{3}230e^{-j90°}\ \text{V}; V_{31} = \sqrt{3}230e^{j150°}\ \text{V}$$

The phase currents for the delta-connected case are

$$I_{12} = \frac{V_{12}}{Z} = 19.92e^{-j15°}\ \text{A}$$

$$I_{23} = \frac{V_{23}}{Z} = 19.92e^{-j135°}\ \text{A}$$

$$I_{31} = \frac{V_{31}}{Z} = 19.92e^{j105°}\ \text{A}$$

The RMS value of the current in the loads is the square root of three times higher if they are delta connected than if they are star connected (compare, for instance, I_{12} with I_{1Y}).

The line currents would be

$$I_{1\Delta} = I_{12} - I_{31} = 34.50e^{-j45°}\ \text{A}$$

$$I_{2\Delta} = I_{23} - I_{12} = 34.50e^{-j165°}\ \text{A}$$

$$I_{3\Delta} = I_{31} - I_{23} = 34.50e^{j75°}\ \text{A}$$

The RMS value of the line current is three times higher if they are delta connected than if they are star connected (compare, for instance, $I_{1\Delta}$ with I_{1Y}).

We take the opportunity to prove Eq. 4.31. In fact

$$\phi_Y = \text{angle}(V_1) - \text{angle}(I_1) = 0 + 45 = 45° = \phi$$

$$\phi_\Delta = \text{angle}(V_{12}) - \text{angle}(I_{12}) = 30 + 15 = 45° = \phi$$

In what regards to the active power delivered to the loads, it is

$$P_Y = 3V_1 I_{1Y}\cos\phi = \sqrt{3}V_{12}I_{1Y}\cos\phi = 5610.9\,\text{W}$$

$$P_\Delta = 3V_{12}I_{12}\cos\phi = \sqrt{3}V_{12}I_{1\Delta}\cos\phi = 16,833\,\text{W}$$

We conclude that the active power delivered to the loads is three times higher if they are delta connected than if they are star connected. Of course, this holds true if the three loads are balanced and equal in both connections.

5   Per Unit (PU) System

In electrical systems, the four fundamental quantities are voltage, current, impedance, and power. In per unit (pu) notation, the physical quantity is expressed as a fraction of a reference (base) value. Let us take a general quantity X expressed in its own SI units. The pu value is given by

$$X_{pu} = \frac{X}{X_{base}} \tag{5.1}$$

The base value is a reference value for the magnitude of the quantity. As a consequence, it is a real value. The quantity X is a quantity expressed in SI units: it can be a magnitude, a phasor, or a complex number, or even an instantaneous value.

Given the basic relationships between the four fundamental quantities, it is only necessary to specify the base values for two of the four quantities; the other two will follow directly. It is a common practice to specify the base power and the base voltage. In general, these two base values can be arbitrarily specified, but it is also a common practice to specify the base values as the nominal (nameplate) values of the quantities.

For a single-phase system, the nominal values are the single-phase power and the phase-to-neutral (phase) voltage; for a three-phase system, the nominal values are the three-phase power and the phase-to-phase (line) voltage.

The known relationships between the fundamental quantities also apply in the pu system

$$\text{single} - \text{phase} \begin{cases} S_b = V_b I_b \\ V_b = Z_b I_b \end{cases} \tag{5.2}$$

$$\text{three} - \text{phase} \begin{cases} S_b = \sqrt{3}V_bI_b \\ \frac{V_b}{\sqrt{3}} = Z_bI_b \end{cases} \tag{5.3}$$

We highlight that for the single-phase system, the base power is the single-phase power, and the base voltage is the phase voltage, whereas for the three-phase system, the base power is the three-phase power, and the base voltage is the line voltage.

As so, for a single-phase system, we specify the base power (S_b) and the base voltage (V_b). The corresponding base current (I_b) and base impedance (Z_b) are

$$I_b = \frac{S_b}{V_b} ; Z_b = \frac{V_b}{I_b} = \frac{V_b^2}{S_b} \tag{5.4}$$

For a three-phase system, the base three-phase power (S_b) and the baseline voltage (V_b) should be specified. The corresponding base current (I_b) and base impedance (Z_b) are

$$I_b = \frac{S_b}{\sqrt{3}V_b} \tag{5.5}$$

$$Z_b = \frac{\frac{V_b}{\sqrt{3}}}{I_b} = \frac{V_b^2}{S_b} \tag{5.6}$$

A word is worth clarifying the base impedance. If the system is three-phase and the loads are star connected (this is the usual situation), the phase voltage is applied to each load, and the single-phase power is delivered to each one. Defining the base voltage as the line voltage and the base power as the three-phase power, as usually in a three-phase system, yields

$$Z_b = \frac{V_b^2}{S_b} = \frac{\left(\frac{V_b}{\sqrt{3}}\right)^2}{\left(\frac{S_b}{3}\right)} = \frac{V_b^2}{S_b} \tag{5.7}$$

If the system is three-phase and the loads are delta connected, the line voltage is applied to each load, and the single-phase power is delivered. Again, defining the base voltage as the line voltage and the base power as the three-phase power leads to

$$Z_b = \frac{V_b^2}{S_b} = \frac{V_b^2}{\left(\frac{S_b}{3}\right)} = 3\frac{V_b^2}{S_b} \tag{5.8}$$

One important conclusion is that the usual $\sqrt{3}$ factor, that appears in the three-phase system in SI units, disappears in the pu system. Let us see why

$$\text{single} - \text{phase} \left\{ S_{pu} = \frac{(VI)_{SI}}{V_b I_b} = V_{pu} I_{pu} \right. \tag{5.9}$$

$$\text{three} - \text{phase} \left\{ S_{pu} = \frac{(\sqrt{3}VI)_{SI}}{\sqrt{3}V_b I_b} = V_{pu} I_{pu} \right. \tag{5.10}$$

This means that the power is computed using the same equation, whether it is a single-phase or a three-phase system.

For a connected system, it is obvious that the same bases should be used for the whole network, such that the normal circuit theorems and equations also apply to per unit values. As such, a single base power and a single base voltage (usually the nominal values) should be specified for a connected circuit.

Example 5—1

Solve again Example 4—3 using the pu system. As the system is balanced, solve for phase 1 only.

Let us begin by defining the base (three-phase) power and base (line) voltage

$$S_b = 10\,\text{kVA}; V_b = \sqrt{3}230\,\text{V}$$

Then, we compute, the base current and the base impedances

$$I_b = \frac{S_b}{\sqrt{3}V_b} = 14.49\,\text{A}$$

$$Z_{bY} = \frac{V_b^2}{S_b} = 15.87\,\Omega$$

$$Z_{b\Delta} = 3\frac{V_b^2}{S_b} = 47.61\,\Omega$$

The voltage and the impedances in pu are

$$V = \frac{\sqrt{3}230}{\sqrt{3}230}e^{j0} = 1\,\text{pu}$$

$$Z_Y = \frac{20}{15.87}e^{j45°} = 1.26e^{j45°}\,\text{pu}$$

$$Z_\Delta = \frac{20}{47.61}e^{j45°} = 0.42e^{j45°}\,\text{pu}$$

Note that pu conversion affects the magnitudes, not the angles.

Consequently, the line currents are

$$I_Y = \frac{V}{Z_Y} = 0.79e^{-j45°} \, \text{pu} \Rightarrow I_Y = 0.79 \times 14.43e^{-j45°} = 11.5e^{-j45°} \, \text{A}$$

$$I_\Delta = \frac{V}{Z_\Delta} = 2.38e^{-j45°} \, \text{pu} \Rightarrow I_\Delta = 2.38 \times 14.43e^{-j45°} = 34.5e^{-j45°} \, \text{A}$$

The active power delivered to the loads is (compare with Example 4—3 results)

$$P_Y = VI_Y cos\phi = 0.57 \, \text{pu} \Rightarrow P_Y = 0.56 \times 10 \, \text{kW} = 5.6 \, \text{kW}$$

$$P_\Delta = VI_\Delta cos\phi = 1.7 \, \text{pu} \Rightarrow P_\Delta = 1.68 \times 10 \, \text{kW} = 16.8 \, \text{kW}$$

In a transformer, two circuits are not directly connected but magnetically coupled. The voltages of the windings are in the ratio of turns and currents in inverse ratio.[4] For the coupled circuit, we should then choose the same base power and base voltages in the ratio of turns. Therefore, in a circuit with transformers, all the nominal voltage levels are base voltages, and a single base power should be chosen. For instance, if we have a 400 V/15 kV transformer, both 400 V and 15 kV are base voltages.

The main advantages of the pu system may be summarized as follows:

- Normally we are dealing with numbers near unity rather than over a wide range.
- Provides a more meaningful comparison of parameters of machines with different ratings.
- As the pu values of parameters of a rotating machine or a transformer normally fall within a certain range, a typical value can be used if such parameters are not provided.
- Calculations involving transformers are much easier because there is no need to refer circuit quantities to one or the other side of the transformer.
- The pu values clearly represent the relative values of the circuit quantities. Many of the ubiquitous constants are eliminated.

6   AC Circuit Application: The Induction Motor

The induction motor is perhaps the most common type of electric motor in the world because it is simple, robust, reliable, and cheap. An electric motor converts electrical energy, that it receives from the grid, into mechanical energy (energy associated with the rotation of a shaft, for instance), that it delivers to a load.

---

[4]In fact, the voltages of the windings are almost in the ratio of turns and currents in inverse ratio, because the transformers are not ideal transformers.

An induction motor has 2 main parts: the stator and rotor. The stator is the stationary part (Fig. 37) and the rotor, which sits inside the stator, is the rotating part (Fig. 38). There is also a small gap between the rotor and stator, known as airgap. The value of the radial airgap may vary from 0.5 to 2 mm.

6.1   Basic Operating Principle

Suppose that a magnetic pole is moving at an angular speed ω_s in relation to a closed winding. In these conditions, a current i will flow in the winding. The interaction between the current and the magnetic field created by the rotating pole causes a force. This force tries to keep the relative position of the winding in

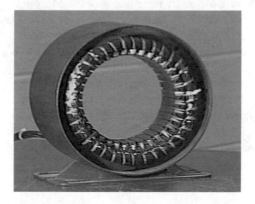

Fig. 37 Stator of an induction motor

Fig. 38 Short-circuited (squirrel cage) rotor of an induction motor

relation to the changing magnetic field. Consequently, the winding will rotate at an angular speed $\omega_r<\omega_s$, because there are forces that oppose the movement, friction forces, for instance. You can think of the rotor frantically trying to "catch up" with the rotating magnetic field in an effort to eliminate the difference in motion between them.

If $\omega_r=\omega_s$, the winding would not "see" any magnetic field variation, as so, no induced EMF, nor induced current would be produced, and no force would drag the winding. As in practice, $\omega_r<\omega_s$, it follows that there is a slip between the rotating winding (which rotates at ω_r) and the rotating pole (that rotates at synchronous speed, ω_s). This is why the induction motor is also called an asynchronous motor.

This is the basic operating principle of an induction motor. In a real induction motor, the poles do not actually rotate. What rotates is a magnetic field distribution, which is identical to the one that would be obtained if the poles were rotating. This is the so-called Rotating Magnetic Field and is obtained from the grid AC balanced and symmetrical power supply to the three stator coils 120° spatially phase shifted. Actually, there is a ring of grid-connected coils arranged around the outside (making up the stator), which are specially designed to produce a rotating magnetic field. Inside the stator, there is a loop of wires assembled in a squirrel cage made of metal bars and interconnections or some other freely rotating metal part that can conduct electricity. This is the most used design of induction motors, the so-called squirrel cage induction motor, or short-circuited rotor induction motor.

The electrical currents that flow in the rotor are obtained by induction; no galvanic connection exists between stator and rotor. The coupling is purely magnetic; the rotating magnetic field drags the rotor.

6.2   Deriving the Equivalent Circuit

For quantitative predictions about the behaviour of the induction machine, under various operating conditions, it is convenient to represent it as an equivalent circuit under a sinusoidal steady state. Since the operation is balanced, a single-phase equivalent circuit is sufficient for most purposes.

We will begin by defining the induction motor slip as

$$s = \frac{\omega_s - \omega_r}{\omega_s} \qquad (6.1)$$

where ω_s is the synchronous speed and ω_r is the rotor speed. When $s=1$, the rotor is blocked; when $s=0$, the motor is in no-load condition (no current is induced).

Let us now look at the stator circuit of an induction motor. It should contain inductances to account for the leakage (flux that does not link stator and rotor) and mutual (magnetization) fluxes and a resistance to account for the resistance of the stator winding.

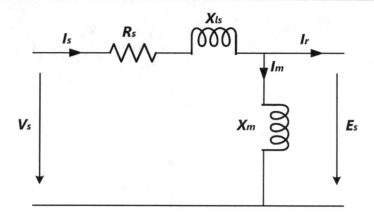

Fig. 39 Stator equivalent circuit of an induction motor

Figure 39 shows the equivalent circuit of one phase of the stator of the induction motor. The elements R_s and $X_{ls} = \omega_s L_{ls}$ are the stator winding resistance and leakage reactance, and X_m is the magnetizing reactance.

Using complex amplitudes, we can write

$$V_s = (R_s + j\omega_s L_{ls})I_s + E_s \tag{6.2}$$

where V_s is the stator voltage, I_s is the stator current, and E_s is the EMF induced in the stator coil by the mutual flux.

We now need to add the rotor to the equivalent circuit. The issue is that the rotor is rotating. The first consequence of this is that the relative angular speed of the stator field and rotor coil is $s\omega_s = \omega_s - \omega_r$ instead of ω_s. Let us recall that (remember Eq. 2.2)

$$E_s = k\omega_s\varphi \tag{6.3}$$

Therefore

$$E_r = ks\omega_s\varphi = sE_s \tag{6.4}$$

i.e., the EMF induced in the rotor will be $E_r = sE_s$.

A second consequence is that since the frequency of the rotor currents is $s\omega_s$, the leakage rotor reactance will have a value of $s\omega_s L_{lr}$.

These two consequences yield the rotor equivalent circuit of Fig. 40.

Again, using phasor notation, we can write (note that the rotor is short-circuited in our induction motor)

$$0 = (R_r + js\omega_s L_{lr})I_r - sE_s \tag{6.5}$$

Fig. 40 Rotor equivalent
circuit of an induction motor

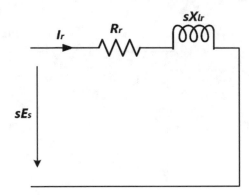

The elements R_r and $X_{lr} = \omega_s L_{lr}$ are efficiency induction motor has the following leakage reactance and I_r is the rotor current.

Dividing by the slip, one gets

$$E_s = \left(\frac{R_r}{s} + j\omega_s L_{lr} \right) I_r \tag{6.6}$$

which leads to the rotor equivalent circuit of Fig. 41.

Now, we can connect the stator circuit (Fig. 39) with the rotor circuit (Fig. 41). Before that, it is instructive to split the resistance into two separate components. For convenience, we can write

$$\frac{R_r}{s} = R_r + \left(\frac{1-s}{s} \right) R_r \tag{6.7}$$

and the induction motor equivalent circuit is depicted in Fig. 42 (for simplification, the EMF is represented by E).

Fig. 41 Rotor equivalent
circuit of an induction motor
reduced to the stator by
frequency

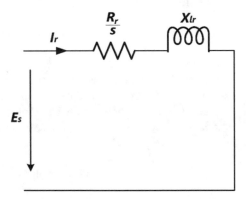

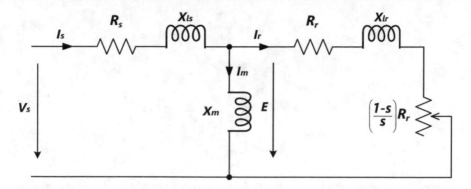

Fig. 42 Equivalent circuit of an induction motor

A couple of remarks should be made regarding the equivalent circuit of Fig. 42:

- The rotor quantities and parameters are referred to the stator, meaning that they are affected by the turns ratio of the stator and rotor. This inconvenience disappears if the circuit is solved in pu: in pu the turns ratio is 1. Nevertheless, the stator and rotor turns are usually the same, therefore, the turns ratio is frequently 1.
- The core (iron) losses can be represented by a resistance in parallel with the magnetizing reactance. In the deducted equivalent circuit, we have neglected these losses.
- An induction motor has also mechanical losses, due to friction and windage.[5] If considered, the useful power output (P_{out}) is computed by subtracting these losses from the mechanical power delivered in the shaft (P_{mec}).

6.3   Mechanical Power and Efficiency

From the equivalent circuit, one can see that the dissipation in R_s represents the stator losses. Therefore, the power absorption indicated by the rotor part of the circuit in Fig. 42 must represent all other means of power consumption, namely the actual mechanical output, friction and windage losses, and the rotor copper losses. Since the dissipation in R_r is rotor copper losses, the power dissipation in $\left(\frac{1-s}{s}\right)R_r$ is the total of the remaining (mechanical power plus friction and windage losses). In standard terminology (SI units; please note that I_r is the RMS value of the rotor current)

---

[5]Windage losses refer to the losses sustained by a machine due to the resistance offered by air to the rotation of the shaft.

$$P_{ag} = 3\frac{R_r}{s}I_r^2 \tag{6.8}$$

$$P_{Lr} = 3R_rI_r^2 \tag{6.9}$$

$$P_{mec} = 3R_r\left(\frac{1-s}{s}\right)I_r^2 = P_{ag} - P_{LR} \tag{6.10}$$

$$P_{out} = P_{mec} - P_{Lfw} \tag{6.11}$$

where, P_{ag} is the air-gap power, P_{Lr} is the rotor copper loss, P_{mec} is the mechanical power output, P_{out} is the useful power output and P_{Lfw} and the friction and windage losses. Out of the power P_{ag} transferred at the airgap, a fraction s is dissipated in the rotor, and $(1-s)$ is delivered as output at the shaft. If there are no mechanical losses, this represents the power delivered in the shaft and available to the load.

The input electrical power is given by

$$P_{in} = 3\left(R_sI_s^2 + \frac{R_r}{s}I_r^2\right) = 3V_sI_s\cos\phi \tag{6.12}$$

and the efficiency is

$$\eta = \frac{P_{out}}{P_{in}} = \frac{3\frac{1-s}{s}R_rI_r^2 - P_{Lfw}}{3\left(R_sI_s^2 + \frac{R_r}{s}I_r^2\right)} \tag{6.13}$$

The absorbed reactive power is

$$Q_{in} = 3\left(X_mI_m^2 + X_{ls}I_s^2 + X_{lr}I_r^2\right) = 3V_sI_s\sin\phi \tag{6.14}$$

Example 6—1
A 50 Hz, 460 V, 5 HP, 0.86 pf, 90% efficiency induction motor has the following equivalent circuit parameters: $R_s = 1.21\ \Omega$; $R_r = 0.742\ \Omega$; $X_{lr} = 2.41\ \Omega$; $X_{ls} = 3.10\ \Omega$; $X_m = 65.6\ \Omega$. Find the starting and no-load currents for this machine.

1 HP (horsepower) is a power unit equivalent to 746 W. This motor has a rated power output of 3.73 kW. The rated current is

$$I_N = \frac{P_N}{\sqrt{3}V_N\cos(\phi_N)\eta} = 6.05\ A$$

At starting, the rotor speed is zero, meaning that the slip is 1. From an equivalent impedance, as seen from the stator point-of-view, the stator impedance is in series with the parallel of the magnetizing impedance and rotor impedance. As so, we can write

$$Z_{eq} = (R_s + jX_{ls}) + \frac{jX_m(R_r + jX_{lr})}{jX_m + (R_r + jX_{lr})} = 1.90 + j5.43\,\Omega = 5.75e^{j70.72°}\,\Omega$$

Taking the equivalent circuit of Fig. 42 into consideration, the absorbed current is

$$I_{st} = \frac{V_s}{Z_{eq}} = \frac{\frac{460}{\sqrt{3}}e^{j0}}{5.75e^{j70.72°}} = 46.15e^{-j70.72°}\,A$$

The starting current is 7.6 times higher than the nominal current.

In no-load, the slip is zero, because the rotor speed is equal to the synchronous speed (neglecting mechanical losses). Therefore, the rotor part equivalent circuit is an open circuit, and the equivalent impedance is

$$Z_{eq2} = R_s + jX_{ls} + jX_m = 1.21 + j68.70\,\Omega = 68.71e^{j88.89°}\,\Omega$$

The no-load current is given by

$$I_{nl} = \frac{V_s}{Z_{eq2}} = \frac{\frac{460}{\sqrt{3}}e^{j0}}{68.71e^{j88.89°}} = 3.87e^{-j88.89°}\,A$$

Example 6—2

A 460 V, 25 hp, 50 Hz, Y-connected induction motor has the following impedances in ohms per phase referred to the stator circuit: $R_s=0.641\Omega$; $R_r=0.332\Omega$; $X_{ls}=1.106\Omega$; $X_{lr}=0.464\Omega$; $X_m=26.3\Omega$. The total rotational losses are 1100 W and are assumed to be constant. The core loss is lumped in with the rotational losses. For a rotor slip of 2.2 percent at the rated voltage and rated frequency, find the motor's efficiency.

First, let us remark that the rated power output of this motor is 25 hp = 18.65 kW.

To compute the efficiency, we have to compute the output power and the input power. Let us start with the latter.

The equivalent impedance as seen from the stator terminals is given by

$$Z_{eq} = (R_s + jX_{ls}) + \frac{jX_m\left(\frac{R_r}{s} + jX_{lr}\right)}{jX_m + \left(\frac{R_r}{s} + jX_{lr}\right)} = 11.70 + j7.80\,\Omega = 14.06e^{j33.67°}\,\Omega$$

This enables the computation of the stator current as

$$I_s = \frac{V_s}{Z_{eq}} = 18.89e^{-j33.67°} \text{ A}$$

The input electrical power is

$$P_{in} = 3V_sI_s\cos\phi = 3 \times \frac{460}{\sqrt{3}} \times 18.89 \times \cos(33.67) = 12,526 \text{ W}$$

As for the mechanical power computation, there are several options. We will present two paths

A simpler path immediately arises after Eq. 6.12, where can write

$$I_r^2 = \frac{\frac{P_{in}}{3} - R_sI_s^2}{\frac{R_r}{s}} = 261.51 \text{ A}^2$$

and the mechanical power and output power are, respectively

$$P_{mec} = 3R_r\left(\frac{1-s}{s}\right)I_r^2 = 11,579 \text{ W}$$

$$P_{out} = P_{mec} - P_{Lfw} = 10,479 \text{ W}$$

At this operating point, the output useful power is 10.479 kW, meaning that it is not operating at rated power (18.65 kW). It is roughly half-load.

The efficiency is given by

$$\eta = \frac{P_{out}}{P_{in}} = 83.66\%$$

A more complicated path, but that can be useful in solving other exercises, is related to the rotor current computation.

We may begin by computing the EMF as (see Eq. 6.2)

$$E = V_s - (R_s + jX_{ls})I_s = 243.92 - j10.67 \text{ V} = 244.15e^{-j2.51°} \text{ V}$$

The magnetizing current is

$$I_m = \frac{E}{jX_m} = 9.28e^{-j92.51°} \text{ A}$$

Applying KCL, we obtain the rotor current, which is

$$I_r = I_s - I_m = 16.13 - j1.20 \text{ A} = 16.17e^{-j4.25°} \text{ A}$$

Still another way to obtain the rotor current is to apply Eq. 6.6

$$I_r = \frac{E}{\frac{R_r}{s} + jX_{lr}} = 16.17e^{-j4.25°}\,\text{A}.$$

6.4 Rotor Angular Speed

We have seen that a symmetrical and balanced system of currents feeding three coils 120° phase shifted can produce a rotating magnetic field with angular speed $\omega_s = 2\pi f = 100\pi$, for a 50 Hz frequency. This holds true if the induction motor has 2 poles (1 pair of poles). Usually, common induction motors have 4 poles (2 pairs).

In general, for an induction motor with p pairs of poles, the angular speed of the rotating magnetic field (also called synchronous speed) is

$$\omega_p = \frac{\omega_s}{p} = \frac{2\pi f}{p}\ \text{rad/s} \tag{6.15}$$

In rpm (revolutions per minute), the synchronous speed is

$$N_p = \frac{60\omega_s}{p2\pi}\ \text{rpm} \tag{6.16}$$

For instance, for an induction machine with 2 pairs of poles, the angular speed of the rotating magnetic field is

$$\omega_p = \frac{\omega_s}{p} = \frac{2\pi 50}{2} = 50\pi\ \text{rad/s} \tag{6.17}$$

$$N_p = \frac{60\omega_s}{p2\pi} = \frac{60 \times 50}{2} = 1500\ \text{rpm} \tag{6.18}$$

This means that the generalized equation for the slip is

$$s = \frac{\omega_p - \omega_r}{\omega_p} = \frac{N_p - N_r}{N_p} \tag{6.19}$$

Example 6—3
The nominal slip of a four-pole induction motor is 6%. Find the nominal revolutions.

From Eq. 6.19, one gets

$$N_r = N_p(1 - s) = 1500(1 - 0.06) = 1410\,\text{rpm}.$$

6.5 Operation as a Generator

We have seen that if a symmetrical and balanced AC supply is connected to the stator three 120° phase-shifted coils, a rotating magnetic field is produced in the stator which pulls the rotor to run behind it, while the machine is acting as a motor.

Now, suppose that the rotor is accelerated to the synchronous speed by means of a prime mover. Under these circumstances, the slip will be zero, and hence no EMF is induced, the rotor current will become zero and no output mechanical power is produced.

If the rotor is made to rotate at a speed higher than the synchronous speed, the slip becomes negative. A rotor current is generated in the opposite direction, due to the rotor conductors cutting the stator rotating magnetic field, therefore, producing a rotating magnetic field in the rotor. This causes a stator voltage that pushes current flowing out of the stator winding. Thus, the machine is now working as an induction generator.

The induction machine, whether it is operating as a motor or as a generator, always takes reactive power from the AC power supply.[6] When it is in generator operating mode, it supplies active power back into the grid. Reactive power is needed for producing the rotating magnetic field.

7 Conclusions

This book is about electricity production from renewables. This means that some electrical engineering concepts are required to understand how electricity is produced, transformed, and conditioned in renewable-based power plants. One of these concepts is related to electrical circuits, because electrical engineering, and more precisely, power systems engineering, is critically based on them.

Electrical circuits are a tremendously complex topic, requiring solid backgrounds on many subjects, mainly electromagnetism. In electrical engineering courses, several subjects are dedicated to the study of electrical circuits. Here, we have compressed everything into one chapter. The scientific deepness has been somehow sacrificed and our aim was to provide the main concepts related to the electrical circuits in the simpler and straightforward way we could.

We began by presenting how a sinusoidal voltage is obtained from the rotation of a winding inside a magnetic field. This allowed for the introduction of the three basic electrical circuits—resistive, inductive, and capacitive. It was proved that if more complex circuits were to be studied using this theory, the task would be almost unfeasible. The notion of the phasor, a complex number, was introduced to facilitate the computations involved in solving electrical circuits. With this method, the study of complex circuits became much more straightforward.

---

[6]Just take a look at the equivalent circuit. Remember that the coils absorb reactive power, which must be supplied by the AC power system.

In power systems engineering, the concept of power is fundamental. The notion of complex power and its components, active and reactive power, was derived using phasor theory. Also, the need for reactive power compensation was shown and the ways to perform it using capacitors were presented.

Power systems are three phased and not single phased. The simpler theory of balanced three-phase AC circuits was presented, involving the definition of phase and line voltage and the study of the different ways of connecting the loads (star and delta). Also, the important notion of three-phase power was addressed.

Another technique aimed at simplifying the calculations is the per unit (pu) system. This units' system allows for dealing with numbers near the unity, provides a meaningful comparison of parameters of machines with different ratings, and facilitates the computations involving transformers.

A common method used in power systems to study the behaviour of electrical machines is to derive an equivalent electrical circuit that represents the operating modes of the machine. We have exemplified this method by presenting a steady-state model of the induction (asynchronous) machine, extensively used as a motor in the industry. The previous knowledge about electrical circuits proved to be very useful in the task of writing the equations that govern the electrical behaviour of the induction motor.

8 Proposed Exercises

Problem EC1

The working voltage of a factory is 230 V, 50 Hz. The factory comprises 5 motors, each with a capacity of 7 kW, an efficiency of 85%, and a power factor equal to 0.7 ind. The lighting installation is composed of forty 65 W lamps with a power factor equal to 0.5 ind. Compute:

1. The current absorbed by the motors, the lamps, the total current, and the power factor of the factory.
2. The capacitance of the capacitor that corrects the power factor of the lighting installation to 0.92 ind.

Solution

(1) The electrical power absorbed by each motor is

$$P_{el1} = \frac{P_{mec}}{\eta} = 8.2353 \, \text{kW}$$

Current absorbed by the motors, the lightning, and total

$$I_1 = I_1 e^{-j\phi_1} = \frac{5P_{ell}}{Vpf_1} e^{-jacos(pf_1)} = 255.7545 e^{-j45.5730°} \text{ A}$$

$$I_2 = I_2 e^{-j\phi_2} = \frac{40P_{el2}}{Vpf_2} e^{-jacos(pf_2)} = 22.6087 e^{-j60°} \text{ A}$$

$$I = I_1 + I_2 = Ie^{-j\phi} = 277.7074 e^{-j46.7352°} \text{ A}$$

$$pf = \cos\phi = \cos 46.7352° = 0.6854 \text{ ind}$$

(2) The vectorial diagram after the installation of the compensation capacitor is in Fig. 43.

$$I_{C2} = I_2 \sin\phi_2 - I_{2comp} \sin\phi_{2comp}$$

$$I_{2comp} = I_{2comp} e^{-j\phi_{2comp}} = \frac{40P_{el2}}{Vpf_{2comp}} e^{-jacos(pf_{comp})} = 12.2873 e^{-j23.0739°} \text{ A}$$

$$I_{C2} = 22.6087 \sin(60°) - 12.2873 \sin(23.0739°) = 14.7641 \text{A}$$

$$I_{C2} = j\omega C_2 V$$

$$C_2 = \frac{I_{C2}}{2\pi 50 V} = 204.33 \mu\text{F}$$

Problem EC2

A three-phase, 4-wire feeder with a voltage of 400/230 V feeds the following loads:

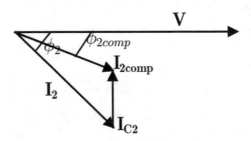

Fig. 43 Power factor compensation diagram from Problem EC1

- An electric oven, which may be modelled as a resistor, draws 10 kW and is connected between phase 1 and the neutral.
- Between phase 2 and the neutral, six lamps that draw 6 kW with a power factor of 0.5 ind are installed.
- A single-phase induction motor is connected between phase 2 and phase 3. The motor draws an active power equal to 7 kW with a power factor equal to 0.7 ind.
- Between phase 1 and phase 3 a 14 kvar capacitor is connected.

Compute

1.   The current absorbed by each load.
2.   The currents in the 4 wires.

Solution
A schematic of the installation is in Fig. 44.
This is an unbalanced feeder, as usual in the distribution system.
The phase-to-neutral and phase-to-phase voltages are

$$V_1 = 230e^{j0}\text{V}$$

$$V_2 = 230e^{-j120°}\text{V}$$

$$V_3 = 230e^{j120°}\text{V}$$

$$V_1 - V_2 = V_{12} = \sqrt{3}230e^{j30°}\text{V}$$

$$V_2 - V_3 = V_{23} = \sqrt{3}230e^{-j90°}\text{V}$$

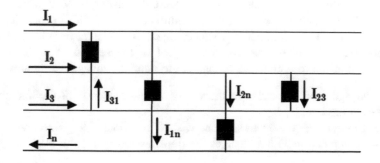

Fig. 44 Schematic of the installation of Problem EC2

$$V_3 - V_1 = V_{31} = \sqrt{3}230e^{j150°}\,\text{V}$$

(1) The requested currents are shown in the figure

$$I_{1n} = \left(\frac{P_{1n}+j0}{V_1}\right)^* = 43.4783e^{j0}\,\text{A}$$

$$I_{2n} = \left(\frac{P_{2n}+jP_{2n}\tan[\text{acos}(pf_{2n})]}{V_2}\right)^* = 52.1739e^{j180°}\,\text{A}$$

$$I_{23} = \left(\frac{P_{23}+jP_{23}\tan[\text{acos}(pf_{23})]}{V_{23}}\right)^* = 25.1022e^{-j135.5730°}\,\text{A}$$

$$I_{31} = \left(\frac{jQ_{31}}{V_{31}}\right)^* = 35.1431e^{-j120°}\,\text{A}$$

We recall that $S = P+jQ = P(1+j\tan\phi)$ and $Q_{31} = -14\,\text{kvar}$ because the capacitor supplies reactive power.

(2) According to the directions shown in the figure, the 4-wire currents are

$$I_1 = I_1n - I_{31} = 68.2155e^{j26.4974°}\,\text{A}$$

$$I_2 = I_{2n} + I_{23} = 72.2692e^{-j165.9281°}\,\text{A}$$

$$I_3 = I_{31} - I_{23} = 12.8682e^{-j88.4191°}\,\text{A}$$

$$I_n = I_1 + I_2 + I_3 = 8.6957e^{j180°}\,\text{A}$$

Problem EC3

A 60 kV, 30 km, overhead three-phase line must transmit 10 MW with a cos ϕ equal to 0.8 ind. The line has 70 mm^2 aluminium conductors with a resistivity at 40° C equal to 28.3 Ω mm^2/km and a reactance equal to 0.4 Ω/km per phase. Assume that the value of the reactance is independent of the section of the conductor. Assume that at the reception the voltage is the nominal voltage. Compute:

1. The voltage drop in percent of the voltage at the reception of the line and the losses in the line.
2. The section of the conductors, if one wishes to maintain the same transmission losses computed in 1) but changes the transmission voltage to 3 kV.

Solution

(1) The voltage drop in a transmission line is defined as the difference between the RMS voltages at both ends of the line (sending and receiving). In percentage, we have

$$\Delta V = \frac{V_e - V_r}{V_r} \times 100\%$$

This distribution line may be considered a short line and we can use the R-L model. We neglect the capacitive effects, and the single-phase equivalent diagram is in Fig. 45.

We consider

$$V_r = \frac{60 \times 10^3}{\sqrt{3}} e^{j0} = 34.6410\,\text{kV}$$

The sending voltage is

$$V_e = (R_L + jX_L)I + V_r$$

$$I = \left(\frac{P_r + jP_r \tan[\text{acos}(pf)]}{3V_r}\right)^* = 120.2813 e^{-j36.8699°}\,\text{A}$$

$$R_L = \rho \frac{l}{s} = 12.1286\,\Omega;\, X_L = 12\,\Omega$$

$$V_e = 36.6752 e^{j0.4365°}\,\text{kV}$$

$$\Delta V = \frac{V_e - V_r}{V_r} \times 100\% = 5.87\%$$

The losses are

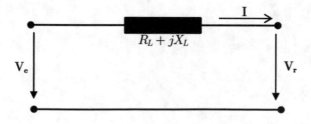

Fig. 45 Single-phase equivalent diagram from Problem EC3

$$P_L = 3R_L I^2 = 0.5264\,\text{MW} = 5.26\%$$

(2) If the transmission voltage was 3 kV, i.e., 20 times smaller, the line current would be 20 times higher, i.e., 2.4056 kA. The losses, which depend on the square of the current, would be 400 times higher. This shows the impossibility to use the same conductors to transmit the required power.

To keep the same level of losses, the resistance should be 400 times lower, i.e., we would need to increase the cross-section by 400 times. The new cross-section would be 280 cm^2. It is apparent that we cannot transmit 10 MW at 3 kV voltage level.

Problem EC4
In an electrical outlet of 230 V, 50 Hz, a resistance of 20 Ω, a coil of 60 mH, and a capacitor of 100 µF are connected in parallel. Compute the current drawn by the circuit.

Solution
We consider $V = 220e^{j0}$ V.

$$I_R = \frac{V}{R} = 11.5e^{j0}\,\text{A}$$

$$I_L = \frac{V}{j\omega L} = 12.2019e^{-j90°}\,\text{A}$$

$$I_C = j\omega c V = 7.2257e^{j90°}\,\text{A}$$

$$I = I_R + I_L + I_C = 12.53e^{-j23.399°}\,\text{A}$$

Problem EC5

A three-phase electrical oven is heated by resistances ($R = 10\,\Omega$). The oven is fed by a three-phase feeder, 230–400 V at the oven terminals. For both star and delta-connected systems, compute:

1. The RMS voltage in each resistance.
2. RMS current in the feeder.
3. The active power supplied to the oven.
4. Power factor.
5. Energy consumed by the oven for 8 h.

Solution
See Table 2.

Problem EC6
A 10 kW, 4 poles, 400 V, 50 Hz star-connected asynchronous three-phase motor
has the following parameters:

$$R_s = 0.5\,\Omega; R_r = 0.8\,\Omega; X_s = 2.4\,\Omega; X_r = 2.4\,\Omega; X_m = 40.9\,\Omega$$

For a certain load, the motor is fed at rated voltage and frequency and shows an
efficiency of 94%. Neglect the iron losses and mechanical losses and compute:

1. The slip.
2. The current is drawn from the grid.

Use the simplified equivalent diagram of Fig. 46.

Solution
(1) The equivalent diagram of the figure is a simplified representation of the
induction motor. Neglecting the iron losses corresponds to make $G_m = \infty$;
neglecting the mechanical losses corresponds to make $P_{Lfw} = 0$.

$$P_{mec} = P_{out} = 3R_r \left(\frac{1-s}{s} \right) I_r^2$$

$$P_{in} = 3(R_s + R_r)I_r^2 + P_{mec}$$

Table 2 Solution of Problem EC5

	Star-connected	Delta-connected
Voltage (V)	$V = 230$	$V = 400$
Current (A)	$I = \frac{V}{R} = 23$	$I = \frac{\sqrt{3}V}{R} = 69.3$
Power (kW)	$P = 3RI^2 = 15.87$	$P = 3R\left(\frac{V}{R}\right)^2 = 48$
Power factor	$pf = 1$	$pf = 1$
Energy (kWh)	$E = P\Delta t = 126.96$	$E = P\Delta t = 384$

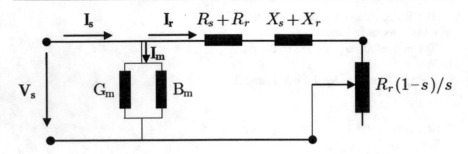

Fig. 46 Equivalent diagram from Problem EC6

$$\eta = \frac{P_{out}}{P_{in}} = \frac{R_r \frac{1-s}{s}}{\frac{R_r}{s} + R_s}$$

$$s = \frac{R_r(1-\eta)}{\eta R_s + R_r} = 3.78\%$$

(2) $V_s = \frac{400}{\sqrt{3}} e^{j0}$ Vbecause the equivalent diagram is a single-phase representation.

$$I_s = I_m + I_r = 12.8684 e^{-j37.8574°}\,\text{A}$$

$$I_m = jB_m V_s = \frac{V_s}{jX_m} = -j5.6465\text{A}$$

$$I_r = \frac{V_s}{\left(\frac{R_r}{s} + R_s\right) + j(X_r + X_s)} = 10.4065 e^{-j12.4915°}\,\text{A}.$$

Economic Assessment of Renewable Energy Projects

3

Abstract

The success of Renewable Energy Sources (RES) is strongly dependent on the economics of these projects. RES are currently the preferred source of electricity production mainly because producing electricity from RES is cheaper than the alternatives. The proper assessment of the economic viability of RES projects is a matter of utmost importance. In this chapter, we are concerned with two main aspects of the economic assessment of RES: the electricity production cost and the indexes used to evaluate the economic interest of the projects. Economic models are proposed to compute the Levelized Cost Of Energy (LCOE), the indicator that is used to study the discounted production cost. Economic decisions are based on economic assessment indexes, namely the Net Present Value (NPV) and Internal Rate of Return (IRR). These indicators, together with other less-used ones, are introduced and their application is discussed in this text. To compute the said economic indicators, energy production is necessary. It is more practical that energy production is made independent of the installed capacity. This is achieved by means of the utilization factor and capacity factor. The importance of these technical indicators is highlighted in this chapter.

1 Introduction

Renewable Energy Sources (RES) are nowadays a common element of the landscape. As of 2018, 95% of the new generating facilities installed in Europe were renewable-based. This is a reality mainly because RES are cheaper than conventional fossil fuel-fired power plants, a conclusion which is based on the assessment of the economics of the projects.

© The Author(s), under exclusive license to Springer Nature Switzerland AG 2022   105
R. Castro, *Electricity Production from Renewables*,
https://doi.org/10.1007/978-3-030-82416-7_3

Where different technical solutions are possible or where various investment opportunities are offered, it is also necessary to evaluate the available projects, so that one can decide the ones to be carried out and the ones to be disregarded.

The proper assessment of the economic viability of investments in renewable electricity generation facilities was a necessary condition for the development of the RES new technologies to be made robustly and convincingly.

This justifies the introduction of this chapter on economic evaluation criteria for projects related to the installation of electricity production units using renewable resources. Nevertheless, the reader is warned that only some limited aspects of energy economics are discussed here. We shall focus on the main subjects that usually concern engineers who are called to provide a view on the economic viability of investments in renewable-based production of electricity. Therefore, two main aspects are approached in this chapter: the electricity production cost and the economic assessment indexes.

The electricity production cost that is usually used to compare different technologies is the Levelized Cost Of Energy (LCOE). The LCOE is the discounted average cost, i.e., the average cost considering that all relevant expenses are discounted to the present time. A detailed model is developed to compute the LCOE. As the data to feed this model may turn difficult to know beforehand, the usually used simplified model is offered.

The LCOE does not consider the revenues. When the overall economic viability of a project is to be assessed, the revenues must be considered, thus leading to the definition of the Net Present Value (NPV) and Internal Rate of Return (IRR). These are the two main indicators that, together with other less-used indicators, are exposed in the following text.

Also, in this chapter, the concepts of utilization factor and capacity factors are introduced. These technical indexes are very important to compare the production of different technologies, independently of the respective installed capacity.

Economic and financial assessment can be carried out at constant prices, when the effects of inflation are ignored, or at current prices if these effects are accounted for. In periods of controlled inflation and considering that inflation equally affects revenues and expenses, an analysis at constant prices can be carried on. This option is followed in the following text.

2 Utilization Factor and Capacity Factor

Before we go further, it is worth clarifying the meaning of two important factors related to the power plant operation that will be used throughout this text: the capacity factor (C_f) and the annual utilization factor (h_a).

$$C_f = \frac{P_{avg}}{P} \qquad (2.1)$$

$$h_a = \frac{E_a}{P} = \frac{P_{avg}8760}{P} = 8760 C_f \qquad (2.2)$$

In the above equations, P_{avg} is the average power, P is the maximum or installed or rated capacity, and E_a is the annual produced electricity.

The annual utilization factor is measured in "hours" and represents the number of hours that a power generation installation would operate at rated capacity to produce the very same energy that it has produced operating under its own generation diagram, during the whole year. An example may help in understanding the meaning of the annual utilization factor.

Let us assume the generation diagrams (representation of generating power as a function of time) depicted in Fig. 1.

The annual energy production is given by the area below the generation diagram $(E_{a1} = 4\,\text{GWh}; E_{a2} = 2\,\text{GWh})$. The installed capacity is the same $(P_1 = P_2 = 1\,\text{MW})$. However, it can be seen that to produce the same energy, power generation installation 1 would need to operate for 4000 h and installation 2 for 2000 h at rated power, instead of the 8760 h that they have actually operated at. Power plant 1 is being better exploited than power plant 2 because the rated capacity is being used for more equivalent time.

Another perspective on the utilization factor is that it is the number of MWh each MW of installed capacity can produce. Of course, the maximum value for the utilization factor is 8760 h (capacity factor equal to one), meaning that each MW of installed capacity can produce 8760 MWh of electricity at maximum. As seen before, the capacity factor is the utilization factor in a percentage of the yearly hours.

Of course, the higher the utilization factor, the better. The ideal would be a utilization factor of 8760 h. This is theoretically possible in fossil fuel power plants, whose output power depends on a controllable resource. Theoretically, a coal-fired power plant could have a unitary capacity factor provided we supply coal during the whole year. For power plants based on renewable resources, this is not possible, because the output power depends on an uncontrollable resource.

Depending on the renewable resource conditions, the utilization factor of a wind power plant typically varies between 2000 and 2500 h, if located onshore. For offshore installations, where the wind speed is higher, the utilization factor may reach 3000 or 4000 h. In what concerns solar photovoltaic power plants, the output power depends on the solar power and the temperature. In Central Europe, the utilization factor is not higher than 1000 h, while in Southern Europe it reaches 2000 h.

The utilization factor is a very important factor because it allows for a comparison between the yield of different generation sources, independently of the installed capacity. We will use this index a lot in the sequence.

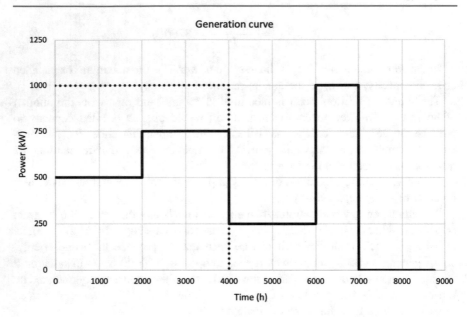

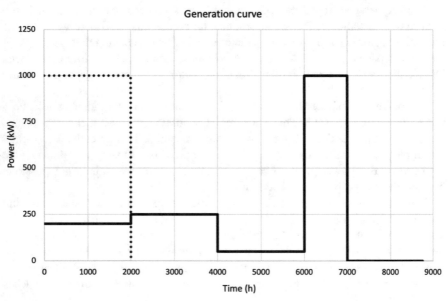

Fig. 1 Example of two-generation diagrams: solid line: actual generation diagram; dashed line: equivalent generation diagram always at rated capacity operation

3   Electricity Average Cost

3.1   Annual Average Cost

To calculate the average annual cost, i.e., the cost of each unit of electricity produced in one specific year, the annual overall costs C_a (€) are divided by the annual electricity production E_a (MWh).

The cost computed in this way will change from one year to another and, as such, is not adequate to evaluate the global economic relevance of a particular electric power source. Yet, it is important to judge the economics of a project in a particular year.

The average annual cost c_a(€/MWh) can, in general, be stated as

$$c_a = \frac{C_a}{E_a} = \frac{\left(i' + c_0\right)I_t}{E_a} \tag{3.1}$$

in which

- i': annual fixed capital cost (%);
- c_0: other annual fixed costs, (e.g., O&M) (%);
- I_t: total investment (€).

It should be remarked that we are assuming that the renewable energy project does not present variable costs, related to fuel and environmental costs. Also, we are assuming that the other costs are fixed costs.

Let us divide the numerator and the denominator by the power plant rated capacity, P:

$$c_a = \frac{\left(i' + c_0\right)I_{01}}{h_a} \tag{3.2}$$

We recall that the utilization factor, h_a, is given by

$$h_a = \frac{E_a}{P} \tag{3.3}$$

and that I_{01} is the unitary investment (€/MW), i.e., the investment per unit of capacity.

3.2   Discount Rate

The annual average cost, which was introduced in the previous paragraph, may be used to monitor the project economics on a year-by-year basis, but it is not a proper criterion to evaluate the economic interest of projects. There are cases in which the average annual cost of a given project is the lowest, but the same project is not most economically interesting when analysed in an integrated perspective over a lifetime.

A known difficulty in the economic evaluation of projects results from the fact that cash inflows and outflows are staggered in time according to the most varied sequences. It is common knowledge that it is not indifferent to pay (or receive) a certain amount of money today or to pay (or receive) the same amount after a few years. The use of the discount rate allows us to overcome this difficulty.

Between paying a certain amount immediately or paying it within 10 years, the natural choice is for the payment after 10 years. It is not the hope that the lender disappears in between that justifies the option; furthermore, the explanation does not rely on thinking that in the long term the same amount eroded by inflation corresponds to a much smaller real value.

The amount to be paid in the future may be invested during this period, after which the actual accumulated amount may be much higher than the amount that has to be paid. The money invested over time will give a real income, which justifies the option for the forward payment.

It should be emphasized that this reasoning is done with a constant price model, in which inflation is absent. Income obtained, thanks to inflation, is illusory, because the inflated currency loses purchasing power: the "profit" obtained in a devalued currency may correspond to a real loss.

Let F_0 (€) be an amount of money available in the present time (t = 0). If this amount of money is invested during t years, the total accumulated amount after t years will be F_t, which is obtained by

$$F_t = F_0(1+a)(1+a)\ldots(1+a) = F_0(1+a)^t \qquad (3.4)$$

a (%) being the annual real yield of capital, known as the discount rate.

We can conclude that a payment, F_0, made today is equivalent to a (larger) payment made after t years. Conversely, a payment, F_t, made within t years amounts to a (lower) payment, F_0, made today, being

$$F_0 = \frac{F_t}{(1+a)^t} \qquad (3.5)$$

It is said that F_0 is the present (or discounted) value of a payment made in time t, and F_t is the future value. The rate at which one can convert payments (or incomes) made at different times to the present time is called the discount rate.

It can then be stated that discounting is a concept associated with an arithmetic process that allows converting an amount of money referred to a given date to the equivalent amount on another date (usually the present time). Thus, values distributed over different time instants can be converted to discounted values at the present time, and, provided that they are expressed in the same units, can be added.

The foregoing also shows that the concept of discount rate is linked to the concept of the real rate of return of an investment, also known as the opportunity cost of capital. Thus, the discount rate is nothing more than the minimum profitability that the investor requires to invest in a given project.

Usually, enterprise sponsoring organizations have their own guidelines for setting the discount rate, which should reflect the minimum rate of return available for the capital invested and the risk associated with the investment option.

For the private sector, the discount rate must be determined by examining known similar transactions, because the project will be implemented in a market environment. On the other hand, governmental organizations may use the reference discount rate recommended by the state banking institutes, for the economic activity sector in which the project is to be deployed.

Finally, it should be noted that the discount rate does not coincide—except in a perfect market, which does not exist—with the bank interest rate, although the two rates are somewhat related. For instance, if the investment is financed through a loan (which is a common practice) at a fixed interest rate, this will influence the required rate of return and, therefore, the discount rate.

3.3   Levelized Cost Of Energy (LCOE)

The average annual cost is significant for each year. However, it is not significant if the evaluation period extends from the time of the investment decision till the end of the power plant lifetime.

To obtain the discounted average cost, usually known as Levelized Cost Of Energy (LCOE), the different cost parcels (e.g., investment, O&M, and others) and the total electricity production are separately discounted over the lifetime of the renewable power plant or during a given analysis time period. Discounting means computing the present value.

Let us denote each discounted cost parcel by c_{di}(€) and the total discounted electricity production by E_d (MWh). So, the LCOE (€/MWh) will be given by

$$LCOE = \frac{\sum_{i=1}^{n_c} c_{di}}{E_d} \tag{3.6}$$

where n_c is the number of cost parcels.

Discounting consists of calculating the amounts corresponding to payments and revenues made on different dates as if they were all made at time $t = 0$ (for example, the time at which the economic evaluation of the project is being carried out).

A general model may admit that both money inflows (energy sales) and money outflows (investment, O&M, …) are erratically distributed during the analysis timeframe. However, in this paper, we will assume that

- Expenses are due on the first day of the year during which they are paid.
- Revenues enter on the last day of the year during which they are actually received.

We will know in detail each cost parcel and the discounted electricity production as follows.

3.3.1   Discounted Cost Parcel 1—Investment Cost

A fairly general model may consider that the investment is distributed over N years before $t = 0$ and $n - 1$ years after $t = 0$ (n is project lifetime). Given these conditions, the discounted investment cost, c_{d1}, is

$$c_{d1} = I_{td} = \sum_{j=-N}^{n-1} \frac{I_j}{(1+a)^j} \tag{3.7}$$

where a is the discount rate (%), I_j (€) is the investment in year j, and I_{td} (€) is the total discounted investment.

Of course, if the investment is fully concentrated at the initial time instant ($t = 0$), the investment cost does not require any discounting and is given by (I_t is the total investment):

$$c_{d1} = I_t \tag{3.8}$$

The capacity of renewable power plants is relatively small, and the equipment deployment is a rapid procedure. As such, consideration of this hypothesis usually involves a minimal error.

3.3.2   Discounted Cost Parcel 2—O&M Cost

We shall consider that the O&M costs are fixed costs and as such they may be introduced as a percentage of the total investment. We recall that the total investment is a fixed cost as it depends on the capacity of the power plant. Variable costs (fuel costs and CO_2 emissions costs) are not considered as we are dealing with renewable-based power plants.

The discounted O&M costs are due during the power plant lifetime (which begins in $t = 1$) and may be computed as

$$c_{d2} = I_t \sum_{j=1}^{n} \frac{c_{omj}}{(1+a)^j} \tag{3.9}$$

where c_{omj} (%) are the O&M costs in year j, given as a percentage of the total investment I_t.

3.3.3 Discounted Electricity Production

The electricity production may be discounted as follows:

$$E_d = \sum_{j=1}^{n} \frac{E_{aj}}{(1+a)^j} = P \sum_{j=1}^{n} \frac{h_{aj}}{(1+a)^j} \tag{3.10}$$

where E_d (MWh) is the discounted electricity production, E_{aj} (MWh) is the electricity production in year j, P (MW) is the power plant rated capacity, and h_{aj} (h) is the power plant utilization factor in year j.

According to Eq. 3.6, the LCOE is given by

$$LCOE = \frac{c_{d1} + c_{d2}}{E_d} = \frac{\sum_{j=-N}^{n-1} \frac{I_j}{(1+a)^j} + I_t \sum_{j=1}^{n} \frac{c_{omj}}{(1+a)^j}}{P \sum_{j=1}^{n} \frac{h_{aj}}{(1+a)^j}} \tag{3.11}$$

3.4 LCOE Simplified Model

Let us assume that

- The total investment is concentrated at the initial instant, $t = 0$, and is denoted by I_t.
- The annual utilization factor is constant throughout the power plant lifetime and equals h_a.
- O&M expenses are constant over the power plant lifetime and equal c_{om}.

We shall define the factors k_a and i as (note that the sum of the series is given by the indicated analytical expression)

$$k_a = \sum_{j=1}^{n} \frac{1}{(1+a)^j} = \frac{(1+a)^n - 1}{a(1+a)^n} ; i = \frac{1}{k_a} = \frac{a(1+a)^n}{(1+a)^n - 1} \tag{3.12}$$

k_a

is known as the discount factor.

Given these conditions, Eq. 3.11 becomes

$$LCOE = \frac{I_t + c_{om} I_t k_a}{E_d k_a} = \frac{I_t(i + c_{om})}{E_d} \tag{3.13}$$

Or, if one divides by the power plant rated capacity, the LCOE may be computed through

$$LCOE = \frac{I_{01}(i + c_{om})}{h_a} \tag{3.14}$$

Example 3—1
The capacity factor of a 10 MW wind park is 28.54%. The investment is 1.2 M€/ MW, the expected lifetime is 20 years, and the annual O&M costs are 1.5%. Compute the LCOE (€/MWh) for a 5% discount rate.

Solution:

The solution is given by Eq. 3.14:

$$LCOE = \frac{I_{01}(i + c_{om})}{h_a} = \frac{1.2 \times 10^6 (0.0802 + 0.0150)}{0.2854 \times 8760} = 45.71\,€/\text{Mwh}$$

4 Economic Assessment Indexes

The evaluation criteria for profitability that are commonly used to measure the economic interest of projects may appear to be objective, but in reality, they are not. They count for sure on future expenses and revenues, and the future is, as we know, more or less uncertain. Thus, when the parameters that determine the evaluation (costs, revenues, equipment lifetime, O&M costs, and others) are admitted as reliable, this results more from the mental attitude of the evaluator than from objective evidence. As a consequence, it is more correct to state that the project's economic assessment is to be obtained based on a forecast of the required data.

In what follows, we assume that cash outflows occur irregularly from $t = 0$ to $t = n - 1$ and that revenues are also obtained irregularly from $t = 1$ to $t = n$. The previous convention is kept for the dates on which expenses and revenues are due.

Net Present Value (NPV) and Internal Rate of Return (IRR) are the most used economic assessment indexes for the evaluation of investment projects in dispersed renewable power plants.

4.1 NPV—Net Present Value

4.1.1 NPV General Model

NPV is the difference between discounted cash inflows and outflows, called the cash-flows, during the project's lifetime (n years):

$$NPV = \left(\sum_{j=1}^{n} \frac{R_j}{(1+a)^j} + \frac{V_S}{(1+a)^n} \right) - \left(\sum_{j=0}^{n-1} \frac{I_j}{(1+a)^j} + I_t \sum_{j=1}^{n} \frac{c_{omj}}{(1+a)^j} \right)$$

$$(4.1)$$

where R_j is the revenue coming from the electricity sales in year j and V_S is the salvage value due in $t = n$.

A positive NPV is a sign of the economic viability of the project. It means that the results achieved cover the initial investment, as well as the minimum remuneration required by the investor (represented by the discount rate), and also generate a financial surplus. A zero NPV means full recovery of the initial investment, plus the minimum income required by investors and no more than that, so the profitability of a project with these characteristics is uncertain. A negative NPV is a clear indication of the economic nonviability of the project.

It is interesting to note that the higher the discount rate considered in the NPV calculation, the lower NPV will be obtained since a higher return on the project investment is required.

4.1.2 NPV Simplified Model

If the simplified model assumptions, introduced earlier in this paper, hold true, and furthermore the salvage value may be neglected, and the annual revenue is constant and equal to R during the project's lifetime, then Eq. 4.1 becomes

$$NPV = (R - e_{om})k_a - I_t = R_N k_a - I_t \tag{4.2}$$

where e_{om} are the total annual O&M expenses, and R_N is the annual net revenue.

4.2 IRR—Internal Rate of Return

4.2.1 IRR General Model

The Internal Rate of Return (IRR) is the discount rate that cancels the NPV. Then, from Eq. 4.1, the results that the IRR (%) will satisfy are

$$\left(\sum_{j=1}^{n} \frac{R_j}{(1+IRR)^j} + \frac{V_S}{(1+IRR)^n} \right) = \left(\sum_{j=0}^{n-1} \frac{I_j}{(1+IRR)^j} + I_t \sum_{j=1}^{n} \frac{C_{omj}}{(1+IRR)^j} \right)$$

$$\tag{4.3}$$

The IRR assessment immediately places the interest of the project on the financial market evaluation scale, which is not the case with the NPV.

An IRR greater than the discount rate considered in the NPV computation means that the project can generate a rate of return higher than the opportunity cost of capital. We are therefore facing an economically viable project. The opposite situation means that the required minimum profitability is not achieved.

4.2.2 IRR Simplified Model

In the general case, the IRR computation from Eq. 4.3 can be solved using iterative methods, which makes the IRR calculation a complicated task. This scenario is somewhat attenuated in the simplified model conditions. The equation to be solved is as follows, IRR being the unknown:

$$R_N \frac{(1+IRR)^n - 1}{IRR(1+IRR)^n} = I_t \qquad (4.4)$$

where R_N is the annual net annual revenue, supposed constant over the project's lifetime:

$$R_N = R - e_{om} \qquad (4.5)$$

It is apparent that Eq. 4.4 is easier to solve, although it does not dispense the use of iterative methods, for example, a simple Gauss method can be applied. For this purpose, Eq. 4.4 can be written in a form suitable to apply the method (take note that (k) is the order number of the iteration):

$$IRR^{(k+1)} = \frac{R_N}{I_t} \frac{\left(1+IRR^{(k)}\right)^n - 1}{\left(1+IRR^{(k)}\right)^n} \qquad (4.6)$$

Usually, convergence, with a small error (let's say 0.01), is obtained in 4–5 iterations. To obtain a faster convergence, a Newton-type method can be used, but it is much more complicated to implement.

4.2.3 IRR Approximate Computation

Often, in practice, an expedited IRR calculation is required for a fast estimate. For this purpose, it is usual to use an approximate IRR computation employing a linear interpolation. It has already been mentioned that NPV decreases with the discount rate increase. Figure 2 illustrates the non-linear, typical variation of the NPV with the discount rate.

As depicted in the figure, the IRR is the discount rate that cancels the NPV. The IRR value can be obtained, approximately, by linearizing the section of the curve around the point of nullification. For this purpose, two NPV values are

Fig. 2 Variation of NPV with the discount rate

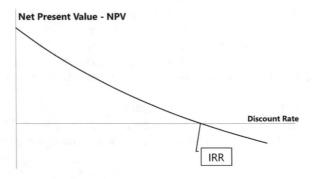

Net Present Value - NPV

Discount Rate

IRR

calculated, one positive (NPV_1) and one negative (NPV_2), corresponding to the discount rates a_1 and a_2, respectively. It is easy to verify that the line passing through these two points has a zero at the abscissa point:

$$IRR \approx a_1 - (a_2 - a_1) \frac{NPV_1}{NPV_2 - NPV_1} \tag{4.7}$$

Example 4—1

Retake Example 3—1 and compute the NPV and IRR, considering that the Feed-In Tariff is 50 €/MWh.

Solution:

Considering the given data, we will use the NPV simplified model (Eq. 4.2):

$$
\begin{aligned}
NPV &= (R - e_{om})k_a - I_t \\
&= \left(50 \times 10 \times 0.2854 \times 8760 - 0.0150 \times 1.2 \times 10^6 \times 10\right) \times 12.4622 \\
&\quad - 1.2 \times 10^6 \times 10 = 1{,}335{,}213 \text{ €}
\end{aligned}
$$

As far as the IRR is concerned if the simplified model (Eq. 4.4) is used,

$$IRR^{(k+1)} = \frac{R_N}{I_t} \frac{\left(1 + IRR^{(k)}\right)^n - 1}{\left(1 + IRR^{(k)}\right)^n} = \frac{1{,}070{,}052}{12{,}000{,}000} \frac{\left(1 + IRR^{(k)}\right)^{20} - 1}{\left(1 + IRR^{(k)}\right)^{20}}$$

Iteration	IRR (%)
0	10.00
1	7.59
2	6.85
3	6.55
4	6.41
5	6.34
6	6.31
7	6.29
8	6.29

The result is IRR = 6.29%.
Using the IRR approximate computation (Eq. 4.7)

$$IRR \approx a_1 - (a_2 - a_1)\frac{NPV_1}{NPV_2 - NPV_1}$$
$$= 0.05 - (0.07 - 0.05)\frac{1,335,213}{-663,854 - 1,335,213} = 0.0634$$

Note that for $a_2 = 7\%$, $NPV_2 = -663,854$ €. The other required point is $a_1 = 7\%$, $NPV_1 = 1,335,213$ €.
The result is IRR = 6.34%, which is a very good approximation of the real IRR.

4.3   Payback Period

The payback period is the length of time the investment takes to be recovered supposing the net revenue is constant and equal to the value it has in year 1. It is measured in years. The following formula applies:

$$PB = \frac{I_t}{R_1 - e_{om1}} \tag{4.8}$$

where I_t is the total investment, R_1 is the revenue in year 1, and e_{om1} are the O&M expenses in year 1.

The payback period is a coarse index, but its application is very simple: no discount is performed, and it is assumed the net revenue is constant year after year.

4.4   Discounted Payback Period

The discounted payback is a more thorough way of measuring the payback time of an investment. The discounted payback is the number of years the investment takes to be recouped, considering the discount.

Considering that the investment is concentrated in $t = 0$, the Discounted Payback Period (DPP) is computed using the equation:

$$\sum_{j=1}^{DPP} \frac{R_{Nj}}{(1+a)^j} = I_t \tag{4.9}$$

where R_{Nj} is the net revenue in year j.

Developing and assuming the conditions of the simplified model:

$$I_t = R_N \frac{(1+a)^{DPP} - 1}{a(1+a)^{DPP}} = R_N \left(\frac{1}{a} - \frac{1}{a(1+a)^{DPP}} \right) \tag{4.10}$$

$$\frac{1}{(1+a)^{DPP}} = 1 - \frac{aI_t}{R_N} \tag{4.11}$$

$$(1+a)^{DPP} = \frac{R_N}{R_N - aI_t} \tag{4.12}$$

and finally

$$DPP = \frac{\ln\left(\frac{R_N}{R_N - aI_t}\right)}{\ln(1+a)} \tag{4.13}$$

4.5 Return on Investment

The Return On Investment (ROI) is defined as

$$ROI = \frac{\sum_{j=1}^{n} \frac{R_{Nj}}{(1+a)^j}}{\sum_{j=0}^{n-1} \frac{I_j}{(1+a)^j}} \tag{4.14}$$

and in the simplified model conditions, it becomes

$$ROI = \frac{R_N k_a}{I_t} \tag{4.15}$$

ROI is a measure of the effective profitability of the project per unit of capital invested. $ROI = 1$ means that for per each unit of capital invested (discounted), precisely one unit of return (discounted) is obtained. Therefore, $ROI = 1$ is equivalent to $NPV = 0$.

Example 4—2
Retake Example 4—1 and compute the PB, DPP, and ROI of the project.

Solution:

The solution to this problem is just a matter of applying Eq. 4.8, Eq. 4.13, and Eq. 4.15, respectively. We recall that the simplified model conditions hold.

$$PB = 11.2144 \, \text{year}$$

$$DPP = 16.8603 \text{ year}$$

$$ROI = 1.1113$$

Note the difference between PB and DPP is significant due to the discounting applied to the DPP. Also, note that for $n = DPP$, it is $NPV = 0$.

5 Conclusions

The interest in properly evaluating the economic interest of Renewable Energy Sources (RES) projects is evident. This assessment is important for both investors and policymakers. Investors require a solid economic grounding to bet in RES projects to the detriment of other investment options. Policymakers rely on the economic perspectives of each renewable technology to decide which needs to be supported and which should be left alone to compete in the markets.

In this chapter, we have provided the basics of the economic assessment of RES projects. The objective was not to deliver a detailed economics course, but instead to provide the basic assessment tools that allow the engineers to follow the economics of a RES project.

The general framework that runs through the closing chapter is that the value of money is not timeless but instead depends on the moment in which it is traded. Thus, it has become clear that it is necessary to discount, to the present time, all monetary transactions carried out at different moments in time, to provide a common and fair basis for comparison. This is the rationale behind the discount rate, which was defined in this chapter and is the basis of the considered economic models. Another view on the discount rate is it is the minimum rate of return demanded by an investor to invest in a given project.

The economic analysis of an investment requires estimates for the energy production of the technology being assessed. To facilitate the comparison between the energy production of different technologies, it is usual to use as a technical indicator the ratio of the produced energy to the installed capacity. This is known as the utilization factor and the capacity factor is the utilization factor as a percentage of the yearly 8760 h. These indicators are very useful because they allow us to compare the energy production of different technologies regardless of the installed capacity.

The discounted average production cost, also known as Levelized Cost Of Energy (LCOE), i.e., the production cost of each MWh of energy, properly discounted and integrated along the total period of analysis, was introduced in this chapter. This economic indicator is usually used to compare the economic interest of different generating technologies. This is the indicator that allows the conclusion that Photovoltaics (PV) and wind are currently the cheapest ways of producing electricity.

To quantify the economic viability of a project, two economic assessment indexes are mainly used: Net Present Value (NPV) and the Internal Rate of Return (IRR). NPV computes the balance between discounted, i.e., referred to the present time, revenues, and expenses along the project lifetime. IRR is the discount rate that turns the NPV equal to zero, i.e., it is the real rate of return of the project. In this chapter, these indexes were presented and the decision support information they provide was analysed. Moreover, other related indexes less used in the economic analysis of projects were introduced.

In all cases, the reader has been provided with simplified models. In the preliminary phases of the project in which these economic analyses are carried out, it is proven difficult to correctly feed the detailed models, therefore, the need for simplified models. The simplified models are a commonly used tool that facilitates the dialogue between engineers and economists.

6 Proposed Exercises

Problem EA1
The total investment in a small hydro project equals 3 M€. The annual O&M costs equal 75 k€ and, in the average year, the installation produces 3.75 GWh. Compute the average sales price of energy that turns the NPV (VAL) equal to 0 for a time frame of 10 years and a discount rate of 7%.

Solution

$$\text{NPV} = R_d - C_d = p_s E_a k_a - c_{om} k_a - I_T = 0$$

$$p_s = \frac{\frac{I_T}{k_a} + c_{om}}{E_a} = 133.90 \text{ €/MWh}$$

$$k_a = \frac{(1+a)^n - 1}{a(1+a)^n} = 7.024$$

Problem EA2
A Small Hydropower Plant (SHP), with an installed capacity of 500 kW, starts its operation at $t=1$. The total investment costs were 940 k€ with the following yearly distribution (year; k€): (−2; 120), (−1; 600), and (0; 220). The average net income of selling energy to the grid equals 67.5 €/MWh, in the first 15 years, and half this value between year 16 and year 30. The lifetime of the SHP equals 30 years. Do not consider O&M costs. Assuming a discount rate of 7% and that the net income is equal every year of the two periods detailed above, compute

1. The minimum utilization of the installed capacity that renders the investment interesting.
2. For 3000 h, annual utilization of the installed capacity:

a. The NPV for a period of 15 years and a discount rate of 7%.
b. The IRR for a 15-year lifetime of the project.

Solution

(1)

Profitability boundary condition: NPV $= 0$

$$NPV = R_d - C_d = Ph_a k_a^{(15)} \left(p_{s1} + p_{s2} \frac{1}{(1+a)^{15}} \right) - I_{Td} = 0$$

$$k_a^{(15)} = 9.1079$$

$$I_{Td} = \sum_{j=-2}^{0} \frac{I_j}{(1+a)^j} = 999.388 \text{ €}$$

$$h_a = 2752 \text{ h}$$

(2)

$$NPV = R_d - C_d = Ph_a k_a^{(15)} p_{s1} - I_{Td} = -77{,}212 \text{ €}$$

$$IRR \approx a_1 - (a_2 - a_1) \frac{NPV_1}{NPV_2 - NPV_1} = 5.94\%$$

$$a_2 = 5\% \xrightarrow{yields} NPV_2 = 68{,}640 \text{ €}$$

Problem EA3
Consider a wind park with 10 MW installed capacity and an annual utilization of the installed capacity of 2500 h. The specific investment equals 1200 €/kW and the expected lifetime of the project is 20 years. It is foreseeable that, in year 10, a 50% increase in the installed capacity will be performed. This increase in capacity shall start operation in year 11. The tariff paid by the grid for the electric energy delivered shall be equal to 75 €/MWh in the first 15 years. In and after year 16, the tariff paid reduces to 50% of the value of the tariff in year 15. The O&M costs represent 1.5% of the investment costs. After 20 years, the salvage value of the equipment that began operation in year 1 is zero and the salvage value of the reinforcement equipment equals half the investment carried in year 10. Assuming a discount rate of 7%, compute the NPV of the project.

Solution

$$NPV = R_d - C_d = 4{,}618{,}129 \; €$$

$$R_d = p_{s1}P_N h_a k_a^{(10)} + p_{s1}1.5P_N h_a k_a^{(5)} \frac{1}{(1+a)^{10}} + p_{s2}1.5P_N h_a k_a^{(5)} \frac{1}{(1+a)^{15}}$$

$$+ \frac{I_{01}0.5P_N}{2} \frac{1}{(1+a)^{20}} = 21{,}896{,}486 \; €$$

$$C_d = I_{01}P_N + I_{01}0.5P_N \frac{1}{(1+a)^{10}} + c_{o\&M}I_{01}P_N k_a^{(10)} + c_{o\&M}I_{01}1.5P_N k_a^{(10)} \frac{1}{(1+a)^{10}}$$

$$= 17{,}278{,}357 \; €$$

Problem EA4

A 10 MW wind park has an average annual utilization of 2000 h. The investment cost is 1.2 M€/MW and it is foreseen that the wind park operates for 20 years. The annual O&M expenses are 1.5%. In years 5, 10, and 15, major repairs are expected, each one in the amount of 10% of the initial investment. Grid injected energy is to be paid at 75 €/MWh during the first 15 years of operation. From year 16 on, the gross income is reduced to 50% of the value it had in year 15. The salvage value after the lifetime expectancy is 20% of the initial investment. Assume a discount rate of 7% and compute the NPV and the IRR.

Solution

$$NPV = R_d - C_d = -410{,}809 \; €$$

$$R_d = p_{s1}P_N h_a k_a^{(15)} + p_{s2}P_N h_a k_a^{(5)} \frac{1}{(1+a)^{15}} + 20\% \cdot I_{01}P_N \frac{1}{(1+a)^{20}}$$

$$C_d = I_{01}P_N + c_{o\&M}I_{01}P_N k_a^{(20)} + 10\% \cdot I_{01}P_N \left(\frac{1}{(1+a)^5} + \frac{1}{(1+a)^{10}} + \frac{1}{(1+a)^{15}} \right)$$

$$IRR = a_1 - (a_2 - a_1)\frac{NPV_1}{NPV_2 - NPV_1} = 6.55\%; a_2 = 6\%, NPV_2 = 502{,}872 \; €$$

Solar Power

4

Abstract

PV power is a mature technology that will lead to further development of Renewable Energy Sources (RES). Moreover, an impressive drop-down in PV investment costs has been witnessed in recent years, making PV a strong competitor in the RES market. The main aim of this chapter is to study PV systems. To compute the DC power output of a PV module, two models based on the respective equivalent electrical circuit are proposed—the simpler 1-diode and 3-parameter model and the more complex 1-diode and 5-parameter model. The possible options for the layout of utility-scale PV parks, concerning the location of both the Maximum Power Point Trackers (MPPT) and the inverters, are reviewed. An inverter model based on its efficiency is proposed, which, together with the output power model, allows for the electricity produced to be computed. Solar tracking systems and floating PV systems are presented. Both the conventional silicon and thin film conventional technologies are explained, together with the new emerging technologies. In the final part, Concentrating Solar Power (CSP) technologies are presented and discussed.

1 Introduction

In each hour, the sun delivers to the Earth the same energy that is used in human activities for one year, i.e., about 4.6×10^{20} J $= 1.28 \times 10^5$ TWh.[1] A tiny parcel of

---

[1]George Crabtree, Nathan Lewis, Solar Energy Conversion, 2007.

© The Author(s), under exclusive license to Springer Nature Switzerland AG 2022
R. Castro, *Electricity Production from Renewables*,
https://doi.org/10.1007/978-3-030-82416-7_4

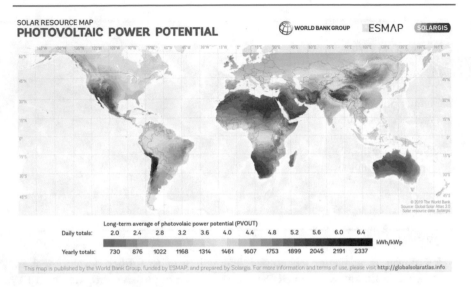

Fig. 1 PV power potential. *Source* Solargis, https://solargis.com/

this huge input energy can be converted into electricity. Figure 1 shows the world's potential for electricity production using Photovoltaic (PV) systems.

The measuring index is the utilization factor introduced in Chap. 3. We recall that the utilization factor is the number of hours the PV system should operate always at rated power[2] to produce the same electricity it produced for one year. The defining equation is

$$h_a = \frac{E_a}{P_p} \tag{1.1}$$

where E_a is the yearly produced energy and P_p is the peak power. The units are $\frac{kWh}{kW_p} = h$.

There are many regions in the world where the utilization factors are higher than 2000 h, which is a remarkable index for a PV system, given that there is no sun at night. In global terms, Europe is perhaps the continent with a lower solar resource, except for the southern countries (Portugal and Spain) where the solar power is significant (about 2000 h utilization factor).

Today, it is commonly accepted that Renewable Energy Source (RES) further development will significantly rely on PV generators. This will take advantage of the impressive cost decrease that has been witnessed in recent years. Based on

---

[2]We will see later that the equivalent concept of rated power applicable to PV systems is peak power. Peak power is measured in Wp.

projects completed in 2019, the global weighted average Levelized Cost of Energy (LCOE) of large-scale solar PV plants is down by 89% since 2009.[3]

In the following chapter, sun power is addressed, in particular, the electricity production from PV utility-scale parks. We begin by introducing some definitions and the indexes most used to assess the performance of PV systems. An explanation of the operating principle of a PV cell, which dismisses the turbine and the generator used in all other electricity production technologies, is offered. Current PV technology uses a silicon crystal, adequately prepared, to directly produce a DC current. To connect a PV module to the grid, a DC/AC converter, called an inverter, is required.

Using the common approach in electrical engineering, an equivalent circuit is derived to represent the behaviour of a PV cell. The output power of a PV module depends on the irradiance and the module temperature. Two models that can compute the output DC power of a PV module based on an equivalent circuit are presented— the simpler 1-diode and 3-parameter model and the more complex 1-diode and 5-parameter model. A common feature of both models is that their parameters can be computed-based solely on readily available datasheet information.

As mentioned, inverters are required to interface the PV production system with the grid. Another required equipment is the Maximum Power Point Tracker (MPPT) that ensures the operation of the PV modules at maximum power given the irradiance and temperature conditions. The main layouts used to dispose of both the MPPT and the inverters inside the PV park are reviewed.

Once the injected AC power into the grid is computed, based on the modules' DC power and the efficiency of the MPPT and the inverter, the electricity produced is easily computed based on the known relationship between average power and time interval. The input data required to compute the electricity production are presented.

The efficiency of PV panels is low, therefore ways of improving the overall efficiency of the conversion are being tested. Examples are solar tracking systems and floating PV, i.e., installing the PV systems above the water to take advantage of the cooling effect.

A review of the technologies, both the conventional and the emerging, is proposed. PV power is produced using today's standard silicon modules. This technology is currently mature and well-known. Current monocrystalline silicon modules can reach maximum efficiency of about 18%, proving that the investment in research paid off. Alternative thin-film technology is currently losing market share, mainly because the real-world efficiencies are not what has been promised.

Moreover, emerging technologies are being intensively researched and are expected to reach the commercial phase soon. Examples of new technologies are Gratzel, organic, and Perovskite cells, where new materials are being tested. The manufacturers are already offering bifacial modules that will increase the efficiency of the conversion.

The option to concentrate the solar beams is on the table, either in association with multijunction PV cells or as a separate and different technology, known as Concentrating Solar Power (CSP).

---

[3]Lazard's Levelized Cost of Energy Analysis—Version 13.0, www.lazard.com.

Building Integrated Photovoltaics (BIPV) are nowadays a common element of the rural and urban landscapes. The four main options for building integration of PV modules are on sloped roofs, flat roofs, facades, and shading systems. BIPV are part of the building's envelope, acting like a material that generates power. PV power is the preferred technology for micro-generation applications in PV buildings, where it can be cost-effective. Also, Near Zero Energy Buildings (NZEB), which make intensive use of PV power, are becoming common.

The last part of the chapter deals with other applications of solar power, as is the case of CSP that does not make use of PV technology. The main CSP technologies —parabolic troughs and solar tower—are introduced.

As anticipated, the present text will approach solely the high-power applications of PV power, known as utility-scale PV parks. Medium power applications, used in rural electrification and micro-generation installed near the end-users, and low power applications, used in watches and pocket calculators, battery chargers, road signals, parking meters, etc., are out of the scope of this text.

2   Basic Concepts

2.1   Some Definitions

The study of PV power requires some prior definitions to provide the context. Here, they are

- Irradiance—The solar power; it is represented by G and is measured in W/m^2.
- Irradiation (Insolation)—The solar energy; it is represented by H_i and is measured in Wh/m^2.
- Peak Power—The peak power; it is represented by P_p and is measured in W_p. It is the PV module electrical output DC power under Standard Test Conditions (STC).
- Standard Test Conditions—The internationally agreed conditions by the manufacturers to perform the PV modules factory tests. They are defined as $G^r = 1000$ W/m^2 and module temperature $\theta_m^r = 25\,^{\circ}\mathrm{C} \Leftrightarrow T^r = 273 + 25 = 298$ K. In the sequence, the quantities referenced by the superscript r are considered to refer to STC. Also, the Greek letter θ is used to represent the temperature in $^{\circ}$C, and T is used to represent the temperature in K.

2.2   Performance Indexes

The efficiency at STC is defined as (A is the module area)

$$\eta^r = \frac{P_p}{G^r A} \tag{2.1}$$

The efficiency at other irradiance (G) and module temperature (T) conditions is

$$\eta = \frac{P_{DC}(G,T)}{GA} \tag{2.2}$$

where $P_{DC}(G,T)$ is the PV module DC output power for the given irradiance and temperature conditions.

The utilization of peak power, or utilization factor, is

$$h_a = Y_F = \frac{E_a}{P_p} \tag{2.3}$$

where E_a is the annual electricity yield injected in the grid by the PV module. For instance, in Portugal, this index changes between 1500 h (in the northern part) to more than 1800 h (in the southern part).

The reference index is

$$Y_r = \frac{H_i}{G^r} \tag{2.4}$$

The performance ratio is defined as

$$PR = \frac{Y_F}{Y_r} = \frac{E_a}{\eta^r A H_i} \tag{2.5}$$

that is the ratio between the electricity injected in the grid and the one that would have been produced by the PV module if the efficiency was constant and equal to the efficiency at STC. Typical values for PR lay between 0.7 and 0.8.

2.3 PV System Equipment

A typical PV system is comprised of the equipment shown in Fig. 2.

The MPPT is a Maximum Power Point Tracker that optimizes the operating point of the PV module and extracts the maximum possible power from it, given the conditions of irradiance and temperature. The inverter is a DC/AC converter, which is required because the operating principle of PV modules implies Direct Current (DC) power is produced. However, the power system operates in Alternating Current (AC), thus an inverter is needed. Both devices are described further in this text.

The transformer is a device that transforms the AC output voltage to the adequate level to be injected into the grid. Details about DC, AC, and transformers can be found in Chap. 2.

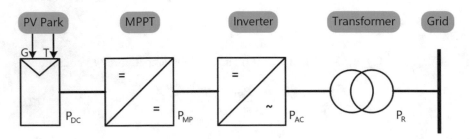

Fig. 2 PV system equipment

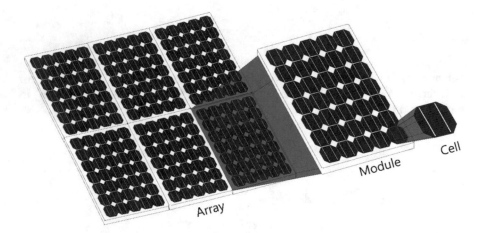

Fig. 3 PV cells, modules, and arrays

To increase their utility (the output voltage of a single PV cell is typically 0.5 V), several individual PV cells are interconnected together in a sealed, weatherproof package called a Module. For example, it is common to have modules composed of 72 PV cells connected in series. The characteristic parameters of PV modules are given in the datasheets. To achieve the desired voltage and current, modules are wired in series and parallel into what is called a PV Array. The flexibility of the modular PV system allows designers to create solar power systems that can meet a wide variety of electrical needs. Figure 3 shows a PV cell, a module, and an array.

Many PV arrays connected in series and parallel, together with the inverters and transformer, compose a utility-scale PV park.

2.4 PV Cell Operating Principle

All other generation technologies, such as hydro, coal, combined cycle, and wind, use the same operating principle based on a turbine coupled with a generator. PV technology shows a completely different approach.

First, we require a material with the appropriate characteristics. Silicon is the most used material for PV conversion. Silicon has 14 electrons, with 4 electrons in the valence band. To achieve the desired objective of having 8 electrons in the valence band, each atom of silicon makes 4 bonds with 4 neighbour atoms, sharing one electron with each one of the neighbours, as seen in Fig. 4. These bonds are called covalent bonds. In this way, the valence band is full, and we have a stable connection.

The particles that compose solar radiation are photons. When the silicon crystal is exposed to sunlight, photons with enough energy, higher than the silicon energy gap band, 1.12 eV, can displace electrons to the conduction band and originate hole–electron pairs.

If nothing else was done, the free electrons would return to the valence band after a while, because no electric field would keep them in the conduction band.

So, we need to create an electric field to keep the free electrons in the conduction band. This is achieved by doping the silicon. The doping process consists of injecting, on one side of the crystal, a material with one less electron than silicon, for instance, boron. This way, a positive zone has been created, the silicon type

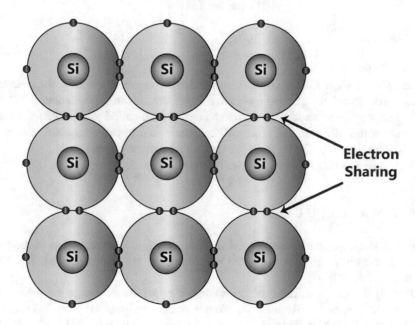

Fig. 4 Silicon covalent bonds

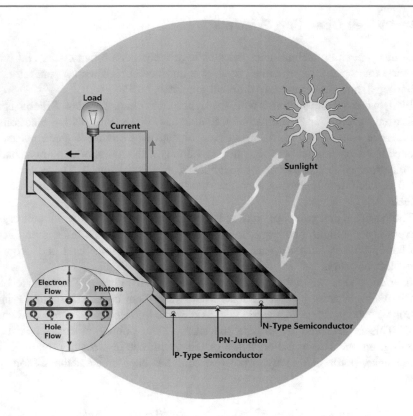

Fig. 5 PV effect

p zone. On the other side, we inject a material with one more electron, for instance, phosphorus, therefore, creating a negative zone, the silicon type *n* zone.

In this way, we created an electrical field, the so-called *p–n junction*. In the presence of this electric field, holes will be accelerated to the + terminal, electrons will be accelerated to the — terminal, and a DC current is produced.

A picture showing this process can be seen in Fig. 5.

2.5   A Note on PV Costs

It is well known that PV costs experienced an amazing cost reduction in the past years. Just to have an idea about the magnitude of the cost reduction, Fig. 6 is offered.

This figure shows an impressive investment cost reduction over the past years for all PV segments: Residential, Commercial, Utility-Scale Fixed Tilt, and Utility-Scale One-Axis Tracker. The figures are for the year 2017. Today, the

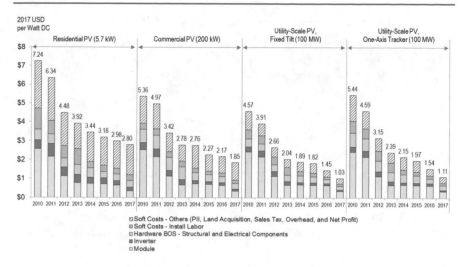

Fig. 6 PV cost reduction. *Source* NREL, www.nrel.gov, 2017

investment cost is even lower, the last Figs. (2020) pointing to 0.6 €/Wp for utility-scale PV parks.

3   PV Performance Models

In this section, the theoretical backgrounds of the commonly used PV performance model are presented. These models are simplified Fast Estimate (FE) and the intermediate 1D + 3P (1 Diode + 3 Parameters).

3.1   Data Provided by Manufacturers in Datasheets

Manufacturers of PV modules provide datasheets detailing the characteristics of their products. Table 1 shows the data commonly provided by the manufacturers. Data given at STC is mandatory. Data given at Normal Operating Conditions (NOC) is optional. NOC are defined as irradiance = 800 W/m^2 and ambient temperature = 20 °C.

3.2   Simplified Model—Fast Estimate (FE)

To establish a simple model to describe the behaviour of a PV module, it is assumed that the DC power output depends linearly on the irradiance, G, and the temperature

Table 1 Data commonly provided by manufacturers in PV module datasheets

Symbol	Unit	Description
$P^r_{MP} = P_p$	Wp	Peak power—Maximum DC power output @STC
V^r_{MP}	V	Output voltage at maximum power @STC
I^r_{MP}	A	Output current at maximum power @STC
V^r_{oc}	V	Open-circuit voltage @STC
I^r_{sc}	A	Short-circuit current @STC
$NOCT$	°C	Normal Operating Conditions Temperature (NOCT)—Module temperature in Normal Operating Conditions (NOC)
μ_{Isc}	%/°C	Temperature coefficient of the short-circuit current
μ_{Voc}	%/°C	Temperature coefficient of open-circuit voltage
μ_{Pp}	%/°C	Peak power temperature coefficient
N_s		Number of cells connected in series in the module
P^{NOC}_{MP}	W	Maximum DC power output under NOC
V^{NOC}_{MP}	V	Voltage at maximum power under NOC
I^{NOC}_{MP}	A	Current at maximum power under NOC
V^{NOC}_{oc}	V	Open-circuit voltage under NOC
I^{NOC}_{sc}	A	Short-circuit current under NOC

correction is included through the peak power temperature coefficient, μ_{Pp}, usually supplied in the manufacturer's datasheets.

$$P(G, T) = \frac{G}{G^r} P_p \left[1 + \mu_{Pp}(T - T^r) \right] \tag{3.1}$$

The advantage of this model is its simplicity; the disadvantage is that it does not allow access to other quantities of interest, such as the maximum power voltage, the maximum power current, the open-circuit voltage, and the short-circuit current.

3.3   Intermediate Model—1 Diode and 3 Parameters (1D + 3P)

We highlight that we are seeking a model that can compute the PV module DC output power for any given irradiance and temperature conditions.

With this objective in mind, a more sophisticated model can be constructed by assuming that the PV module can be described by the equivalent electric circuit shown in Fig. 7.

The p–n junction acts as a large diode. The fundamental property of a diode is that it conducts electric current in only one direction. When the applied voltage is

Fig. 7   Equivalent circuit of a
PV module—1D + 3P model

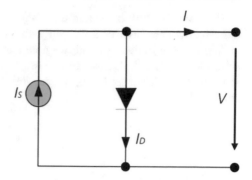

positive and greater than a certain minimum, then the current flows through the
diode. If the voltage is negative, then the diode does not conduct current,[4] at least
until a certain voltage is reached. Figure 8 helps in understanding the process.

Current I_D is the current across a diode, whose equation is given by Eq. (3.2)
and the graphics is displayed in Fig. 8.

$$I_D = I_0\left(e^{\frac{V}{mV_T}} - 1\right) \tag{3.2}$$

where I_0—Diode's inverse saturation current; V—Terminal voltage; m—Diode's
ideality factor (ideal diode: $m = 1 \times N_s$; real diode: $m > 1 \times N_s$, N_s being the
number of cells connected in series in a module). The inverse saturation current is
the very small inverse current (magnitude order of μA) flowing across the diode
when the voltage is negative.

Still regarding Eq. (3.2), V_T is the thermal voltage in V, given by

$$V_T(T) = \frac{K}{q}T \tag{3.3}$$

where K is the Boltzmann constant ($K = 1.38 \times 10^{-23}$ J/K), T is the absolute
temperature in K, and q is the electron's electrical charge ($q = 1.6 \times 10^{-19}$ C). For
the STC temperature, it is $V_T^r = 0.0257$ V.

Current I_S is represented by a current source and is the current generated by the
beam of light radiation photons, upon reaching the active surface of the module
(photovoltaic effect); this unidirectional current is constant for a given incident
irradiance. As mentioned, the p–n junction operates as a diode crossed by a unidirectional internal current I_D, which depends on the terminal voltage V.

Referring to Fig. 7, current I is

$$I = I_s - I_D = I_s - I_0\left(e^{\frac{V}{mV_T}} - 1\right) \tag{3.4}$$

---

[4]In fact, a very small current, called inverse saturation current, flows.

Our first task will be to find out if the electric circuit represented in Fig. 7 can describe the behaviour of a PV cell with an appropriate degree of accuracy.

With this objective in mind, we note that the source current, I_S, is equal to the short-circuit current, I_{sc}, as can be seen from Eq. (3.4), by setting $V = 0$. In the same equation, if we set $I = 0$, we can derive an equation for the inverse saturation current, I_0, as

$$I_0 = \frac{I_{sc}}{e^{\frac{V_{oc}}{mV_T}} - 1} \tag{3.5}$$

where V_{oc} is the open-circuit voltage.

Replacing Eq. (3.5) in (3.4) and taking into consideration that $e^{\frac{V_{oc}}{mV_T}} \gg 1$; $e^{\frac{V}{mV_T}} \gg 1$, we obtain

$$I = I_{sc}\left(1 - e^{\frac{V-V_{oc}}{mV_T}}\right) \tag{3.6}$$

For a given PV cell with an area of 0.01 m², we have measured the following values: $G = 430$ W/m², $T = 273 + 25 = 298$ K, $I_{sc}=1.28$ A, and $V_{oc} = 0.56$ V. Then, we applied different voltage values and measured the corresponding current. The experimental I-V graphic is shown in Fig. 9 by the dark dots.

Then we proceeded to the simulation. By replacing T, I_{sc}, and V_{oc} in Eq. (3.6), we can give values to V and compute the corresponding I. We note that m is still unknown, and till now we have not found a way to compute it. As such, let us for the time being impose $m = 1$,[5] which represents an ideal diode in the circuit of Fig. 7.

The obtained results are shown in Fig. 9 by the fine dashed line. We were unable to reproduce the experimental results, even when we tried $m = 2$, as shown by the thick dashed line in Fig. 9. Nevertheless, the shape of the curves is fairly similar, which indicates that there exists an m value, which allows for the experimental curve to be reproduced through simulation. This validates our 1D + 3P model.

Let us go on with the computation of the DC power. For a given irradiance and module temperature, the electrical DC output power, P, is

$$P = VI = V\left[I_{sc} - I_0\left(e^{\frac{V}{mV_T}} - 1\right)\right] \tag{3.7}$$

The maximum power is obtained when $dP/dV = 0$, which leads to

$$I_{sc} + I_0\left(1 - e^{\frac{V}{mV_T}} - \frac{V}{mV_T}e^{\frac{V}{mV_T}}\right) = 0$$

---

[5]We are considering only one cell.

$$e^{\frac{V}{mV_T}} = \frac{\frac{I_{sc}}{I_0} + 1}{\frac{V}{mV_T} + 1} \qquad (3.8)$$

The determination of the maximum power is of utmost importance because PV modules are equipped with a power electronic device, the Maximum Power Point Tracker (MPPT), which ensures the module operates at maximum power given the existent irradiance and temperature conditions. As such, we keep our objective in mind, which is to be able to compute the maximum PV module DC output power, given the module temperature and irradiance.

The solution to Eq. (3.8) is $V = V_{MP}$, the maximum power voltage, and the corresponding current is $I = I_{MP}$, the maximum power current, respectively, given by

$$V_{MP} = mV_T ln\left(\frac{\frac{I_{sc}}{I_0} + 1}{\frac{V_{MP}}{mV_T} + 1}\right) \qquad (3.9)$$

$$I_{MP} = I_{sc} - I_0\left(e^{\frac{V_{MP}}{mV_T}} - 1\right) \qquad (3.10)$$

The maximum power, hereafter denoted by P_{DC} to highlight it is DC power, is $P_{MP} = V_{MP}I_{MP} = P_{DC}$.

As Eq. (3.9) is a non-linear equation, its solution requires iterative methods. If Gauss–Seidel is used, the required iterative equation to be solved is (k is the iteration number)

$$V_{MP}^{(k+1)} = mV_T ln\left(\frac{\frac{I_{sc}}{I_0} + 1}{\frac{V_{MP}^{(k)}}{mV_T} + 1}\right) \qquad (3.11)$$

To solve Eq. (3.11), we need a starting guess, $V_{MP}^{(0)}$, and the knowledge of the three parameters m, I_0, and I_{sc}. We recall that the thermal voltage is known because it depends solely on the module temperature, which we assume is known.

A proper starting guess is $V_{MP}^{(0)} = V_{MP}^r$. We will now proceed with the determination of the three parameters of the model at STC.

First, let us write the fundamental Eq. (3.4) at the short-circuit ($V = 0$), open-circuit ($I = 0$), and maximum power ($V = V_{MP}; I = I_{MP}$) points, respectively, and at STC:

$$I_{sc}^r = I_s^r \qquad (3.12)$$

$$0 = I_{sc}^r - I_0^r\left(e^{\frac{V_{oc}^r}{m^r V_T^r}} - 1\right) \qquad (3.13)$$

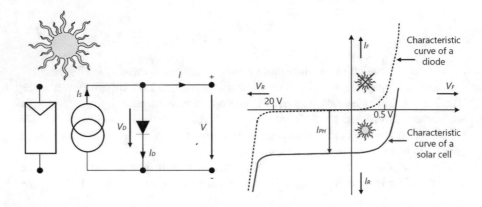

Fig. 8 Sun power source + diode

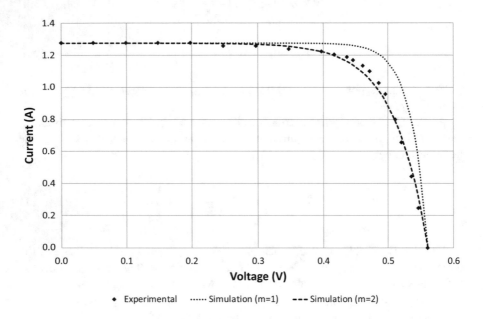

Fig. 9 1D + 3P model validation against experimental results

$$I^r_{MP} = I^r_{sc} - I^r_0 \left(e^{\frac{V^r_{OC}}{m^r V^r_T}} - 1 \right) \tag{3.14}$$

We immediately conclude that Eq. (3.12) portrays the first equation we are looking for. The parameter I^r_{sc} is directly given in the module datasheet, as seen in Table 1.

From Eq. (3.13), we obtain the second parameter:

$$I_0^r = \frac{I_{sc}^r}{e^{\frac{V_{oc}^r}{m^r V_T^r}} - 1}$$

(3.15)

where V_{oc}^r is the open-circuit voltage at STC.

Finally, the third parameter is obtained by replacing Eq. (3.15) into Eq. (3.14), thereby leading to

$$m^r = \frac{V_{MP}^r - V_{oc}^r}{V_T^r ln\left(1 - \frac{I_{MP}^r}{I_{sc}^r}\right)}$$

(3.16)

It is important to highlight that the 3 parameters of the model can be computed solely based on datasheet open information, as can be verified in Table 1.

Now, the conditions are met to compute the maximum power voltage, given by Eq. (3.11), at STC, the obtained result being

$$V_{MP}^{r}{}^{(k+1)} = m^r V_T^r ln\left(\frac{\frac{I_{sc}^r}{I_0^r} + 1}{\frac{V_{MP}^{r}{}^{(k)}}{m^r V_T^r} + 1}\right)$$

(3.17)

Of course, this has no interest at all, because the STC maximum power voltage is given at the module datasheet, and therefore is known. What would be interesting is to find the maximum power voltage for irradiance and temperature conditions different from STC.

Before that, it would be interesting to see how an I-V and a P-V curve look like for STC. The relevant equations are Eq. (3.4) and Eq. (3.7) that we recover here written for STC.

$$I = I_{sc}^r - I_0^r \left(e^{\frac{V}{m^r V_T^r}} - 1\right)$$

(3.18)

$$P = VI$$

(3.19)

Based on datasheet information, we can compute the 3 parameters of the 1D + 3P model and give values to V and compute the corresponding I. The obtained result is shown in Fig. 10.

One can observe the peak power, i.e., the I-V point of maximum power under STC. The peak power, P_p, is defined as the PV cell DC output power at STC ($G^r = 1000$ W/m^2 and $T^r = 298$ K (25°C)). In fact, we have

$$P_p = V_{MP}^r I_{MP}^r = P_{DC}^r$$

(3.20)

One quantity that is usually defined to account for the performance of a PV module is the Fill Factor at STC, defined as

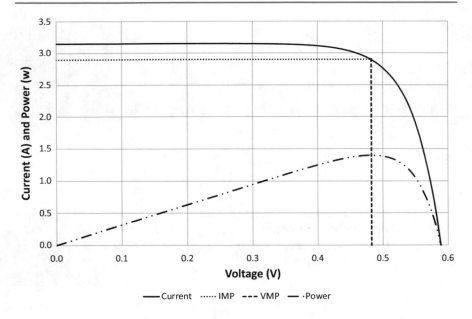

Fig. 10 I-V and P-V curves @STC

$$FF^r = \frac{P_p}{V^r_{oc}I^r_{sc}} \qquad (3.21)$$

The maximum voltage of a PV module is V^r_{oc} and the maximum current is I^r_{sc}. However, at both of these operating points, the power from the PV module is zero. So, the product $V^r_{oc}I^r_{sc}$ is impossible to attain. The fill factor is defined as the ratio between two areas, as shown in Fig. 11.

Fig. 11 Fill Factor of a PV cell

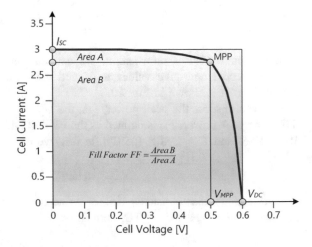

The fill factor is always less than one, the higher the fill factor, the best. The typical FF of a PV silicon module is between 0.7 and 0.8.

Influence of irradiance and temperature

The influence of the irradiance and module temperature in the 1-diode and 3-parameter model is included by making the following approximations:

- the ideality factor is constant $m^r = m$;
- the module temperature influence is accounted for in the inverse saturation current, $I_0 = I_0(T)$;
- the variation of the irradiance is incorporated in the short-circuit current, $I_{sc} = I_{sc}(G)$.

These approximations are justified by experimental evidence.

Therefore, for any temperature and irradiance given conditions, Eq. (3.11) can be written as

$$V_{MP}^{(k+1)}(G, T) = mV_T(T)ln\left(\frac{\frac{I_{sc}(G)}{I_0(T)}+1}{\frac{V_{MP}^{(k)}}{mV_T(T)}+1}\right) \tag{3.22}$$

It remains to be explained how we are going to incorporate the influence of the temperature on the inverse saturation current and the influence of the irradiance on the short-circuit current.

Let us begin with the former. It falls outside the scope of this course, but it is possible to demonstrate that the simplest model accounts for the inverse saturation current dependence on the temperature by

$$I_0(T) = DT^3 e^{\frac{-N_s\varepsilon}{mV_T(T)}} \tag{3.23}$$

where D is a constant, $\varepsilon = 1.12$ eV is the silicon bandgap, and N_s is the number of series-connected cells in a PV module. The value of D is not relevant, because we can write Eq. (3.23) at STC as

$$I_0^r = DT^{r3} e^{\frac{-N_s\varepsilon}{mV_T^r}} \tag{3.24}$$

Dividing Eq. (3.23) by Eq. (3.24), we obtain the relationship we are looking for:

$$I_0(T) = I_0^r \left(\frac{T}{T^r}\right)^3 e^{\frac{N_s\varepsilon}{m}\left(\frac{1}{V_T^r}-\frac{1}{V_T(T)}\right)} \tag{3.25}$$

We can plot the I-V curve for different temperatures, keeping the irradiance at its STC value ($G^r = 1000$ W/m^2). This is achieved by using fundamental Eq. (3.4), written at the appropriate operating point, as follows:

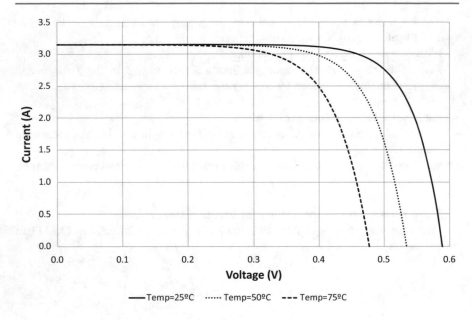

Fig. 12 I-V curves for different temperatures; $G = G^r$

$$I = I_{sc}^r - I_0(T)\left(e^{\frac{V}{mV_T(T)}} - 1\right) \tag{3.26}$$

The obtained result is depicted in Fig. 12.

We can conclude that the output power drops as the temperature increases, which is a known drawback of PV modules.

As for the influence of the irradiance on the short-circuit current, we are going to use the simplest model, which states that the short-circuit current is linearly dependent on the irradiance:

$$I_{sc}(G) = I_{sc}^r \frac{G}{G^r} \tag{3.27}$$

We highlight that the irradiance influences the inverse saturation current, and the temperature influences the short-circuit current, but we are neglecting these variations, as experimental evidence shows that they are minor.

In a similar way that we have done for the temperature, we can compute the I-V curves for different irradiances, keeping the temperature at its STC value (T^r=298 K), as shown in Fig. 13.

The used equation is now as follows:

$$I = I_{sc}(G) - I_0^r\left(e^{\frac{V}{mV_T^r}} - 1\right) \tag{3.28}$$

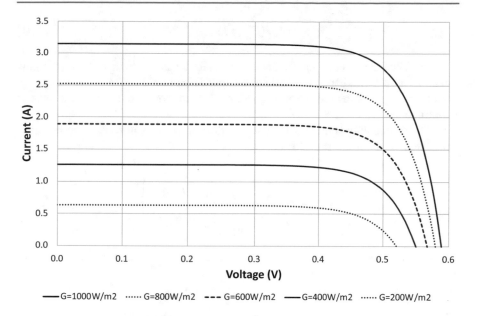

Fig. 13 I-V curves for different irradiances; $T = T^r$

It is concluded that the output power increases with the irradiance, as would be expected. Higher solar power would give rise to more PV output power.

We are now able to compute the maximum power voltage for any irradiance and temperature conditions, iteratively solving Eq. (3.22), taking into account Eqs. (3.16), (3.3), (3.27), and (3.25). After $V_{MP}(G, T)$ is obtained, the maximum power current is computed through (see Eq. (3.10), which is derived from the fundamental Eq. (3.4))

$$I_{MP}(G,T) = I_{sc}(G) - I_0(T)\left(e^{\frac{V_{MP}(G,T)}{mV_T(T)}} - 1\right)$$

(3.29)

The DC power output is finally computed by multiplying Eq. (3.22) by Eq. (3.29).

Simplified computation process of *PDC*.

Let us look again at Eq. (3.29). We notice that it can be written as follows:

$$V_{MP}(G,T) = mV_T(T)ln\left(\frac{I_{sc}(G) - I_{MP}(G,T)}{I_0(T)}\right)$$

(3.30)

The issue in this equation is that I_{MP} depends on V_{MP}. If the aim is to simplify the computation process, we can introduce a simplification to overcome this problem. As so, let us assume the maximum power current changes linearly with the irradiance, as we have assumed for the short-circuit current:

$$I_{MP}(G) = I^r_{MP}\frac{G}{G^r} \tag{3.31}$$

Experimental evidence shows that considering the short-circuit current changing linearly with the irradiance is a fairly good approximation. For the maximum power current, the approximation is not so good, but we are going to use it for the sake of simplification.

This considerably simplifies the computation process, as now V_{MP} can be easily calculated, without the need for iterations.

$$V_{MP} = mV_T ln\left(\frac{\frac{G}{G^r}(I_{sc} - I_{MP})}{I^r_0\left(\frac{T}{T^r}\right)^3 e^{\frac{N_s\varepsilon}{m}\left(\frac{1}{V^r_T} - \frac{1}{V_T}\right)}}\right) \tag{3.32}$$

The DC power output is obtained from the multiplication of algebraic Eqs. (3.31) and (3.32).

Example 3—1
The datasheet of a PV module is known and given in Table 2.
For the Normal Operating Conditions, the datasheet indicates $P_{DC} = 72.3$ W.
Compute the output power for NOC and the respective error using

---

- *The 1D + 3P model;*
- *The 1D + 3P model with the Simplified Computation (SC) of PDC;*
- *The Fast Estimate (FE).*

Table 2 Manufacturer datasheet (Example 1)

Monocrystalline Silicon			
Peak power	P_p	100.3	Wp
Maximum power current	I^r_{MP}	5.9	A
Maximum power voltage	V^r_{MP}	17.0	V
Short-circuit current	I^r_{sc}	6.5	A
Open-circuit voltage	V^r_{oc}	21.0	V
NOCT	NOCT	45	°C
Temperature coefficient P_p	μ_{Pp}	− 0.45	%/°C
Temperature coefficient I_{sc}	μ_{Isc}	2.80×10^{-3}	A/°C
Temperature coefficient V_{oc}	μ_{Voc}	$- 7.60 \times 10^{-2}$	V/°C
Number of cells in series	N_s	36	
Length	C	1.316	m
Width	L	0.660	m

First, let us recall that the NOC are $G^{NOC} = 800$ W/m^2 and $T^{NOC} = NOCT = 318$ K.

The 3 parameters of the model are computed for STC and are given by

$$m^r = \frac{V_{MP}^r - V_{oc}^r}{V_T^r ln\left(1 - \frac{I_{MP}^r}{I_{sc}^r}\right)} = 65.38$$

$$I_0^r = \frac{I_{sc}^r}{e^{\frac{V_{oc}^r}{m^r V_T^r}} - 1} = 2.40 \times 10^{-5} \text{A}$$

$$I_{sc}^r = 6.50 \text{ A}$$

The influence of irradiance and temperature is accounted for in the short-circuit current and inverse saturation current, respectively:

$$I_{sc}(G^{NOC}) = I_{sc}^r \frac{G^{NOC}}{G^r} = 5.2 \text{ A}$$

$$I_0(T^{NOC}) = I_0^r \left(\frac{T^{NOC}}{T^r}\right)^3 e^{\frac{N_{se}}{m}\left(\frac{1}{V_T^r} - \frac{1}{V_T(T^{NOC})}\right)} = 1.32 \times 10^{-4} \text{ A}$$

Also, the thermal voltage depends on the temperature:

$$V_T(T^{NOC}) = \frac{K}{q} T^{NOC} = 0.0274 \text{ V}$$

1. First, we have to compute the maximum power voltage at NOC using the Gauss–Seidel iterative process:

$$V_{MP}^{(k+1)}(G^{NOC}, T^{NOC}) = mV_T(T^{NOC})ln\left(\frac{\frac{I_{sc}(G^{NOC})}{I_0(T^{NOC})} + 1}{\frac{V_{MP}^{(k)}}{mV_T(T^{NOC})} + 1}\right)$$

Here are the steps of the iterative procedure:

$$V_{MP}^{(0)} = V_{MP}^r = 17 \text{ V}$$

$$V_{MP}^{(1)} = 14.74 \text{ V}$$

$$V_{MP}^{(2)} = 14.97 \text{ V}$$

$$V_{MP}^{(3)} = 14.95 \text{ V}$$

$$V_{MP}^{(4)} = 14.95 \text{ V}$$

We stop the iterative process when the difference between V_{MP} computed in two consecutive iterations is lower than a pre-defined tolerance ϵ_1 (typically $\epsilon_1 = 10^{-2}$).

$$\left| V_{MP}^{(k)} - V_{MP}^{(k-1)} \right| < \epsilon_1$$

After the fourth iteration, the final result is obtained:

$$V_{MP}^{(4)} = 14.95 \text{ V} = V_{MP}(G^{NOC}, T^{NOC})$$

Now, we can compute the maximum power current as

$$I_{MP}(G^{NOC}, T^{NOC}) = I_{sc}(G^{NOC}) - I_0(T^{NOC}) \left(e^{\frac{V_{MP}(G^{NOC}, T^{NOC})}{mV_T(T^{NOC})}} - 1 \right) = 4.64 \text{ A}$$

and the DC output power is

$$P_{DC}(G^{NOC}, T^{NOC}) = V_{MP}(G^{NOC}, T^{NOC}) I_{MP}(G^{NOC}, T^{NOC}) = 69.43 \text{ W}$$

The error relative to the datasheet value is

$$\text{error} = \frac{P_{DC}(G^{NOC}, T^{NOC}) - P_{DC}}{P_{DC}} = -3.97\%$$

2. Using the Simplified Computation (SC) of PDC, we assume that

$$I_{MP}(G^{NOC}) = I_{MP}^r \frac{G^{NOC}}{G^r} = 4.72 \text{ A}$$

The maximum power voltage can therefore be computed as

$$V_{MP}(G^{NOC}, T^{NOC}) = mV_T(T^{NOC}) \ln \left(\frac{I_{sc}(G^{NOC}) - I_{MP}(G^{NOC})}{I_0(T^{NOC})} \right) = 14.69 \text{ V}$$

The final result achieved by this SC process is

$$P_{DC}(G^{NOC}, T^{NOC}) = V_{MP}(G^{NOC}, T^{NOC}) I_{MP}(G^{NOC}) = 69.32 \text{ W}$$

$$\text{error} = -4.12\%$$

3. If the FE method is used, the final result is

$$P(G^{NOC}, T^{NOC}) = \frac{G^{NOC}}{G^r} P_p \left[1 + \mu_{P_p}(T^{NOC} - T^r) \right] = 73.02\,\text{W}$$

$$\text{error} = 0.99\%$$

The FE method achieved the best estimation. The FE method usually performs well at NOC because the peak power temperature coefficient is computed by the manufacturers using this test at NOC. For other operating conditions, the estimates are not so good. A further problem is that FE does not provide any information about other relevant quantities, such as voltages and currents.

3.4 Detailed Model—1 Diode and 5 Parameters (1D + 5P)

The equivalent circuit of this model is represented in Fig. 14. It accounts for several losses that exist in a PV module.

Once again, the objective is to be able to compute the 5 model parameters—I_s^r, I_0^r, m, R_s e R_{sh}—which are based solely on the manufacturer's datasheet, namely V_{oc}^r, I_{sc}^r, V_{MP}^r, I_{MP}^r, μ_{Isc} and μ_{Voc}.

In this case, current I can be written as

$$I = I_s - I_D - I_{sh} = I_s - I_0 \left(e^{\frac{V + R_s I}{mV_T}} - 1 \right) - \frac{V + R_s I}{R_{sh}} \tag{3.33}$$

If we write Eq. (3.33) corresponding to the short-circuit, open-circuit, and maximum power operating points, at STC, one can conclude that we have 3

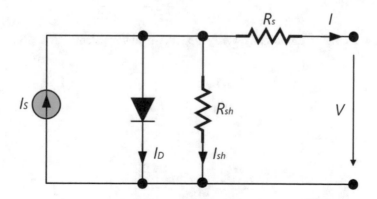

Fig. 14 Equivalent circuit of a PV module—1D + 5P model

equations and 5 unknowns. However, we can express 2 unknowns (I_0^r and I_s^r) as a function of the remaining 3 (m, R_s, and R_{sh}) using the former three equations. The result is

$$I_s^r = I_0^r e^{\frac{V_{oc}^r}{mV_T^r}} + \frac{V_{oc}^r}{R_{sh}} \tag{3.34}$$

$$I_0^r = \left(I_{sc}^r - \frac{V_{oc}^r - R_s I_{sc}^r}{R_{sh}} \right) e^{\frac{-V_{oc}^r}{mV_T^r}} \tag{3.35}$$

If we replace Eq. (3.34) and Eq. (3.35) in the maximum power point, we then obtain the first equation as

$$I_{MP}^r = I_{sc}^r - \frac{V_{MP}^r + R_s I_{MP}^r - R_s I_{sc}^r}{R_{sh}} - \left(I_{sc}^r - \frac{V_{oc}^r - R_s I_{sc}^r}{R_{sh}} \right) e^{\frac{V_{MP}^r + R_s I_{MP}^r - V_{oc}^r}{mV_T^r}} \tag{3.36}$$

In the maximum power point situation, we have

$$\left. \frac{dP}{dV} \right|_{\substack{V = V_{MP}^r \\ I = I_{MP}^r}} = 0 = \left. \frac{d(VI)}{dV} \right|_{\substack{V = V_{MP}^r \\ I = I_{MP}^r}} = \left. \left(I + \frac{dI}{dV} V \right) \right|_{\substack{V = V_{MP}^r \\ I = I_{MP}^r}} \tag{3.37}$$

This allows obtaining the second equation as

$$\left. \frac{dP}{dV} \right|_{\substack{V = V_{MP}^r \\ I = I_{MP}^r}} = I_{MP}^r + \frac{-\frac{\left(R_{sh} I_{sc}^r - V_{oc}^r + R_s I_{sc}^r \right) e^{\frac{V_{MP}^r + R_s I_{MP}^r - V_{oc}^r}{mV_T^r}}}{mV_T^r R_{sh}} - \frac{1}{R_{sh}}}{1 + \frac{R_s \left(R_{sh} I_{sc}^r - V_{oc}^r + R_s I_{sc}^r \right) e^{\frac{V_{MP}^r + R_s I_{MP}^r - V_{oc}^r}{mV_T^r}}}{mV_T^r R_{sh}} + \frac{R_s}{R_{sh}}} V_{MP}^r = 0 \tag{3.38}$$

It is possible to empirically verify that the shunt resistance, R_{sh}, follows the relationship:

$$\left. \frac{dI}{dV} \right|_{I = I_{sc}^r} = -\frac{1}{R_{sh}} \tag{3.39}$$

which results in

$$\left. \frac{dI}{dV} \right|_{I = I_{sc}^r} = \frac{-\frac{\left(R_{sh} I_{sc}^r - V_{oc}^r + R_s I_{sc}^r \right) e^{\frac{R_s I_{sc}^r - V_{oc}^r}{mV_T^r}}}{mV_T^r R_{sh}} - \frac{1}{R_{sh}}}{1 + \frac{R_s \left(R_{sh} I_{sc}^r - V_{oc}^r + R_s I_{sc}^r \right) e^{\frac{R_s I_{sc}^r - V_{oc}^r}{mV_T^r}}}{mV_T^r R_{sh}} + \frac{R_s}{R_{sh}}} = -\frac{1}{R_{sh}} \tag{3.40}$$

In Eq. (3.40) is displayed the third equation that is needed. The solution of the system of equations—Eqs. (3.36), (3.38), and (3.40)—allows obtaining m, R_s, and R_{sh} at STC. Then, we use Eqs. (3.34) and (3.35) to find I_0^r and I_s^r.

This model considers that m, R_s, and R_{sh} are constant, meaning that they do not depend on either irradiance or module temperature. The dependence on irradiance and module temperature is taken by the short-circuit current and the open-circuit voltage. Regarding the variation of the temperature coefficients, μ_{Isc} and μ_{Voc}, with the module temperature, given in the datasheets will be

$$I_{sc}(G,T) = I_{sc}(G)\left[1 + \mu_{I_{sc}}(T - T^r)\right]$$

$$V_{oc}(G,T) = V_{oc}(G)\left[1 + \mu_{V_{oc}}(T - T^r)\right] \tag{3.41}$$

To take into account the dependence of the short-circuit current with the irradiance, a linear variation is assumed:

$$I_{sc}(G) = \frac{G}{G^r}I_{sc}^r \tag{3.42}$$

As for the dependence of the open-circuit voltage with the irradiance, Eq. (3.33) is written at the open-circuit point, which leads to

$$V_{oc}(G) = mV_T^r ln\left(\frac{\frac{G}{G^r}I_s R_{sh} - V_{oc}(G)}{I_0^r R_{sh}}\right) \tag{3.43}$$

In Eq. (3.43), it was admitted that I_S current changes linearly with the irradiance and I_0 current is not much affected by the irradiance with its value being equal to the value at STC.

The remaining parameters of the model (I_0 and I_s) depend on irradiance and module temperature according to (see Eq. (3.34) and Eq. (3.35))

$$I_0(G,T) = \left(I_{sc}(G,T) - \frac{V_{oc}(G,T) - R_s I_{sc}(G,T)}{R_{sh}}\right)e^{\frac{-V_{oc}(G,T)}{mV_T(T)}} \tag{3.44}$$

$$I_s(G,T) = I_0(G,T)e^{\frac{V_{oc}(G,T)}{mV_T(T)}} + \frac{V_{oc}(G,T)}{R_{sh}} \tag{3.45}$$

The dependence of the five model parameters on the irradiance and module temperature was established. Now, the method to determine the maximum output power will be addressed. This is not trivial, because Eq. (3.33) is transcendental on the current, meaning that the derivative depends on the current itself and the voltage.

To overcome this difficulty, we note that from Eq. (3.37) one can write

$$\left.\frac{dI}{dV}\right|_{\substack{V = V_{MP} \\ I = I_{MP}}} = -\frac{I_{MP}}{V_{MP}} \tag{3.46}$$

Developing Eq. (3.46) leads to

$$\frac{-\dfrac{(R_{sh}I_{sc}-V_{oc}+R_sI_{sc})e^{\frac{V_{MP}+R_sI_{MP}-V_{oc}}{mV_T}}}{mV_TR_{sh}}-\dfrac{1}{R_{sh}}}{1+\dfrac{R_s(R_{sh}I_{sc}-V_{oc}+R_sI_{sc})e^{\frac{V_{MP}+R_sI_{MP}-V_{oc}}{mV_T}}}{mV_TR_{sh}}+\dfrac{R_s}{R_{sh}}} = -\frac{I_{MP}}{V_{MP}} \tag{3.47}$$

The other equation which is required is Eq. (3.36) written for the desired irradiance and module temperature conditions:

$$I_{MP} = I_{sc} - \frac{V_{MP}+R_sI_{MP}-R_sI_{sc}}{R_{sh}} - \left(I_{sc} - \frac{V_{oc}-R_sI_{sc}}{R_{sh}}\right)e^{\frac{V_{MP}+R_sI_{MP}-V_{oc}}{mV_T}} \tag{3.48}$$

Equation (3.47) together with Eq. (3.48) allows finding V_{MP} and I_{MP} and, therefore, $P_{MP} = P_{DC} = V_{MP}I_{MP}$.

4   PV System Components

The main component of a PV system is the PV module, whose models we have already approached in the last section. Other components are the Maximum Power Point Tracker (MPPT) and the inverter, which we will approach in this section.

4.1   MPPT—Maximum Power Point Tracker

As seen before, the PV module DC output power changes with atmospheric conditions (irradiance and temperature) and with the terminal voltage. Let us look back at Fig. 10, where an I-V curve is depicted. In general, any point of the I-V curve is a valid operating point, but only one is desired—the one that corresponds to the maximum DC output power, for the given conditions of irradiance and temperature. In Fig. 10, this point is highlighted as P_p, because that I-V curve refers to STC.

The maximum output power occurs for a particular voltage, known as maximum power voltage, V_{MP}. In the MPPT, this voltage is computed through a dedicated algorithm. The voltage reference value as calculated by the MPPT is the input of a DC/DC converter that adjusts the output voltage to the input voltage of the inverter.

In this way, it is assured that the PV module always operates in optimal conditions, because the maximum possible output power for the given conditions of irradiance and temperature is achieved. This highlights the need for accurate and efficient maximum power computation through appropriate simulation models because PV modules are expected to operate always at the maximum power point, for the existing irradiance and temperature conditions.

Most inverters on the market today perform the MPPT function, meaning that MPPT is incorporated inside the inverter, however, current trends point to a separate MPPT.

4.2 Inverter

PV systems convert solar radiation energy into electricity. The physical phenomena related to this conversion imply that the output of a PV system is Direct Current (DC), as seen before. Most of the PV systems in operation worldwide are connected to the power system, which is operated in Alternate Current (AC). Therefore, the PV system DC output power must be converted into an AC power, with the proper voltage and current characteristics (both in terms of magnitude and frequency) suitable to be injected into the grid. To perform this conversion, an interface device, the inverter, is required. The inverter is a piece of electrical equipment, composed of power electronics devices, whose objective is to convert DC electrical quantities into AC electrical quantities. For this reason, the inverter is also known as DC/AC converter.

Nowadays, the inverters used in PV systems are very complete, namely including in the same device the functions of

- DC/AC conversion (inverter function) with high-quality standards.
- Maximum power tracking (MPPT function), i.e., regulating the voltage to the value corresponding to the maximum power. In modern PV systems, this function tends to be performed by a separated DC/DC converter.
- System protection, namely overload, overvoltage, frequency, interconnection, detection of the island, etc. Islanding occurs whenever the grid is disconnected, and PV systems keep feeding isolated consumers. This operating mode is normally forbidden by the grid codes. Nevertheless, it is technically possible, providing that PV systems are adequately designed for that purpose.

4.2.1 Inverter Configuration

Inverters for PV systems can be classified according to their position on the terrain, as follows:

- Central inverter (Fig. 15a);
- String inverter (Fig. 15b);
- Multi-string inverter (Fig. 15c);
- Module integrated inverter (Fig. 15d).

Central inverter—This was the configuration used in the first grid-connected PV parks: the series- and parallel-connected modules are coupled to a single central

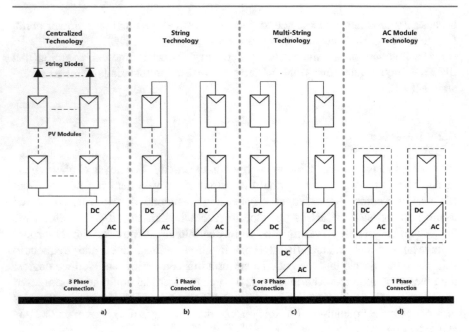

Fig. 15 Types of grid-connected inverters for PV systems: **a** central inverter; **b** string inverter; **c** multi-string inverter; **d** module integrated inverter

inverter, which integrates both the MPPT and DC/AC inverter functions. Each series-connected module string has a diode to prevent current flow circulation between strings (string diode in Fig. 15a). The increase in the voltage as obtained by the series connection is enough to fulfil the input voltage requirements of the inverter. This type of configuration presents some disadvantages: (i) increased losses, due to the centralized location of the inverter (local control of each module optimal voltage is impossible, only module ensemble control is performed); (ii) need for High Voltage (HV) DC cables to allow for the connection of the PV modules to the central inverter; (iii) design with little flexibility; (iv) relatively high price. Mass production was not reached, mainly due to these drawbacks.

String inverter—Nowadays, this is the most used configuration to connect big PV parks to the grid. Each string of series-connected PV modules is coupled to its own inverter, therefore, there is no need for string diodes (Fig. 15b). The nominal capacity of each inverter is $1/n$ of the nominal capacity of the equivalent central inverter, n being the number of strings. In general, the string voltage is enough to feed the inverter, so, no voltage amplification is necessary. If that is not the case, a DC/DC boost converter is required. String inverter configuration shows some advantages: (i) each string has its own MPPT, thus losses are reduced; (ii) reduced costs, due to mass production; (iii) improved overall efficiency.

Multi-string inverter—Each string is coupled to a common DC/DC converter (including the MPPT), which, in turn, feeds a unique inverter with a simplified

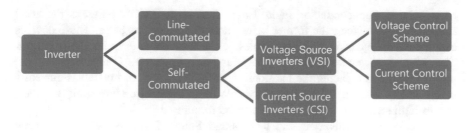

Fig. 16 Inverter types

control system (Fig. 15c). The advantages of the configurations "string inverter" and "module integrated inverter" are combined in this configuration, also known as Smart PV Panel. Each PV module string is individually controlled, therefore, increasing the overall efficiency. Moreover, costs are reduced, and the design has improved flexibility. This technology is currently showing major developments.

Module integrated inverter—This option represents the integration of both the inverter and the PV module into a unique electrical device (Fig. 15d), known as the AC module. The inverter is mounted inside or below the PV module, thus temperature-related constraints must be considered. Module integrated inverters, or micro-inverters, are used in low power applications, typically in the range 200–300 W. They present both MPPT and inverter functions, therefore, it is assumed that each module operates always at the maximum possible power point. This feature is especially important because AC modules are normally used in urban environments where shading effects may occur. Other advantages may be outlined: increased modularity, easy plug-and-play installation, and improved security. The disadvantages of this approach are mainly related to the high per-unit power cost, the central inverter being the best solution as far as this parameter is concerned. Nevertheless, this drawback may be overcome if AC modules tend to become the generalized solution.

4.2.2   Inverter Types

Inverters can be classified accordingly to the way semiconductor commutation is performed, as can be seen in Fig. 16.

Line-commutated inverters use thyristor as the commutation element. Thyristors are semi-commanded devices because they allow for ON-state control, but they are unable to control the time instant in which they go to OFF-state. OFF-state is achieved when current is zero, therefore either grid AC voltage support is needed or auxiliary circuitry with the same objective. Its operating principle implies rectangular waveforms to be obtained, with high harmonic content. High harmonic content and the associated reactive power needs are some of the reasons why modern inverters are no longer equipped with thyristors.

Modern inverters are totally commanded, i.e., self-commutated, both ON-state and OFF-state being controlled. Insulated Gate Bipolar Transistor (IGBT) and

Metal Oxide Semiconductor Field Effect Transistor (MOSFET) are the used semiconductor devices, the former being more used than the latter. These devices can operate at high switching frequencies, let us say, tens or even hundreds of kHz. They allow for accurate control of voltage and current on the AC side, adjusting the power factor and reducing the harmonic content. This type of inverter is currently being widely used in PV systems. Possible electromagnetic compatibility issues, due to high switching frequencies, are to be monitored.

Self-commutated inverters can be Voltage Source Inverters (VSI) or Current Source Inverters (CSI). The main difference between these two types of inverters is related to the DC-side representation: VSI represent the DC side as a voltage source, whereas CSI represent it as a current source. Both of them allow for a constant magnitude and variable frequency waveforms to be obtained on the AC side. In PV systems VSI are normally used because the output quantity of a PV module is a DC voltage. When seen from the AC side, VSI can be operated as a voltage source or as a current source depending on the used control system.

In the voltage control scheme, the target voltage is given as a reference and the objective of the control system is to obtain that very same voltage waveform. Pulse Width Modulation (PWM) technique is normally used: commutation time interval is determined by the comparison between the target sinusoidal waveform and a high-frequency triangular waveform, leading to a constant magnitude and variable-width impulse control signals to be obtained. This type of inverter can be used in off-grid applications.

On the other hand, the target current waveform is given as a reference in the current control scheme, and the output voltage is controlled till the target current waveform is obtained.

VSI with the current control scheme are normally used in grid-connected PV systems. Effective power factor control using relatively simple control circuits and the possibility of controlling the current when the grid is disturbed are the main advantages of this configuration. This type of inverter cannot be used in off-grid applications.

Grid connection through inverters should not spoil power supply quality. Modern inverter harmonic content, which is measured by the Total Harmonic Distortion (THD), is usually less than 3%. In many cases, this is a figure that is better than the figure corresponding to the public grid power supply, because nowadays there is a significant number of electronic devices connected to the grid that damages power quality.

4.2.3 Inverter Efficiency

An important aspect is the overall efficiency of the interface device between the PV modules and the grid. The efficiency of the inverter is given by

$$\eta_{inv} = \frac{P_{AC}}{P_{DC}} \tag{4.1}$$

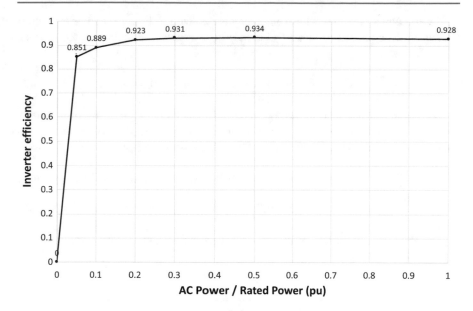

Fig. 17 Typical efficiency curve of grid-connected PV inverters

In Eq. (4.1), P_{AC} and P_{DC} are the active powers in the AC and DC sides, respectively.

Inverters are high-performance devices in the DC/AC conversion. Peak efficiency is usually between 85 and 96% for power values close to the rated power. When the output power is considerably lower than the rated power, for instance, in cloudy sky conditions, in the start-up, or at sunset, or sunrise, the efficiency significantly drops.

Typical inverter efficiencies as a function of the AC output power in the percentage of the rated power can be seen in Fig. 17.

The simpler inverter models, known as efficiency models, consider that the inverter losses can be approximated by a quadratic expression that depends on the input DC power, as shown in Eq. (4.2)

$$\eta_{inv} = \frac{P_{AC}}{P_{DC}} = \frac{P_{DC} - \left(a + bP_{DC} + cP_{DC}^2\right)}{P_{DC}} \tag{4.2}$$

We remark that Eq. (4.2) is written in terms of the input DC power, because this is the quantity that comes out from PV performance models. This formulation includes a constant parcel to model self-consumption; a linear parcel to describe the voltage drop along the semiconductors; and a quadratic parcel to account for Joule losses.

For the typical inverter whose efficiency is depicted in Fig. 17, numerical values for Eq. (4.2), considering that PDC is in pu (base power = inverter rated power), are as follows:

$$a = 6.0878 \times 10^{-3}; b = 0.0473; c = 0.0164 \qquad (4.3)$$

When the performance of several inverters is to be compared, uniform metrics are convenient. Therefore, the European Efficiency has been defined as a weighted average of the inverter efficiency at different load operating points:

$$\eta_E = 0.03\eta_{5\%} + 0.06\eta_{10\%} + 0.13\eta_{20\%} + 0.10\eta_{30\%} + 0.48\eta_{50\%} + 0.20\eta_{100\%} \quad (4.4)$$

In Eq. (4.4), η_E is the European Efficiency and $\eta_{i\%}$ is the inverter efficiency at $i\%$ of the rated power.

The European Efficiency of the typical inverter whose efficiency curve is shown in Fig. 17 is 0.9259.

5   Electricity Delivered to the Grid

One of the objectives of this chapter is to provide a methodology to compute the electricity that is injected into the grid by a PV module.

We have seen that the DC power output of a PV module depends upon the irradiance and module temperature. Therefore, we assume that information about these two quantities must be available, through dedicated measurements.

5.1   Input Data

Let us begin with the irradiance (sun power) information, which can be measured using a device called a pyranometer. But before that, let us take a look at the sun irradiation (sun energy) received on a tilted plane. Figure 18 shows the monthly sun irradiation received in each square meter of terrain in Lisbon for different tilted planes. Over one year, the annual total received irradiation (given by the sum of the monthly irradiations) for the different inclinations is the following:

- 0°—1798 kWh/m^2;
- 30°—1958 kWh/m^2;
- 60°—1749 kWh/m^2.

It is apparent that the 30° slope is the one that maximizes the sun irradiation over one year. Therefore, this is the recommended tilt for PV modules installed in the Lisbon zone.

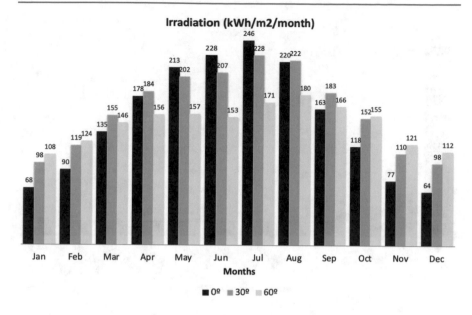

Fig. 18 Monthly sun irradiation received over tilted planes in Lisbon

Figure 19 portrays the monthly irradiance over a 30° tilted plane in Lisbon. Each bar is the monthly average computed based on hourly averages.

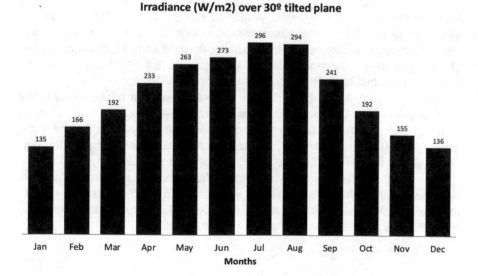

Fig. 19 Monthly average irradiance in Lisbon over a 30° tilted plane

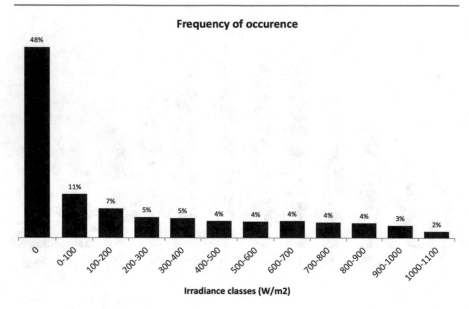

Fig. 20 Irradiance histogram in Lisbon

Figure 20 shows the respective histogram based on hourly average measure-ments and in 100 W/m² bins. One can conclude that each bin occurs more or less with the same frequency.

Irradiance information can be provided on a different time scale. For instance, Fig. 21 shows an example of the irradiance over 5 days of one week with a sampling rate of one minute for the winter and the summer. In this figure, besides global irradiance, diffuse irradiance is also shown. It is apparent that cloudy days make the variation pattern very spiky, deviating from the known bell-shaped pattern of clear sky days. Of course, this has an impact on the PV output power. As expected, peak irradiance is higher in summer than in winter.

The other input that is required to be known as an input of the model is the module temperature. However, this data will be only available after the installation of the system, not before. Before the installation of the physical PV system, the quantity that is available through measurements is the ambient temperature. Fig-ure 22 shows the monthly average ambient temperature in Lisbon.

We should now devise a way of estimating the module temperature, the one that really interests us, from the ambient temperature. To do that, we are going to introduce a simple model (known as the Ross model) that assumes that the module temperature is linearly dependent on the irradiance, that is

$$\theta_m = \theta_a + kG \tag{5.1}$$

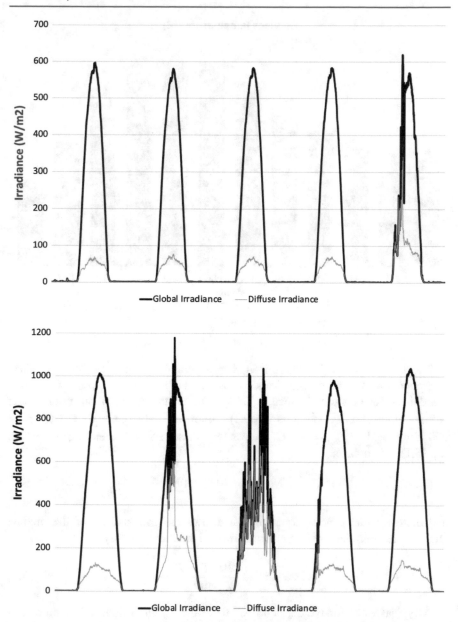

Fig. 21 Example of irradiance over a winter (top) and summer (bottom) 5 days of a week

where θ_m is the module temperature, θ_a is the ambient temperature, and G is the irradiance.

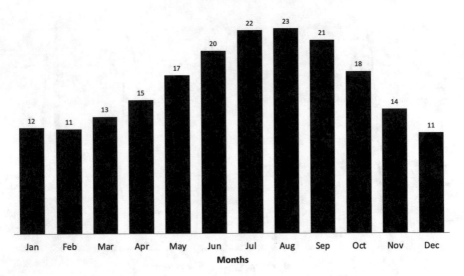

Fig. 22 Monthly ambient temperature in Lisbon

The k constant can be determined if an operating point is known. Datasheet information provides a quantity called NOCT, which stands for Normal Operating Conditions Temperature. As seen in Table 1, this is the module temperature under Normal Operation Conditions (NOC), internationally defined as $G^{NOC} = 800 \, \text{W/m}^2$ and $\theta_{amb}^{NOC} = 20 \, ^\circ\text{C}$. Replacing these known operating points in Eq. (5.1), one obtains

$$NOCT = 20 + 800k \tag{5.2}$$

and therefore, the general equation that allows the computation of the module temperature from the ambient temperature and irradiance is

$$\theta_m = \theta_a + \frac{NOCT - 20}{800} G \tag{5.3}$$

A typical value for NOCT is 45 °C. Considering this value, the module temperature, given the irradiance presented in Fig. 19, and the ambient temperature presented in Fig. 22, is shown in Fig. 23.

We should highlight that this is a simplified model, whose results are estimations. In Fig. 24, we show a comparison of the measured module temperature (in blue) and the module temperature as foreseen by the Ross model (in orange), for one winter day (left) and for a summer day (right). Some differences are noted which show that the Ross model is an approximate model.

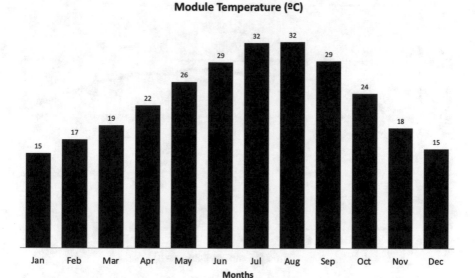

Fig. 23 Monthly module temperature in Lisbon

5.2 Output Annual Electrical Energy

The output annual electrical energy is given by

$$E_a = \sum_{i=1}^{n} \eta_{inv}(P_{DC})_i P_{DC}(G,T)_i \Delta t_i \qquad (5.4)$$

In Eq. (5.4), the quantities are as follows:

E_a—Total PV electricity injected in the grid.
η_{inv}—Efficiency of the power electronics, namely the inverter; in general, it depends on the DC output power level.
n—Total number of time intervals.
P_{DC}—Depending on the desired accuracy, it can be computed using: 1D + 5P model (in general, more accurate); 1D + 3P model with the simplified computation of PDC (intermediate accuracy); Fast Estimate (in general, less accurate).
Δt—Time interval.

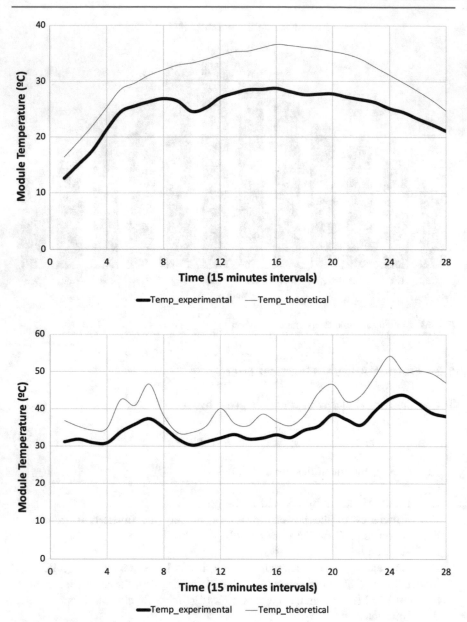

Fig. 24 Comparison of measured module temperature (experimental) with module temperature as foreseen by Ross model (theoretical) for a winter day (top) and a summer day (bottom)

Photovoltaic Solar Electricity Potential in European Countries

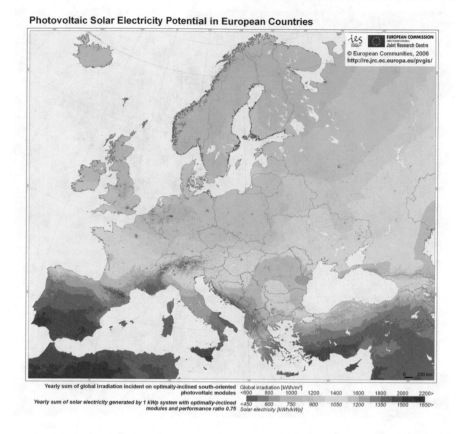

Fig. 25 PV electricity potential in Europe. *Source* Joint Research Centre, European Commission, https://ec.europa.eu/jrc/

6   Improving the Efficiency of Conventional Technologies

In Fig. 25, the European Solar Atlas is depicted, where information regarding the PV electricity potential, namely global irradiation (kWh/m^2) and utilization factor (h), is displayed. We recall that the utilization factor is the number of kWh each kWp can produce (kWh/kWp = h). It can be seen that in Southern Europe, utilization factors as high as 1650 h can be obtained, with optimally inclined modules but without solar tracking systems. These values contrast with Central Europe where no more than 900 h are to be expected.

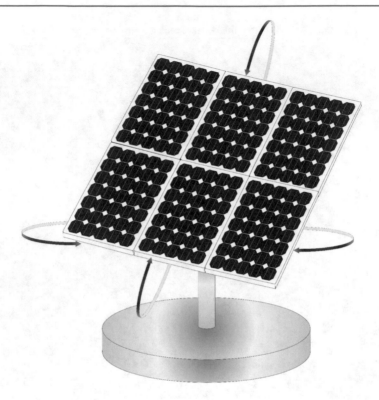

Fig. 26   Dual-axis solar tracker

6.1   Solar Tracking Systems

A typical solar tracking system adjusts the face of the PV panel to align with the sun as it moves across the sky. The aim is to ensure that the sunbeams are always perpendicular to the face of the PV module.

Some single-axis solar trackers rotate on one axis moving back and forth in a single direction. A more sophisticated and efficient tracking system is the dual-axis tracker (Fig. 26) that continually faces the sun because it can move in two different directions, therefore, allowing for maximum solar energy capture.

It is apparent that solar tracking systems increase the utilization factor of a PV park. The drawback is obviously the increased cost of the system. In the past, when the cost of the modules was higher, solar tracking systems were installed in many utility-scale PV parks, therefore, proving their cost-effectiveness. Today, with the impressive cost reduction of the PV modules, which was not followed by the cost reduction of solar tracking systems, the cost-effectiveness is more doubtful.

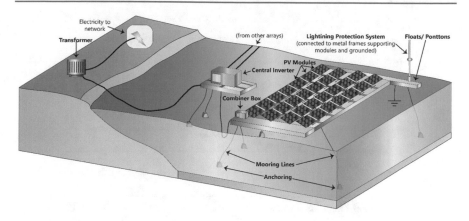

Fig. 27 Schematic representation of an FPV installation

6.2 Floating PV

Another way of increasing solar harvesting is Floating Photovoltaics (FPV), the installation of PV parks in lakes, bays, or reservoirs of hydropower stations. An exponential increase in the number of FPV installations is being witnessed. This is driven by the rapid development of large-scale projects in China in regions that lack available land for solar deployment. In Fig. 27, a schematic view of an FPV park is offered.

The main advantages can be listed as follows.

- Elimination of the need for major site preparation.
- Improved output (due to the cooling effect of water and less dust on panels).
- Reduced evaporation from water reservoirs.
- Use of existing electricity transmission infrastructure at hydropower sites.

The interesting advantage is the higher energy output that can be obtained from FPV, due to the cooling effect of the water. We have learned that the output power increases as the module temperature decreases.

Combining FPV technology with hydropower stations is a very promising solution to take advantage of the hydropower station's electrical infrastructure and reduce the evaporation of the reservoir. The world's first hybrid FPV and hydropower system was installed in 2017 in Portugal (220 kW at the Alto Rabagão Dam).

Capital costs of FPV are still slightly higher, due to the need for floats, moorings, and more resilient electrical components. However, these costs are balanced by a higher expected energy yield of FPV. The benefits are conservatively estimated to be 5% higher, with gains potentially as high as 10–15% in hot climates.

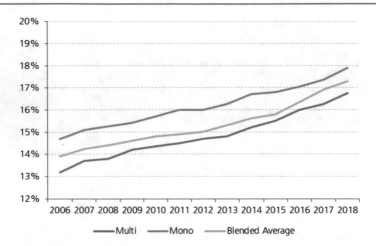

Fig. 28 Evolution of the typical efficiency of crystalline silicon PV cells. *Source* Fraunhofer ISE, www.ise.fraunhofer.de

7 PV Technologies

The PV technologies are normally divided into three categories:

- First generation—crystalline silicon cells.
- Second generation—thin films.
- Third generation—emerging technologies.

In the next sections, we will briefly go through each PV technology.

7.1 First Generation—Crystalline Silicon Cells

Silicon was the material used in the first commercial PV modules and is currently the most used technology by far. Silicon cells can be monocrystalline, using highly pure silicon, or polycrystalline (or multi-crystalline), which uses second-choice silicon. The former is more expensive but capable of achieving higher efficiencies, the typical best efficiency of commercial modules being around 18%. The latter is less efficient (typical best efficiency around 16%) but is cheaper.

An increase in the efficiency of both types of crystalline silicon cells is being witnessed in recent years as seen in Fig. 28.

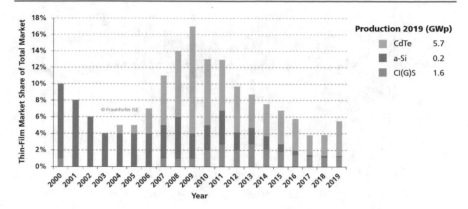

Fig. 29 Evolution of the market share of thin-film technologies. *Source* Fraunhofer ISE, www. ise.fraunhofer.de

7.2   Second Generation—Thin Films

A different PV technology is defended by thin films. Thins films use materials with high light absorption properties. This allows for lightweight solar cells to be made on flexible substrates and therefore for large-scale production. Thin films allow for high efficiencies in a laboratory-controlled environment. However, this is not observed in open-air deployment, i.e., in real-world conditions. Typical module efficiencies range from 10 to 13% depending on the thin-film technology.

Three technologies stand up under the umbrella of the thin film:

- Cadmium Telluride (CdTe), a known drawback is that Cadmium is toxic.
- Amorphous Silicon (a-Si), the less efficient of all the thin-film technologies.
- Copper–Indium–Gallium Selenide (CIGS), showing high promises in the laboratory but not fully confirmed in practice.

The market share for the different thin-film technologies is very volatile, as can be seen in Fig. 29. CdTe cells stand up as the most used thin-film technology since their commercial kick-off back in the early 2000s. However, we must stress that the thin-film market represents currently only 6% of the total PV market.

In Fig. 30, the total market of PV modules is shown.

It can be observed that the total PV market is very volatile with respect to all technologies combined. As of 2019, monocrystalline silicon was the most used technology by far, but if we look back a couple of years ago, the situation was different.

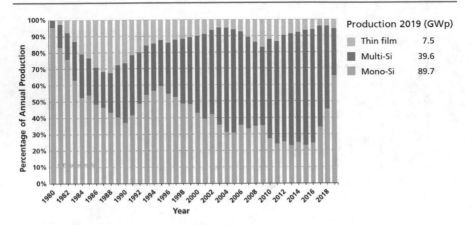

Fig. 30 Evolution of the market share of PV technologies. *Source* Fraunhofer ISE, www.ise. fraunhofer.de

7.3   Third Generation—Emerging Technologies

Several emerging technologies are still in the research phase and yet did not attain commercial deployment. In the sequence, we will introduce the most relevant ones.

An example is the dye-sensitized nanocrystalline cells, also called Gratzel cells, whose base material is titanium dioxide. Their laboratory efficiency reached about 12%, but the first batch produced only achieved an operating efficiency of 5%.

Organic cells, known as OPV—Organic Photovoltaics, use organic pigments as donors and receivers of electrons and holes instead of a p–n junction. These cells are much less efficient than typical silicon cells and are prone to quicker degradation, according to some sources.

The perovskite solar cells employ a perovskite structured compound, normally a hybrid organic–inorganic lead or tin halide-based material. The main issues are related to the overall cost because the electrode material is gold, the short lifespan, and the fact that they contain lead.

In general, the main issue with the mentioned emerging technologies is the low efficiency which is hindering their development.

Bifacial cells are a very promising technology because they allow power production on the back and the front of the cell. The developers claim power gains between 5% and up to 30%, depending on the solar cell technology used, location, and system design.

Finally, let us introduce concentration technologies, known as CPV—Concentration PV. An optical system (Fresnel lenses) is used to concentrate the solar radiation in multiple high-efficient solar PV cells stacked one above the other, called multi-junction cells. Only direct radiation (DNI—Direct Normal Irradiance) can be concentrated, in most cases by a factor of 500. A two-axis solar tracker, and sometimes a cooling system, are needed to maximize the efficiency. In open ground, in normal utilization conditions, efficiencies as high as 35% can be

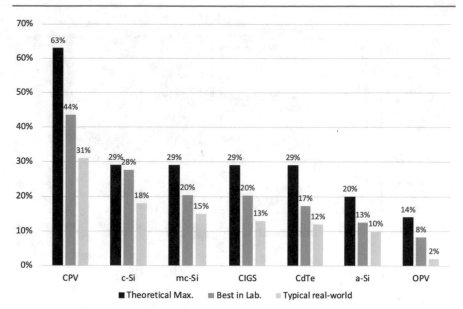

Fig. 31 Comparison of PV technology efficiencies. Adapted from NREL, https://www.nrel.gov, USA

obtained. The main drawback is the prohibitive price of this technology, which is hindering its development.

7.4 Summary of PV Technologies Efficiencies

A comparison of typical efficiencies of the above-mentioned technologies is offered in Fig. 31.

The comparison is interesting because it shows the theoretical maximum efficiency as well as the current typical module efficiency. We can see that CPV is a technology with a wide developing margin, as current technology only uses one-half of the theoretical potential (31% against 63%).

Bearing in mind that the theoretical maximum efficiency of crystalline silicon and thin films is 29%, the current monocrystalline silicon technology (c-Si) efficiency of 18% may be considered reasonable (62% of the theoretical maximum), yet with a margin to further development. For instance, as far as wind power is concerned, the maximum efficiency of current wind generators is about 85% of the maximum theoretical efficiency.

The efficiency of OPV is still too low for this technology may be considered as competitive.

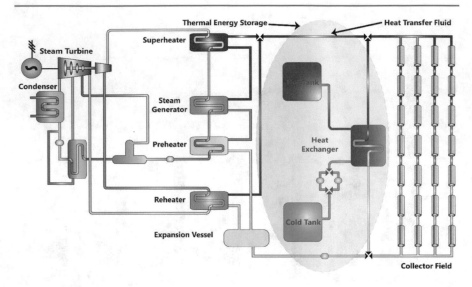

Fig. 32 Diagram of a CSP plant

8   Other Sun Power—Concentrating Solar Power (CSP)

Power available in solar radiation can be used in other ways, rather than in PV systems. A completely different way of using solar power is CSP—Concentrating Solar Power. The operating principle is similar to the conventional thermal power plants: water is overheated in a boiler; steam is produced and is expanded in a steam turbine; electricity is produced via a generator.

The difference is in how steam is produced. In thermal power plants, steam is produced from the combustion of fossil fuel (coal, natural gas…). In CSP installations, sunlight is focused on a receiver to obtain high-temperature heat and produce steam. To focus the sunlight, mirrors or lenses equipped with solar position tracking systems are used.

Figure 32 shows a diagram of a CSP plant.

Many CSP installations have thermal energy storage, therefore, allowing the plant to operate even without sunlight. Usually, a huge metal tank stores the hot liquid, whether in molten salts or in a heat transfer fluid.

Spain (2300 MW, in 2016) and the USA (1700 MW, in 2016) are the countries with more CSP installed. Nevertheless, the costs are still higher than PV, which is hindering its development.

Several different technologies use the CSP concept. The ones that are currently more mature are parabolic troughs and solar towers.

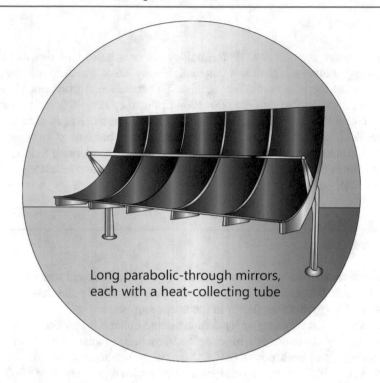

Long parabolic-through mirrors, each with a heat-collecting tube

Fig. 33 Parabolic trough

8.1   Parabolic Troughs

In this CSP technology, sunlight is concentrated using big rectangular mirrors curved in a parabolic shape, each with long heat collector pipes, as shown in Fig. 33.

The system features a one-axis sun tracking system, and the concentration ratio achieves values between 70 and 100. Inside the pipes, there is a heat transfer fluid, which is heated to a high temperature (around 400 °C), capable of generating steam into a heat exchanger. As heat transfer fluid, current plants use some synthetic oil, but alternative concepts include direct steam generation and the use of molten salts as transfer fluid.

Troughs represent the most mature technology and the bulk of current projects; some have significant storage capacities. The solar to electricity conversion can reach an efficiency between 10 and 15% (annual mean value).

8.2   Solar Tower

The solar tower is another CSP technology. It uses a field of distributed flat, sun-tracking mirrors—heliostats—that individually track the sun and focus the sunlight on the top of a tower. We highlight that these are mirrors, not PV arrays.

By concentrating the sunlight 600–1000 times, they can achieve temperatures ranging 600–800 °C. A heat transfer fluid heated in the receiver is used to heat a working fluid, normally molten salt. A cold tank and a hot tank store the working fluid. In a heat exchanger, the solar energy absorbed by the working fluid is used to generate steam to power a conventional turbine. The average efficiency is in the range of 10%.

9   Conclusions

In most countries of the world, Photovoltaics (PV) is the cheapest way of producing electricity. It is believed that PV power will keep on driving the development of Renewable Energy Sources (RES) in the upcoming years.

We have started this chapter by defining some quantities and concepts that are relevant for PV systems. We have defined irradiance (solar power), irradiation (solar energy), and peak power (PV DC output power under Standard Test Conditions (STC)). STC are test conditions defined by irradiance equal to 1000 W/m^2 and module temperature equal to 25 °C. To assess the performance of PV systems some indexes are commonly used, the main ones being the efficiency, the utilization factor, and the performance ratio.

The PV effect, i.e., the production of a DC current from a properly prepared silicon crystal exposed to the sun radiation, was explained, and discussed. A PV system is composed of the PV module, the Maximum Power Point Tracker (MPPT), and the inverter. The MPPT ensures the PV module is always operating at maximum power, given the irradiance and temperature conditions present on-site. The inverter is a power electronics DC/AC converter that is used to interface the DC power output of a PV module power with the AC grid.

To compute the electricity produced by a PV system, the average AC power injected into the grid over a time interval should be multiplied by the respective time interval. To compute the DC power output of a PV module, three models were offered. A fast estimate model that assumes the output power is linearly dependent upon the irradiance and a temperature correction is then applied. This model does not allow other variables of interest, as voltages and currents, to be accessed. Two more sophisticated models, based on equivalent electrical circuits, were proposed. A simpler 1-diode and 3-parameter model and a more complex 1-diode and 5-parameter model. A common feature of the three models is that they can be fed solely with data available in the manufacturer's datasheets.

The output of the modules is DC, so to obtain the required AC power, an efficiency model for the inverter was introduced. This model is simple, nevertheless, it is robust because it is based on data retrieved from commercial inverters. Different topologies are used to dispose of the MPPT and inverters inside a utility-scale PV park. These layouts have been reviewed, and it was concluded that the currently best topology is the string inverter one, which is also the most used.

The efficiency of the PV effect is low; currently, the maximum efficiency of standard monocrystalline silicon modules is about 18%. To increase efficiency, the arrays can be equipped with solar tracking systems. The increase in the energy yield should be confronted with the increasing investment. Floating PV, i.e., the mounting of the PV arrays in lakes, rivers, or reservoirs is acknowledged as increasing the power output due to the cooling effect of the water. We recall that the power output of a PV module increases with the irradiance and decreases with the module temperature.

The technology of PV modules is in constant evolution. Still, first-generation silicon-based modules are by far the preferred technology. It is currently a mature and well-established technology. Thin films promised mass production and to be disruptive, but the promises did not hold mainly because the reported efficiencies fall below what was expected. Meanwhile, new emerging technologies, as dye-sensitized Graetzel, organic and perovskite cells, are being intensively researched and new developments are expected in the upcoming years. Bifacial modules, which allow irradiance to be converted in both sides of the module, have increased efficiency and are already being proposed by several manufacturers. Speaking of increased efficiency, concentration PV technology proposes high efficiencies, but these come with a still high investment cost.

A completely different approach to harvest solar power is Concentrating Solar Power (CSP) that portrays a distinct technology from conventional PV. In this case, the solar irradiance is concentrated into a receiver using mirrors. This allows for high-temperature heat to be produced that ultimately is used to obtain steam. From this point on, the operating principle is equal to conventional thermal power plants, the difference being the steam is produced through a carbon-free process. The two main technologies that use CSP are parabolic troughs and solar towers.

10 Proposed Exercises

Problem PV1

The datasheet of a PV module, as delivered by the manufacturer, is shown in Table 3. Table 4 assembles, under the reference (STC) irradiance, the open-circuit voltage as a function of temperature, as obtained from the I-V curves provided by the manufacturer.

Using the 3-parameter and 1-diode model, compute

(1) The efficiency under STC and the fill factor.

Table 3 PV datasheet for Problem PV1

Monocrystalline silicon		
Peak power	170	Wp
Current at maximum power	4.72	A
Voltage at maximum power	36.0	V
Short-circuit current	5.0	A
Open-circuit voltage	44.2	V
NOCT	47	°C
Current temperature coefficient	3.3×10^{-3}	A/°C
Voltage temperature coefficient	-1.6×10^{-1}	V/°C
Number of cells in series	72	
Length	1.580	m
Width	0.783	m

Table 4 Open-circuit voltage as a function of temperature; G^r=1000 W/m² (Problem PV1)

$\theta(°C)$	$V_{OC}(V)$	$\theta(°C)$	$V_{OC}(V)$
0	48.5	50	39.5
25	44.2	75	35.0

(2) The efficiency for an irradiance equal to 250 W/m² and reference temperature.

(3) The error in the computation of the open-circuit voltage and the output power for a cell temperature of 75 °C and reference irradiance.

Solution
(1)

$$\eta^r = \frac{P_p}{AG^r} = 13.74\%$$

$$FF^r = \frac{P_p}{V_{oc}^r I_{sc}^r} = 0.7689$$

(2)
3 Parameters

$$m^r = m = \frac{V_{MP}^r - V_{oc}^r}{V_T^r \ln\left(1 - \frac{I_{MP}^r}{I_{sc}^r}\right)} = 110.6243$$

$$I_0^r = \frac{I_{sc}^r}{e^{\frac{V_{oc}^r}{mV_T^r}} - 1} = 8.9412 \times 10^{-7} A$$

$$I_{sc}^r = 5.0\,\text{A}$$

Influence of temperature and irradiance

$$I_0(T) = I_0^r \left(\frac{T}{T^r}\right)^3 e^{\frac{N_{s\varepsilon}}{m}\left(\frac{1}{V_T^r} - \frac{1}{V_T(T)}\right)}$$

$$I_{sc}(G) = I_{sc}^r \frac{G}{G^r}$$

Efficiency

$$\eta(G = 250; \theta = \theta^r) = \frac{P(G = 250; \theta = \theta^r)}{250A} = 12.30\%$$

$$P_{DC}(G = 250; \theta = \theta^r) = V_{MP}I_{MP} = 38.0317\,\text{W}$$

$$V_{MP}^{(k+1)} = mV_T(T^r)ln\left[\frac{\left(\frac{I_{sc}(G)}{I_0(T^r)} + 1\right)}{\left(1 + \frac{V_{MP}^{(k)}}{mV_T(T^r)}\right)}\right] = 33.0433\text{V}; V_{MP}^{(0)} = V_{MP}^r$$

$$I_{MP} = I_{sc}(G) - I_0(T^r)\left(e^{\frac{V_{MP}}{mV_T(T^r)}} - 1\right) = 1.1510\,\text{A}$$

(3)

$$V_{oc} = mV_T(T)ln\left(1 + \frac{I_{sc}(G^r)}{I_0(T)}\right) = 36.5442\text{V}; \varepsilon_{V_{oc}} = 4.41\%$$

$$P_{DC}(G = G^r; \theta = 75) = V_{MP}I_{MP} = 130.0377\,\text{W}$$

$$V_{MP}^{(k+1)} = mV_T(T)ln\left[\frac{\left(\frac{I_{sc}(G^r)}{I_0(T)} + 1\right)}{\left(1 + \frac{V_{MP}^{(k)}}{mV_T(T)}\right)}\right] = 28.9805\text{V}; V_{MP}^{(0)} = V_{MP}^r$$

$$I_{MP} = I_{sc}(G^r) - I_0(T)\left(e^{\frac{V_{MP}}{mV_T(T)}} - 1\right) = 4.4871\,\text{A}$$

Problem PV2 The datasheet of a PV module shows the following values: P=150 Wp; V=34 V; I_{sc}=4.8 A; V_{oc}=43.4 V; $NOCT$= 45 °C; N_s=72; c=1.586 m; l=0.769 m. The same datasheet gives the following information: *The relative reduction of module efficiency at an irradiance of 200 W/m² in relation to*

1000 W/m² both at 25°C cell temperature is 7%. Moreover, the datasheet indicates that the open-circuit voltage temperature coefficient is –0.152 V/°C. Compute

(1) The PV module efficiency at STC.
(2) The PV module efficiency as indicated by the manufacturer and both the theoretical DC power and module efficiency for the following conditions: $G=200$ W/m² and $\theta = \theta^r$.
(3) The open-circuit voltage, based on the manufacturer data and foreseen by the theoretical model, for the cell temperature equal to 50°C and the irradiance at STC.

Solution.
(1)

$$\eta^r = \frac{P_p}{AG^r} = 12.30\%$$

(2)

$$\eta_{manuf} = 12.30\% \times 0.93 = 11.44\%$$

$$\eta(G = 200; \theta = \theta^r) = \frac{P(G = 200; \theta = \theta^r)}{200A} = 10.21\%$$

$$P_{DC}(G = 200; \theta = \theta^r) = V_{MP}I_{MP} = 24.8946\,\text{W}$$

$$V_{MP}^{(k+1)} = mV_T(T^r)ln\left[\frac{\left(\frac{I_{sc}(G)}{I_0(T^r)} + 1\right)}{\left(1 + \frac{V_{MP}^{(k)}}{mV_T(T^r)}\right)} \right] = 29.2450\text{V}; V_{MP}^{(0)} = V_{MP}^r$$

$$I_{MP} = I_{sc}(G) - I_0(T^r)\left(e^{\frac{V_{MP}}{mV_T(T^r)}} - 1 \right) = 0.8512\,\text{A}$$

(3)

$$V_{oc}^{manuf} = V_{oc}^r + \mu_{V_{oc}} \times (50 - 25) = 39.6000\,\text{V}$$

$$V_{oc} = mV_T(T)ln\left(1 + \frac{I_{sc}(G^r)}{I_0(T)} \right) = 39.2969\,\text{V}$$

Problem PV3 Table 5 shows the datasheet of a PV module as provided by the respective manufacturer. Table 6 depicts the monthly average irradiance at a certain location. Assume that, in summer months, the PV module temperature is equal to

Table 5 PV datasheet for problem PV3

Polycrystalline silicon		
Peak power	320	Wp
Current at maximum power	8.51	A
Voltage at maximum power	37.62	V
Short-circuit current	9.515	A
Open-circuit voltage	44.84	V
NOCT	47	°C
Current temperature coefficient	−0.05	%/°C
Voltage temperature coefficient	−0.34	%/°C
Power temperature coefficient	−0.45	%/°C
Number of cells in series	72	
Length	1.956	m
Width	0.992	m

Table 6 Average irradiance (Problem PV3)

Month	G (W/m^2)	Month	G (W/m^2)
Jan	99.5	Jul	334.7
Feb	141.4	Aug	297.0
Mar	188.2	Sep	237.5
Apr	259.7	Oct	169.4
May	307.8	Nov	112.5
Jun	325.0	Dec	95.4

the NOCT, and, in winter months, it is equal to the reference temperature. Use the 3-parameter and 1-diode model and adopt 90% as the global efficiency of the set MPPT + inverter. Compute

(1) The electrical energy produced in July and December, and the monthly capacity factor. Use the simplified computation technique.
(2) If the fast estimate was used instead, what would be the associated error in the electrical energy produced in each one of the two referred to months?

Solution
(1)

$$m = \frac{V_{MP}^r - V_{oc}^r}{V_T^r ln\left(1 - \frac{I_{MP}^r}{I_{sc}^r}\right)} = 124.8979$$

$$I_0^r = \frac{I_{sc}^r}{e^{\frac{V_{oc}^r}{mV_T^r}} - 1} = 8.2305 \times 10^{-6} \text{A}$$

$$I_{sc}^r = 9.5150 \text{A}$$

July: $G = 334.70 \frac{W}{m^2}$; $\theta^{NOC} = 47\,^\circ C$

$$P_{DC} = V_{MP} I_{MP} = mV_T(T) \ln \left[\frac{\frac{G}{G^r}(I_{sc}^r - I_{MP}^r)}{I_0^r \left(\frac{T}{T^r}\right)^3 e^{\frac{\varepsilon}{m}\left(\frac{1}{V_T^r} - \frac{1}{V_T(T)}\right)}} \right] \left(\frac{G}{G^r}\right) I_{MP}^r = 85.2616 \text{W}$$

$$E_{July} = \eta_{inv} P_{DC} h_{July} = 57.09 \text{kWh}$$

$$Cf_{July} = \frac{E_{July}}{h_{July} P_p} = 23.98\%$$

December: $G = 95.4 \frac{W}{m^2}$; $\theta^r = 25\,^\circ C$

$$P_{DC} = V_{MP} I_{MP} = mV_T^r \ln \left[\frac{\frac{G}{G^r}(I_{sc}^r - I_{MP}^r)}{I_0^r} \right] \left(\frac{G}{G^r}\right) I_{MP}^r = 24.4149 \text{W}$$

$$E_{Dec} = \eta_{inv} P_{DC} h_{Dec} = 16.35 \text{kWh}$$

$$Cf_{Dec} = \frac{E_{Dec}}{h_{Dec} P_p} = 6.87\%$$

(2)

$$P_{DC}(July) = \frac{G}{G^r} P_p \left[1 + \mu_{Pp}(T - T^r) \right] = 96.50 \text{W}$$

$$P_{DC}(Dec) = 30.53 \text{W}$$

$$\varepsilon_{July} = 13.18\%$$

$$\varepsilon_{Dec} = 25.05\%$$

Problem PV4 Consider a PV module with $NOCT$=44 °C; N_s=60; l=1.649 m; w=0.991 m. The datasheet shows the following values: at STC—P=245 Wp; V=30.2 V; I=8.15 A; I_{sc}=8.65 A; V_{oc}=38.1 V; at Normal Operating Conditions

(NOC)—P=181 W; V=27.5 V; I=6.59 A; I_{sc}=7.03 A; V_{oc}=35.3 V. Using the 3-parameter and 1-diode model, compute

(1) The 3 parameters of the model.
(2) The error made in the computation of the DC power and open-circuit voltage, at NOC. Use the simplified computation technique.
(3) The ratio between the fill factor at NOC and at STC. Use the simplified computation technique.

Solution
(1)

$$m = \frac{\left(V_{MP}^r - V_{oc}^r\right)}{V_T^r ln\left(1 - \frac{I_{MP}^r}{I_{sc}^r}\right)} = 107.7621$$

$$I_{sc}^r = 8.6500A$$

$$I_0^r = \frac{I_{sc}^r}{e^{\frac{V_{oc}^r}{mV_T^r}} - 1} = 9.2510 \times 10^{-6}A$$

(2) Using the Simplified Computation Technique

$$P_{DC}^{NOC} = V_{MP}I_{MP} = mV_T(T) \ln\left[\frac{\frac{G}{G^r}\left(I_{SC}^r - I_{MP}^r\right)}{I_0^r\left(\frac{T}{T_r}\right)^3 e^{\frac{\varepsilon}{m'}\left(\frac{1}{V_T^r} - \frac{1}{V_T(T)}\right)}}\right]\left(\frac{G}{G^r}\right)I_{MP}^r$$

$$= 173.6804 \ W; \quad \varepsilon_{P_{DC}} = -4.04\%$$

(3)

$$FF^r = \frac{P_p}{V_{oc}^r I_{sc}^r} = 0.7434$$

$$FF^{NOC} = \frac{P_{DC}^{NOC}}{V_{oc}^{NOC} I_{sc}^{NOC}} = 0.7162$$

Wind Power

5

Abstract

The conversion of the energy available in the wind into electrical energy is well established as one of the main forms of generating electricity around the world. Among other advantages, wind-based electricity is cheap and complies with environmental constraints. In this chapter, we approach the process of producing electricity using Wind Turbine Generators (WTGs). To be able of performing such computation, two types of information are required. On one side, we need a characterization of the wind profile in the site the WTG is going to be installed. For that purpose, wind speed measurements are required to allow the construction of wind speed histograms. It is better to have the wind speed described by statistical distributions, the Weibull and Rayleigh distributions being the most used. On the other side, information about the conversion machine is required, its features being described by the WTG power curve. With both information in the pocket, estimates for electricity production can be obtained. Other aspects of wind power are also addressed in this text, such as wind forecast, wake effect, power control, and the need for variable speed operation.

1 Introduction

It goes without saying that wind power plays nowadays a major role in the electricity mix in many countries in Europe and in the World. For instance, in Portugal, in 2020, it accounted for roughly 25% of the electricity demand; in Denmark, for 48%, in Ireland for 38%, and in Germany and in the UK for 27%.[1]

---

[1]Source: Wind Europe http://www.windeurope.org/.

Regarding the 2019-year calendar, wind power accounted for 49% of the total new installed capacity in the EU, followed by solar PV with 39%, these two Renewable Energy Sources (RESs) totalling more than 85%. These are very impressive figures that clearly show the importance of wind power in nowadays power generating mix.

Power from the wind is captured using horizontal axis Wind Turbine Generators (WTGs), usually grouped together in the popular wind parks with tens or hundreds of installed MW. Most of the installed capacity is onshore: in the EU, from the grand total of installed 220 GW, only 25 GW (slightly more than 10%) are offshore (2020 figures). Offshore wind power offers tremendous potential, but the associated costs are still higher than onshore. The UK is the country with more offshore wind power installed capacity, reaching a cumulative capacity of almost 10 GW and more than 2400 offshore WTG; Germany follows with 8 GW and 1700 WTG (2020 figures).

WTGs are as tall as more than 100 m and have a turbine composed of three 60 m long blades. The kinetic energy in the wind is used to spin the blades, which turn a shaft connected to the generator, thereby producing electricity.

In Portugal, the average utilization factor usually lies between 2300 and 2500 h, which corresponds to 26% and 28% capacity factor, respectively. This means that, on average, WTG can produce about 27% of the maximum theoretical energy they can produce.

Many studies point to onshore wind power being nowadays one of the cheapest power generating technologies, with a Levelized Cost of Energy (LCOE) around 45 €/MWh. This figure can be easily obtained using typical values for the current investment (about 1 €/W), O&M (1.5%), lifetime (20 years), capacity factor (25%), and discount rate (5%).

WTG capacity has increased over time. In 1985, typical turbines had a rated capacity of 500 kW and a rotor diameter of 15 m. Today's new wind power projects have turbine capacities of about 2–3 MW onshore and 6–8 MW offshore, but there are already commercially available wind turbines with 10 MW capacity, with rotor diameters of up to 180 m.

In the following text, the production of electricity from WTG is lectured. We begin by introducing some basic aspects of wind power conversion, such as the definition of the power available in the wind, the theoretical maximum of the conversion's efficiency (the Betz limit), and the use of Prandtl law to convert wind speeds measured at a measuring height to the wind speeds at rotor height.

The characterization of the wind profile at the site the WTG isgoing to be installed is required to compute the electricity production. Histograms are built after wind speed measurements to describe the frequency of occurrence of each wind speed class. The analysis is enriched if the wind speed is described by statistical distributions, the most used ones are the Weibull and Rayleigh distributions.

On the other side, to compute the electricity production, information about the conversion machine is necessary. This information is provided by the WTG power curve, i.e., the variation of the output electrical power with the input wind speed, which is made available by the manufacturer.

The combined information about the wind profile and the power curve makes it possible to come up with electricity production estimates. Several models to compute the electricity production using WTG are offered, from the more complex to the simpler ones. The specific model to be used depends upon both the quantity and quality of the information the designer has available to perform the calculations.

Besides the computation of electricity production estimates, other wind power aspects are also addressed in this chapter. The wind characteristics and resources are approached by studying the effects of the obstacles in the wind speed decrease, namely the wake effect in wind parks. Also, the importance of wind forecast is highlighted by introducing the basic concepts of the forecasting techniques commonly used.

The need for variable speed operation of the WTG to increase the wind-electricity conversion's efficiency is theoretically demonstrated using the concept of Tip Speed Ratio. Moreover, the power control system of modern WTG is described, the operating principle is lectured, and its main features are highlighted. The chapter ends with a brief analysis of the main forces acting on a WTG blade.

2   Basic Concepts

2.1   Power in the Wind

Consider a volume of air with mass, m (kg), crossing a disk of air with an area, A (m^2) at a constant speed, u (m/s), the thickness of the volume of air being x (m), as in Fig. 1.

The associated kinetic energy is given by

$$E_{kin} = \frac{1}{2}mu^2 = \frac{1}{2}(\rho A x)u^2 \qquad (2.1)$$

where ρ is the air density ($\rho = 1.23$ kg/m^3, $\theta = 15$ °C). The identity $m = \rho A x$ can be easily confirmed by a dimension's analysis.

Fig. 1 Volume of air crossing a section of an air disk at a constant speed

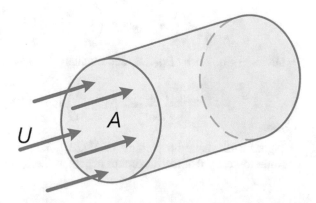

Bearing in mind the relationship between power and energy, the available power in the wind is

$$P_{avail} = \frac{dE_{kin}}{dt} = \frac{1}{2}\left(\rho A u^2\right)\frac{dx}{dt} \qquad (2.2)$$

which can be written as

$$P_{avail} = \frac{1}{2}\rho A u^3 \qquad (2.3)$$

This result shows that the power available in the wind changes with the cube of the wind speed, which is a strong relationship. If the wind speed doubles, the available power is multiplied by 8, but if the area swept by the wind turbine doubles, the available power is also multiplied by 2. On the other side, if the wind speed is reduced to one-half, the available power is only 12.5%.

2.2 Betz Limit

Let us assume that the average wind speed through the rotor of a wind turbine is the average between the wind speed before the turbine (non-perturbed wind speed), u_1, and the wind speed after the turbine (wake wind speed), u_2. The mass flow rate is, therefore,

$$\dot{m} = \frac{dm}{dt} = \rho A \frac{dx}{dt} = \rho A \frac{u_1 + u_2}{2} \qquad (2.4)$$

The power extracted from the wind by the turbine's rotor is

$$P_{ext} = \frac{1}{2}\dot{m}\left(u_1^2 - u_2^2\right) \qquad (2.5)$$

Manipulating Eq. 2.5 and taking into account Eq. 2.4, one can write

$$P_{ext} = \frac{1}{4}\rho A u_1\left(1 + \frac{u_2}{u_1}\right)u_1^2\left(1 - \frac{u_2}{u_1}\right) \qquad (2.6)$$

Dividing Eq. 2.6 by Eq. 2.2, one obtains

$$\frac{P_{ext}}{P_{avail}} = \frac{1}{2}\left(1 + \frac{u_2}{u_1}\right)\left[1 - \left(\frac{u_2}{u_1}\right)^2\right] \qquad (2.7)$$

Figure 2 displays the plotting of Eq. 2.7, i.e., P_{ext}/P_{avail} versus u_2/u_1. It is possible to verify that the maximum occurs at the point (1/3; 16/27), meaning that

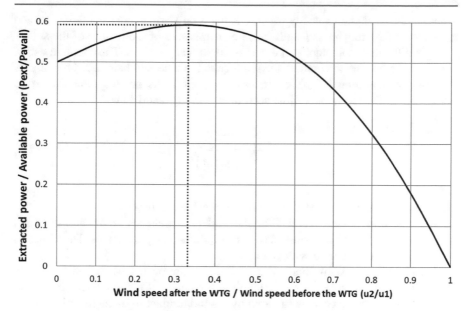

Fig. 2 Extracted power/available power versus wind speed after the WTG/wind speed before the WTG; the point highlighted is the Betz limit = 16/27 = 59.3%

the maximum power that theoretically can be extracted from wind is 59.3% (16/27) of the available power. This is the maximum ratio between the mechanical power output at the ideal turbine's rotor and the wind power input. This value is known as Betz Limit[2] and represents the maximum theoretical efficiency of a wind turbine. Currently, modern wind turbines can extract around 40% of the available power, which is more than 80% of the theoretical maximum.

2.3 Prandtl Law

The atmospheric surface layer, the lowest part of the atmospheric boundary layer, is the region of interest, as far as the wind turbines are concerned. In this region, the wind speed profile is strongly dependent on the topography of the terrain and on the roughness of the soil. The so-called log wind profile is considered to be valid in the surface layer, the wind speed at height h, is given by

$$u(z) = \frac{u_*}{k} ln\left(\frac{z}{z_0}\right) \tag{2.8}$$

---

[2]It was published in 1919, by the German physicist Albert Betz.

where u_* is the friction velocity (m/s), k is the Von Karman constant (~ 0.41), and z_0 is the surface roughness length (m), i.e., a measure of the texture of the soil.

u_* is difficult to compute, as it depends on so many factors. The common use of Eq. 2.8 is to be able to compute the wind speed at a certain height, z, knowing its value at a measuring height, z_R (reference height). As so, the difficulty of u_* computation disappears, as can be seen in Eq. 2.9 (Prandtl Law).

$$\frac{u(z)}{u(z_R)} = \frac{ln\left(\frac{z}{z_0}\right)}{ln\left(\frac{z_R}{z_0}\right)} \tag{2.9}$$

where $u(z_R)$ is the wind speed measured at a reference height.

We note that Eq. 2.9 is valid for flat and homogeneous lands, and does not include topographic effects, obstacles, and roughness length changes. Therefore, its application should be used with caution.

A frequent and useful application of Prandtl law occurs when one intends to determine the wind speed at the turbine's rotor height, which is essential for calculating the energy that can be captured by a wind turbine, from the records of wind speed measuring devices (called anemometers), usually installed at a lower height.

The issue with the use of Eq. 2.9 for this purpose is to determine the proper value of the roughness length, z_0. There are many tables with typical values of z_0. Table 1 illustrates one of the most used tables in wind potential assessment studies, for instance, followed in the preparation of the "European Wind Atlas". As it can be seen, the land is divided into characteristic classes.

Table 1 Roughness length according to land characteristics

Roughness length z_0 (m)	Local terrain type, Landscape, Topography, Vegetation
0.0002	Water surface
0.0024	Complete open terrain with a smooth surface, such as concrete runways in airports and mowed grass
0.03	Open agricultural area without fences and hedgerows and very scattered buildings. Only softly rounded buildings
0.055	Agricultural land with some houses and 8-m-tall sheltering hedgerows within a distance of about 1250 m
0.1	Agricultural land with some houses and 8-m-tall sheltering hedgerows within a distance of about 500 m
0.2	Agricultural land with many houses, shrubs, and plants, or 8 m tall sheltering hedgerows within a distance of about 250 m
0.4	Villages, small towns, agricultural land with many or tall sheltering hedgerows, forests, and very rough and uneven terrain
0.8	Large cities with tall buildings
1.6	Very large cities with tall buildings and skyscrapers

However, this is a very subjective way of working. To overcome this difficulty, two anemometers, mounted at two different heights (z_1 and z_2), can be installed. Then, the unknown is z_0 and Eq. 2.9 becomes

$$\frac{u(z_1)}{u(z_2)} = \frac{ln\left(\frac{z_1}{z_0}\right)}{ln\left(\frac{z_2}{z_0}\right)} \tag{2.10}$$

The solution of Eq. 2.10 is

$$z_0 = exp\left(\frac{u(z_1)ln(z_2) - u(z_2)ln(z_1)}{u(z_1) - u(z_2)}\right) \tag{2.11}$$

This way, the value of z_0 can be computed.

Example 2.1

In a certain place, the average wind speed of 10 m/s was measured at 10 m height. Obtain the variation of the average wind speed with the height for the following surface roughness lengths: $z_0 = 0.01\,\mathrm{m}$, $z_0 = 0.05\,\mathrm{m}$, *and* $z_0 = 0.1\,\mathrm{m}$

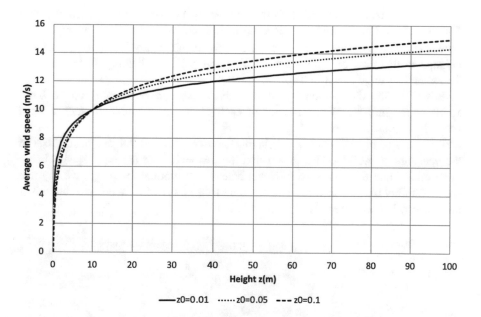

Fig. 3 Variation of the average wind speed with the height for different values of the surface roughness length (Example 2.1)

Solution

Taking $z_R = 10$ m and replacing values in Eq. 2.10, Fig. 3 is obtained.
This characteristic of the wind speed is relevant for the design of WTG. For
example, let us take a 1 MW WTG with a rotor diameter equal to 60 m and a hub
height equal to 60 m. For a surface roughness length of 0.05 m, when the tip of the
blade is in the higher position, the wind speed is 14.1 m/s, whereas when it is in the
lower position it is 12.1 m/s.

3 Wind Data Analysis

Wind data as given by the anemometers and further extrapolated to the rotor height needs preparation to be useful in wind potential assessments. There are two methods used in wind data analysis: the histogram (or method of bins) and statistical analysis.

3.1 Histogram

Data is separated into data intervals (bins or classes) of width 1 m/s, each one with f_j occurrences. The starting point is a file with 8760 values of hourly average wind speeds computed for the rotor height. Then, for instance, a wind speed of 6.5 m/s belongs to the class of 7 m/s; a wind speed of 8.38 m/s belongs to the class of 8 m/s; and so on, as seen in Table 2.

Based on the distribution by wind speed classes, a histogram can be built, counting the number of occurrences of each class. An example of such a wind speed histogram is shown in Fig. 4.

Referring to Fig. 4, wind speeds laying between 7.5 and 8.4 m/s occur 16% of the year (about 1400 h). This means that the probability of the wind speeds being between 7.5 and 8.4 m/s is 16%. Nevertheless, in common language, we say that the probability of the wind speed being equal to 8 m/s is 16%. We know that this is not correct, but we often use this formulation.

Table 2 Example of distribution of hourly average wind speeds by classes

Hour ID	Wind speed (m/s)	Wind class (m/s)
8679	6.5	7
8670	8.38	8
8671	9.75	10
8672	9.76	10
8673	10.98	11
8674	10.64	11
8675	11.76	12
8676	13.72	14
8677	15.42	15
8688	16.94	17

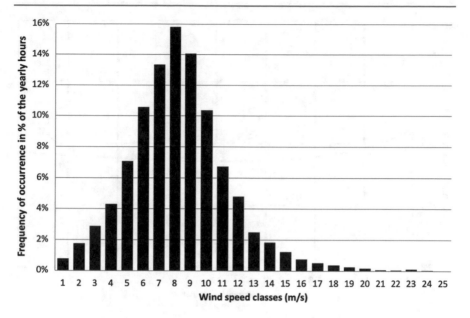

Fig. 4 Example of a wind speed histogram

3.2 Statistical Analysis

Raw wind speed data gain importance if it can be described by analytic functions. Approximation by analytic functions, of probability density type, allows widening the framework of the studies, or projections, or forecasts, that can be made. Weibull distribution is the probability distribution that is normally used to describe wind speed profiles. If a preliminary assessment is intended, the simpler Rayleigh distribution may be used.

3.2.1 Weibull Distribution

The Weibull probability density function (pdf), $f(u)$, is given by

$$f(u) = \frac{k}{c} \left(\frac{u}{c}\right)^{k-1} exp\left\{ -\left[\left(\frac{u}{c}\right)^{k} \right] \right\} \tag{3.1}$$

where u is the wind speed, $c > 0$ (m/s) is a scale parameter, and $k > 0$ is a shape parameter.

Mean annual wind speed and variance

The mean annual wind speed is

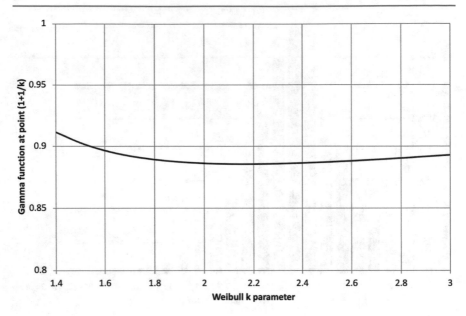

Fig. 5 Gamma function versus Weibull k parameter

$$u_{ma} = \int_0^\infty u f(u)\, du \tag{3.2}$$

The mean annual wind speed and the wind variance can be related to the Weibull parameters k and c through the *Gamma* function Γ^3:

$$u_{ma} = c\Gamma\left(1 + \frac{1}{k}\right) \tag{3.3}$$

$$\sigma^2 = c^2\left[\Gamma\left(1 + \frac{2}{k}\right) - \left(\Gamma\left(1 + \frac{1}{k}\right)\right)^2\right] \tag{3.4}$$

Figure 5 shows the plotting of $\Gamma\left(1 + \frac{1}{k}\right)$ as a function of the Weibull k parameter.

When the Weibull distribution is used to describe the variation of the wind speed, the k parameter usually fluctuates between 1.5 and 2.5. From the plotting of Fig. 5, one can conclude that for $k \in [1.4; 3.0]$, then $\Gamma\left(1 + \frac{1}{k}\right) \approx 0.9$, that is, *Gamma* function computed in the point $\left(1 + \frac{1}{k}\right)$ is always close to 0.9. For $k = 2$, it is $\Gamma\left(1 + \frac{1}{2}\right) = \sqrt{\frac{\pi}{4}} = 0.8862$.

---

[3]The *Gamma* function can be accessed in Excel® with the command EXP(GAMMALN(x)) and in MATLAB® through gamma(x).

Therefore, the following approximation is, in general, valid:

$$u_{ma} = 0.9c \tag{3.5}$$

The close relationship between the mean annual wind speed and Weibull c parameter is apparent from Eq. 3.5.

Variation of k and c

Figure 6 displays the plotting of the Weibull pdf (Eq. 3.1), for 3 different k values, with constant $c = 8\text{m/s}$, as shown in Table 3. It can be seen that the standard deviation is the quantity mainly affected by the k parameter variation. The standard deviation is a measure of wind speed data scattering.

Figure 7 displays the plotting of the Weibull pdf for 3 different c values, with constant $k = 2.3$, as shown in Table 4. It can be seen that the mean annual wind speed is greatly affected by the variation of c parameter, as seen before. However, the variation of the standard deviation is not to be neglected too.

Weibull cdf

The cumulative distribution function (cdf), $F(u)$, is defined, for wind power studies, as the probability of the wind speed to exceed a particular value. Therefore,

$$F(u) = 1 - \int_{-\infty}^{u} f(u)\, du = 1 - \int_{0}^{u} f(u)\, du \tag{3.6}$$

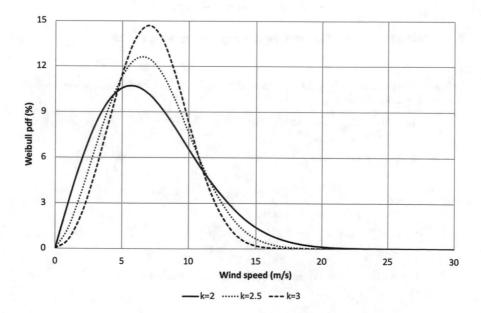

Fig. 6 Weibull pdf versus wind speed, for changing k and constant $c = 8\,\text{m/s}$

Table 3 Mean annual wind speed (u_{ma}) and standard deviation (σ) for a Weibull distribution with constant $c = 8$ m/s and changing k

	$k = 2$	$k = 2.5$	$k = 3$
u_{ma} (m/s)	7.09	7.10	7.14
σ	3.71	3.04	2.60

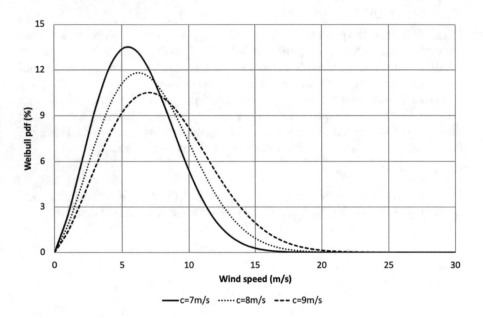

Fig. 7 Weibull pdf versus wind speed, for changing c and constant $k = 2.3$

Table 4 Mean annual wind speed (u_{ma}) and standard deviation (σ) for a Weibull distribution with constant $k = 2.3$ and changing c

	$c = 7$	$c = 8$	$c = 9$
u_{ma} (m/s)	6.20	7.09	7.97
σ	2.86	3.27	3.68

and

$$f(u) = -\frac{dF(u)}{du} \tag{3.7}$$

The application to the Weibull distribution case leads to

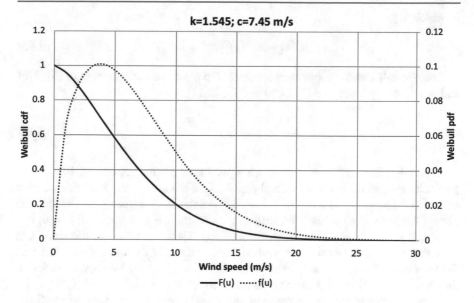

Fig. 8 Weibull pdf (dashed line; right scale) and cdf (solid line; left scale) versus wind speed; $k = 1.545; c = 7.45\,\text{m/s}$

$$F(u) = exp\left[-\left(\frac{u}{c}\right)^k\right] \tag{3.8}$$

In Fig. 8, it is shown the Weibull pdf together with the Weibull cdf for $k = 1.545$ and $c = 7.45\,\text{m/s}$. The cdf values are to be read on the left scale and the pdf values on the right one.

Weibull distribution estimation parameters

A simple method to estimate the k and c parameters of an equivalent Weibull distribution is based on the cdf and is next presented.

Let us begin by applying twice a logarithm to Eq. 3.8, the Weibull cdf. We obtain

$$ln[F(u)] = -\left(\frac{u}{c}\right)^k \tag{3.9}$$

$$ln\{-ln[F(u)]\} = ln\left(\frac{u}{c}\right)^k = kln\,(u) - kln(c) \tag{3.10}$$

It should be noted that Eq. 3.10 is a linear function that can be written in the form:

$$Y = AX + B \tag{3.11}$$

in which:

$$Y = ln\{-ln[F(u)]\}; X = ln(u) \tag{3.12}$$

The Weibull k and c parameters are related to the slope and intercept of Eq. 3.11 and are given by (compare Eq. 3.10 and Eq. 3.11):

$$k = A; c = exp\left(-\frac{B}{A}\right) \tag{3.13}$$

A practical use of this method is when we know a wind speed cdf, $F(u)$, for a particular site, and it is not a Weibull distribution, but something else. So, we give values to the wind speed u and obtain several points (X_i, Y_i) (using Eq. 3.12) in the plane (X, Y). As $F(u)$ is not a Weibull cdf, a straight-line will not be obtained; if it is, of course, a straight-line is directly obtained. Then, we compute the slope and intercept (using the least-squares method, for instance) of the straight-line that best fits the obtained points. Finally, the parameters k and c of the Weibull distribution that best fits the original data are obtained through Eq. 3.13.

A more accurate method to compute the k and c parameters of the Weibull pdf that best fits a known experimental histogram is to use ExcelSolver®. This tool has a built-in optimization algorithm that seeks for the best fit k and c, by minimizing the mean square error between the iteratively computed Weibull pdf and the original histogram. A guess on the initial values is to be provided by the user.

An example of a practical result of the ExcelSolver® application is depicted in Fig. 9. The solid line is the Weibull pdf that best approximates the known experimental histogram (in grey). The parameters of the obtained Weibull distribution are $k = 2.86$ and $c = 8.58$ m/s.

3.2.2 Rayleigh Distribution

To be able to obtain a Weibull distribution, a measured wind speed time series must be available, allowing for an experimental histogram to be produced. If this data is not available, the alternative is to use the simpler Rayleigh distribution that depends only on one parameter—the mean annual wind speed, as shown next.

The Rayleigh distribution is a Weibull distribution in which $k = 2$. So, in Eq. 3.1, it is

$$f(u) = \frac{2}{c}\left(\frac{u}{c}\right)exp\left\{-\left[\left(\frac{u}{c}\right)^2\right]\right\} \tag{3.14}$$

Recalling Eq. 3.3, for $k = 2$:

$$c = \frac{u_{ma}}{\Gamma\left(1 + \frac{1}{2}\right)} = \frac{2}{\sqrt{\pi}}u_{ma} \tag{3.15}$$

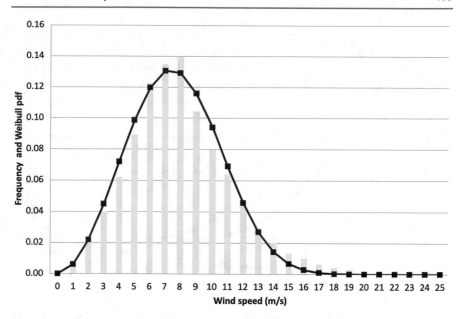

Fig. 9 Experimental histogram (gray bars) and ExcelSolver® computed Weibull pdf approximation (solid line)

Replacing Eq. 3.15 into Eq. 3.14, we finally obtain for the Rayleigh pdf:

$$f(u) = \frac{\pi}{2} \frac{u}{u_{ma}^2} exp\left[-\frac{\pi}{4} \left(\frac{u}{u_{ma}} \right)^2 \right] \qquad (3.16)$$

As can be seen in Eq. 3.16, the Rayleigh distribution depends only on the mean annual speed. As so, it is appropriate to be used to describe the wind speed profile in places where the available data is scarce.

From Eq. 3.6, one obtains the Rayleigh cdf, as

$$F(u) = exp\left[-\frac{\pi}{4} \left(\frac{u}{u_{ma}} \right)^2 \right] \qquad (3.17)$$

Figure 10 displays the Rayleigh pdf (dashed line; right scale) and cdf (solid line; left scale).

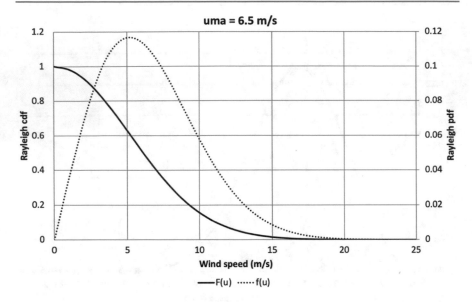

Fig. 10 Rayleigh pdf (dashed line; right scale) and cdf (solid line; left scale) versus wind speed; $u_{ma} = 6.5\,\mathrm{m/s}$

4   The Power Curve

The output power of a WTG depends on the wind speed input following the so-called power curve. Figure 11 shows a typical power curve of a 2 MW WTG.

The WTG are designed to supply the maximum power for a certain wind speed. The maximum power is known as rated power and the corresponding wind speed is known as rated wind speed. In the case of Fig. 11, the rated power is 2 MW and the rated wind speed is 15 m/s. There is no normalized rated wind speed; its value depends on the WTG model, values between 12 and 16 m/s being common.

Analysing the typical power curve shown in Fig. 11, it is possible to conclude that

- For wind speeds lower than a certain value, the so-called cut-in wind speed, the WTG is disconnected, because it is not economical to extract power from the wind, due to the cubic variation law. Usually, the cut-in wind speed is between 3 and 5 m/s, 4 m/s being the cut-in wind speed of the WTG of Fig. 11.
- Then, there is an operating zone, in which the WTG is regulated to capture the maximum power from the wind. In this zone, that goes from the cut-in wind speed to the rated wind speed, the output changes approximately with the cube of the wind speed.

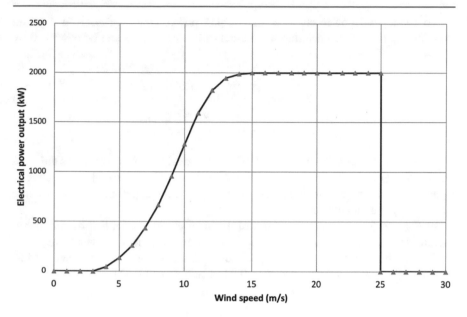

Fig. 11 Typical power curve of a 2 MW WTG

- For wind speeds higher than the rated wind speed, the WTG is regulated to operate at constant rated power, so that the maximum power is not exceeded, and the WTG is not damaged. In this operating zone, the conversion efficiency is artificially lowered by the WTG control system.
- When the wind speed is dangerously high, higher than the so-called cut-off wind speed (usually between 25 and 30 m/s), the WTG is disconnected from the grid, for safety reasons.

All the WTG power curves have the same pattern as depicted in Fig. 11. Of course, the rated power depends on the specific model. Current typical WTG are about 3 to 4 MW rated power, for onshore. For offshore, the typical rated power is higher, about the double of onshore installations. Power curves are available at the manufacturers' internet sites.

4.1   Analytic Equations

It is useful to have the WTG power curve described by an analytic equation. As seen in Fig. 11, the modelling problematic zone is the zone in which the output power changes approximately with the cube of the wind speed; for the remaining operating zones, the modelling is straightforward.

To model the power curve is the output power changing zone, a sigmoid function is often used. Therefore, the model of a power curve can be described by

$$
P_e = \begin{cases} P_e = 0 & u < u_0 \\ \dfrac{P_N}{1 + exp\left(-\frac{u-c_1}{c_2}\right)} & u_0 \le u < u_N \\ P_e = P_N & u_N \le u \le u_{max} \\ P_e = 0 & u > u_{max} \end{cases} \tag{4.1}
$$

In Eq. 4.1, it is as follows: P_e is the WTG electrical output power, P_N is the rated power, u_0, u_N and u_{max} are the cut-in, rated, and cut-off wind speeds, respectively. c_1 and c_2 are fitting constants that are adjusted from the experimental power curve supplied by the manufacturer.

In Fig. 12, we show a comparison of a power curve as supplied by the manufacturer (solid line) and as approximated by a sigmoid function (dotted line).

To find the best fitting c_1 and c_2 parameters, ExcelSolver® may be used. In this case, the sigmoid parameters are $c_1 = 9.0134$ and $c_2 = 1.4400$.

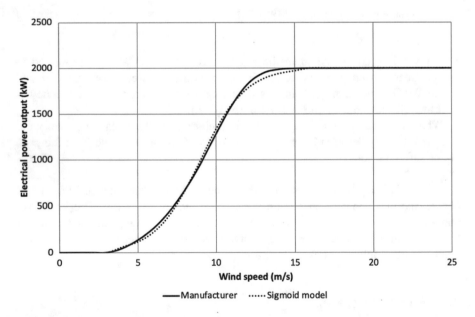

Fig. 12 2 MW WTG power curve: manufacturer (solid line) and sigmoid approximation (dotted line)

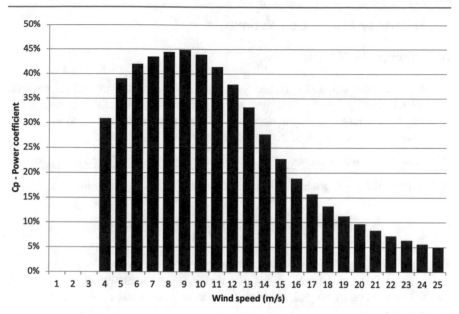

Fig. 13 Power coefficient C_P as a function of the wind speed for a Vestas V90 model

4.2 Power Coefficient—CP

The wind-electric conversion efficiency is called power coefficient and is denoted by C_P, the definition equation being:

$$C_P(u) = \frac{P_e(u)}{P_{avail}(u)} = \frac{P_e(u)}{\frac{1}{2}\rho A u^3} \qquad (4.2)$$

The power coefficient, commonly known as C_P, is, in fact, the WTG efficiency, given by the ratio of the output electric power by the input available power in the wind.

As seen in Eq. 4.2, C_P depends on the wind speed. Given the WTG power curve is available, the power coefficient is very straightforward to compute. Figure 13 displays the graphic of the C_P variation as a function of the wind speed for a Vestas V90[4] WTG.

We note that the WTG efficiency is very low for high wind speeds. This decrease in the efficiency is on-purpose caused by the control system, to achieve rated power limitation for high wind speeds. We will come back to this subject later.

[4]Vestas is a leading WTG manufacturer. Enercon, GoldWind, Siemens, GE are examples of other WTG manufacturers.

5 Electricity Production Estimates

5.1 Complete Models

To be able of computing the electricity production of a WTG, twofold information is required: wind speed profile at rotor height and power curve of the chosen WTG. The way these data are available implies a different appropriate model to use for the estimation of the yearly electricity production of a WTG.

Available data: analytic equations for wind speed pdf and WTG power curve

In this case, the appropriate yearly electricity production equation is

$$E_a = 8760 P_{avg} = 8760 \int_{u_0}^{u_{max}} f(u) P_e(u) du \qquad (5.1)$$

where u_0 and u_{max} are the cut-in and cut-off wind speeds, respectively, $f(u)$ is the analytic equation for the wind speed pdf, and $P_e(u)$ is the analytic equation WTG power curve. Like in Eq. 3.2, we note that the annual average power is given by $\int_{u_0}^{u_{max}} f(u) P_e(u) du$.

An example is provided in the Proposed Exercises section and the end of this chapter.

Available data: wind speed discrete histogram and WTG discrete power curve

If this is the case, then we are not allowed to integrate, and we must sum. Therefore, the yearly electricity production equation is

$$E_a = \sum\nolimits_{u_0}^{u_{max}} f_r(u) P_e(u) \qquad (5.2)$$

where $f_r(u)$ is the wind speed frequency of occurrence (as given by the histogram for each wind speed class) in h/year.

Example 5—1

Consider a WTG with the characteristics shown in Table 5.
The wind speed histogram is known and is shown in Fig. 14. *The WTG power curve is displayed in* Fig. 15.
Compute the best estimate for the annual WTG electricity production.
Solution: For each wind speed, we need to multiply the histogram (number of hours each wind speed occurs) by the power curve (output power for each wind speed) to get the energy produced by each wind speed. The result is displayed in Fig. 16.
To compute the annual electrical energy produced, all the energies produced by the

Table 5 WTG characteristics for Example 5—1			
	Rated power	2310	kW
	Rotor diameter	71	m
	Hub height	80	m

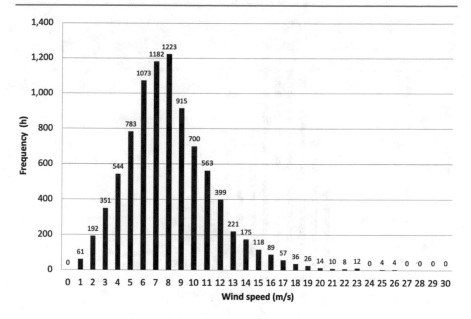

Fig. 14 Wind speed histogram following wind speed measurements (Example 5—1)

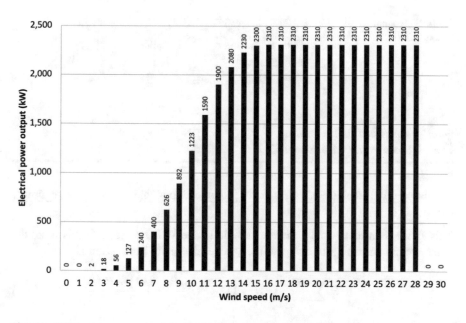

Fig. 15 WTG power curve as given by the manufacturer (Example 5—1)

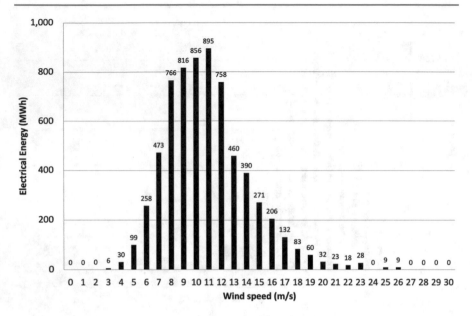

Fig. 16 Electrical energy produced by each wind speed using the method of the histogram (Example 5—1)

Table 6 Example 5—1 results

Annual Energy	6,680	MWh
Utilization factor	2,892	h
Average power	763	kW
Capacity factor	33.01%	

wind speeds must be summed up.
The obtained results are summarized in Table 6.

Available data: analytic equation for wind speed pdf and WTG discrete power curve

This is the data that is more often available in the WTG project. In this case, the most accurate equation to compute the yearly electricity production is

$$E_a = 8760 \sum_{u_0+1}^{u_{max}} \left[(F(i-1) - F(i)) \frac{P_e(i-1) + P_e(i)}{2} \right] \quad (5.3)$$

where $F(i)$ is the wind speed cdf.

We highlight that, for instance, $F(5)-F(6)$ is the probability of the wind speed being between 5 and 6 m/s and $\frac{P_e(5)+P_e(6)}{2}$ is the average output power when the wind speed is between 5 and 6 m/s.

Let us look at an example that will help in better understanding the technique.

Example 5—2

Consider the same WTG of Example 5—1. The best fit Weibull parameters of the histogram (Fig. 17) have been computed leading to $k = 2.86$ and $c = 8.58\,\text{m/s}$. The Weibull cdf is depicted in Fig. 17.

Compute the best estimate for the annual WTG electricity production.

Solution: Given the type of data we have now, the correct model is the one given by Eq. 5.3. Applying this model, the yearly electricity production for each wind speed bin is showed in Fig. 18.

The obtained results are summarized in Table 7.

The results of Tables 6 and 7 are not equal, as would be expected. In fact, when approximating the real histogram by a best fit Weibull distribution an error is made, which is reflected in the results concerning the electricity production.

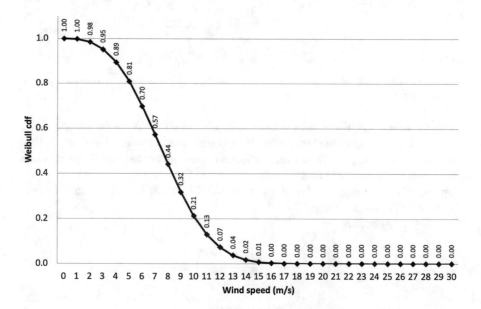

Fig. 17 Wind speed Weibull cdf; $k = 2.86$ and $c = 8.58\,\text{m/s}$ (Example 5—2)

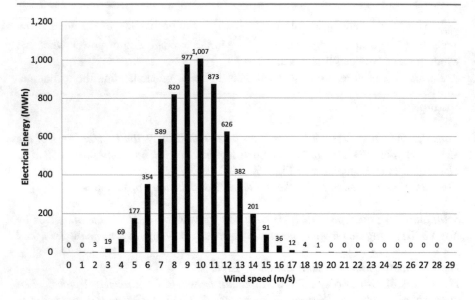

Fig. 18 Electrical energy produced by each wind speed using the method of the Weibull distribution (Example 5—2)

Table 7 Example 5—2 results

Annual energy	6 239	MWh
Utilization factor	2 701	h
Average power	712	kW
Capacity factor	30.83%	

Example 5—3

Consider a 660 kW WTG, with a rotor diameter of 47 m, the rotor height being 40 m. The sigmoid parameters of the power curve are $c_1 = 8.76$ and $c_2 = 1.48$ and the rated wind speed is 15 m/s. The mean annual wind speed at the installation place and at 40 m height is 8.24 m/s.

Compute the electricity produced when the wind speed is between 9 and 11 m/s. Compare different methods.

The relevant functions are

$$P(u) = \left(\frac{660}{1 + \exp\left(-\frac{u - 8.76}{1.48}\right)} \right) \quad u_0 < u < 15$$

$$f(u) = \frac{\pi}{2} \frac{u}{8.24^2} \exp\left[-\frac{\pi}{4} \left(\frac{u}{8.24} \right)^2 \right]$$

$$F(u) = \exp\left[-\frac{\pi}{4} \left(\frac{u}{8.24} \right)^2 \right]$$

As the only wind speed data we have is the mean annual wind speed, a Rayleigh distribution is to be used.

The accurate method is

$$E(9 < u < 11) = 8760 \int_9^{11} P(u)f(u)du = 578.20 \, \text{MWh (step} = 0.01 \, \text{m/s)}$$

This was performed using numerical integration with a step of 0.01 m/s, using Excel®.

With the provided data, the method formulated in Eq. 5.3 can also be used:

$$E(9 < u < 11) = 8760 \left[(F(9) - F(10)) \left(\frac{P(9) + P(10)}{2} \right) + (F(10) - F(11)) \left(\frac{P(10) + P(11)}{2} \right) \right] = 574.34 \, \text{MWh}$$

As can be seen, the accuracy of this method is good, the error being just –0.7%.

5.2   Simplified Models

5.2.1   Johnson's Model

Sometimes, mainly in the early stages of a project, we do not know what specific model of a WTG will be installed. However, we probably know the WTG rated capacity, the cut-in (u_0), rated (u_N) and cut-off (u_{max}) wind speeds (if we do not know these values, typical ones can be assumed) and the wind speed Weibull distribution. In this case, we can use a general-purpose WTG power curve defined as[5]

$$\begin{cases} P_e = 0 & u < u_0 \\ P_e = a + bu^k & u_0 \leq u \leq u_N \\ P_e = P_N & u_N < u \leq u_{max} \\ P_e = 0 & u > u_{max} \end{cases} \tag{5.4}$$

where k is the Weibull parameter and a and b are two constants that can be determined from the boundary conditions:

$$u = u_0 \Rightarrow P_e = 0; u = u_N \Rightarrow P_e = P_N \tag{5.5}$$

The constants a and b can therefore be computed as

$$a = P_N \frac{u_0{}^k}{u_0{}^k - u_N{}^k}; b = P_N \frac{1}{u_N{}^k - u_0{}^k} \tag{5.6}$$

In Fig. 19, we can see a plotting of the manufacturer's (solid line) and Johnson's (dashed line) power curves. We can see that approximation is not perfect, but it is enough to obtain a rough estimation of the annual electricity production, which is valuable in the preliminary phase of the project.

---

[5]Gary L. Johnson, Wind Energy System, Prentice Hall.

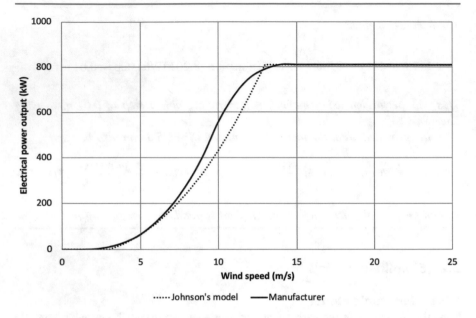

Fig. 19 WTG power curve: Johnson approximation (dashed line) and manufacturer's (solid line)

We aim to compute an estimate for the annual energy produced by a WTG, which is given by

$$E_a = 8760 P_{avg} \tag{5.7}$$

the average power being computed through:

$$P_{avg} = \int_{u_0}^{u_{max}} P_e(u) f(u) du = \int_{u_0}^{u_N} (a + bu^k) f(u) du + P_N \int_{u_N}^{u_{max}} f(u) du \tag{5.8}$$

where $f(u)$ is the Weibull pdf:

$$f(u) = \frac{k}{c} \left(\frac{u}{c}\right)^{k-1} e^{-\left[\left(\frac{u}{c}\right)^k\right]} \tag{5.9}$$

We recall that we assume we know the Weibull distribution parameters. In Eq. 5.8, 2 integrals need to be solved:

$$\int f(u) du; \int u^k f(u) du \tag{5.10}$$

Before proceeding further, we are going to make a change of variable, because this will help in solving the integrals in Eq. 5.10:

$$x = \left(\frac{u}{c}\right)^k$$

(5.11)

and therefore,

$$dx = k\left(\frac{u}{c}\right)^{k-1} d\left(\frac{u}{c}\right)$$

(5.12)

Under these circumstances, we have a first important result for the computation of the 1st integral in Eq. 5.10:

$$\int f(u)du = \int \frac{k}{c}\left(\frac{u}{c}\right)^{k-1} e^{-\left[\left(\frac{u}{c}\right)^k\right]} du = \int e^{-x}dx = -e^{-x}$$

(5.13)

Then, for the 2nd integral of Eq. 5.10, we write

$$\int u^k f(u)du = \int c^k \left(\frac{u}{c}\right)^k f(u)du = c^k \int xe^{-x}dx = -c^k(x+1)e^{-x}$$

(5.14)

which portrays a 2nd important result.

Replacing Eqs. 5.13 and 5.14 in Eq. 5.8, we finally obtain, after some manipulation:

$$P_{avg} = P_N \left(\frac{e^{-\left[\left(\frac{u_0}{c}\right)^k\right]} - e^{-\left[\left(\frac{u_N}{c}\right)^k\right]}}{\left(\frac{u_N}{c}\right)^k - \left(\frac{u_0}{c}\right)^k} - e^{-\left[\left(\frac{u_{max}}{c}\right)^k\right]}\right)$$

(5.15)

An estimate for the annual electricity produced by a general-purpose WTG located in a place where the Weibull parameters are known is (see Eq. 5.7):

$$E_a = 8760 P_N \left(\frac{e^{-\left[\left(\frac{u_0}{c}\right)^k\right]} - e^{-\left[\left(\frac{u_N}{c}\right)^k\right]}}{\left(\frac{u_N}{c}\right)^k - \left(\frac{u_0}{c}\right)^k} - e^{-\left[\left(\frac{u_{max}}{c}\right)^k\right]}\right)$$

(5.16)

5.2.2 Fast Estimate Model

In the very preliminary phases of a WTG project, when only a few data are available, it is common practice to use basic models. This is the case of the fast estimate model that we present hereafter.

The annual electricity production of a WTG is given by

$$E_a = P_{avg}8760 \approx \left(C_p\right)_{avg}(P_{avail})_{avg}8760 \tag{5.17}$$

where $\left(C_p\right)_{avg}$ is the average value of the power coefficient and $(P_{avail})_{avg}$ is the average value of the available power in the wind.

In what concerns the latter, one can write

$$(P_{avail})_{avg} = \left(\frac{1}{2}\rho A u^3\right)_{avg} = \frac{1}{2}\rho A \left(u^3\right)_{avg} \tag{5.18}$$

Our attention will now be focused on the computation of the average of the cube of the wind speed, $(u^3)_{avg}$, which is different from the cube of the average wind speed $\left(u_{avg}\right)^3$.

This model assumes that the wind speed at the installation place is well represented by a Rayleigh distribution. We recall that a Rayleigh distribution is a Weibull distribution in which $k = 2$; we also recall that the annual mean wind speed is $u_{ma} = \int_0^\infty u f(u)\,du$ (Eq. 3.2). Therefore, we can write

$$\left(u^3\right)_{avg} = \int_0^\infty u^3 f_{Rayl}(u)du = \int_0^\infty u^3 \frac{2u}{c^2} exp\left[-\left(\frac{u}{c}\right)^2\right] du$$

$$= \frac{2}{c^2}\int_0^\infty u^4 exp\left[-\left(\frac{u}{c}\right)^2\right] du \tag{5.19}$$

An integrals table shows that

$$\int_0^\infty x^m exp\left(-ax^2\right)dx = \frac{\Gamma\left(\frac{m+1}{2}\right)}{2a^{\frac{m+1}{2}}} \tag{5.20}$$

where $\Gamma\left(\frac{m+1}{2}\right)$ is the value of the *Gamma* function in the point $\frac{m+1}{2}$.

Applying to the case of the integral we want to solve, the parameters assume the following values (compare Eqs. 5.19 and 5.20):

$$m = 4; a = \frac{1}{c^2}; \Gamma\left(\frac{5}{2}\right) = \frac{3}{2}\frac{\sqrt{\pi}}{2}; a^{\frac{5}{2}} = \frac{1}{c^5} \tag{5.21}$$

Replacing Eq. 5.20 in Eq. 5.19 and taking into account the parameters given in Eq. 5.21, we obtain

$$\left(u^3\right)_{avg} = \frac{2}{c^2}\int_0^\infty u^4 exp\left[-\left(\frac{u}{c}\right)^2\right]du = \frac{2\frac{3}{4}\sqrt{\pi}}{c^2\frac{2}{c^5}} = \frac{3}{4}c^3\sqrt{\pi} \tag{5.22}$$

Moreover, recalling that (Eq. 3.3):

$$u_{ma} = c\Gamma\left(1+\frac{1}{k}\right)\underset{\rightarrow}{k=2}u_{ma} = c\frac{\sqrt{\pi}}{2} \tag{5.23}$$

Replacing Eq. 5.23 into Eq. 5.22, one can write

$$\left(u^3\right)_{avg} = \frac{3}{4}\sqrt{\pi}\left(\frac{2u_{ma}}{\sqrt{\pi}}\right)^3 = \frac{3}{4}\frac{8u_{ma}^3}{\pi} = \frac{6}{\pi}u_{ma}^3 \tag{5.24}$$

This is an important result, that we can use by retaking Eq. 5.18, thus leading to

$$\left(P_{avail}\right)_{avg} = \frac{1}{2}\rho A\left(u^3\right)_{avg} = \rho A\frac{3}{\pi}u_{ma}^3 \tag{5.25}$$

We note that the annual electricity production can be written as (see Eq. 5.17):

$$E_a = P_{avg}8760 \approx \left(C_p\right)_{avg}\rho A\frac{3}{\pi}u_{ma}^3 8760 \tag{5.26}$$

As seen before, the power coefficient is not constant, it depends on the wind speed. Modern WTGs show a C_P average value of $\left(C_p\right)_{avg} = 25\%$ (see, for instance, Fig. 13). The air density in standard pressure and temperature conditions is $\rho = 1.23\,kg/m^3$. Moreover, $\frac{3}{\pi} = 0.95$. Therefore, a fast estimate of the annual electricity produced by a WTG is

$$E_a \approx 0.25\rho A\frac{3}{\pi}u_{ma}^3 8760 \approx 0.3Au_{ma}^3 8760 \tag{5.27}$$

where A is the swept area and u_{ma} is the mean annual wind speed at rotor height.

Rewriting Eq. 5.27 as a function of the rotor diameter, RD, and expressing the result in kWh, we finally have

$$E_a \approx 2(RD)^2 u_{ma}^3 \tag{5.28}$$

This is a very simple equation that allows the computation of an estimate of the WTG annual electricity production, given the knowledge of readily available data.

Example 5—4

Retake Example 5—2 data and compute the annual energy yield by using the fast estimate method.

Solution

To use the fast estimate method, we need the mean annual wind speed.
Let us recall that we have approximately: $u_{ma} = 0.9c$ (Eq. 3.5). The data indicate $RD = 71$ m and $c = 8.58$ m/s, therefore $u_{ma} = 7.72$ m/s. Using Eq. 5.28, we obtain $E_a = 4639$ MWh, which deviates from the result obtained with the complete model in –26%.

6 Wind Characteristics and Resources

6.1 Wind Potential and Variation

Figure 20 shows the European wind potential, both onshore and offshore, at 100 m height. It can be seen that the UK holds the highest wind potential in Europe. The offshore potential around the British Islands is very significant.

The mean annual wind speed at 80 m height in onshore Portugal is shown in Fig. 21. For comparison purposes, Fig. 22 depicts the mean annual wind speed at 10 m height. It is apparent why WTG should be high enough so that to take advantage of higher wind speeds.

The offshore wind potential in Portugal is represented in Fig. 23. Significantly higher offshore wind speeds can be observed, with mean annual wind speeds around 9 m/s.

It is common knowledge that wind speed changes a lot with time. In Fig. 24, we can observe the time variation of the hourly average wind speed in one month.

Wind speed suffers inter-annual (30 years are needed to determine long-term values of weather and climate; one year produces long-term seasonal mean wind speeds with 10% accuracy and 90% confidence level), annual (significant variation in seasonal or monthly averaged wind speeds), diurnal (differential heating of the earth surface during the daily radiation cycle), and short-term (turbulence and gusts —variations over intervals of 10 min or less) variations.

6.2 The Effect of Obstacles

The effect of obstacles in the wind flow is very significant in reducing wind speed and thus wind power. This can be seen in Fig. 25, where it is shown that only after 20 times the height of the obstacle (h_s), the wind flow recovers its original free characteristics.

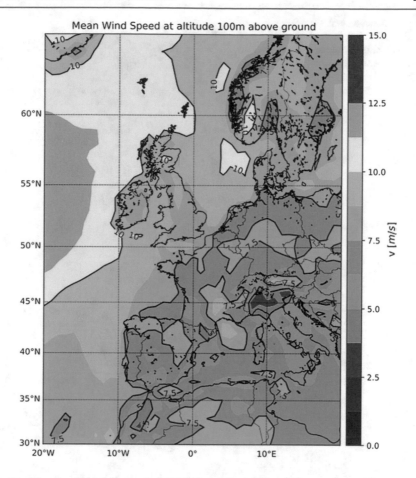

Fig. 20 Mean annual wind speed at 100 m height in Europe; *Source* Airborne Wind Europe, https://airbornewindeurope.org

In a wind park, each WTG is an obstacle to the other ones, due to the downstream wake, portraying what is known as the wake effect. A WTG downwind of another WTG has to cope with wakes with both slower wind speeds and increased turbulence.

The wake effect causes the wind park aggregate production to be lower than the number of WTG times the production of one WTG. Therefore, the wake effect should be minimized by conveniently spacing the WTG apart in a wind park. Recommended practices point to keeping 7 rotor diameters in the wind-stream direction and 4 rotor diameters in the perpendicular direction (see Fig. 26). Even if this layout is followed, a loss of 5% due to the wake effect is to be expected.

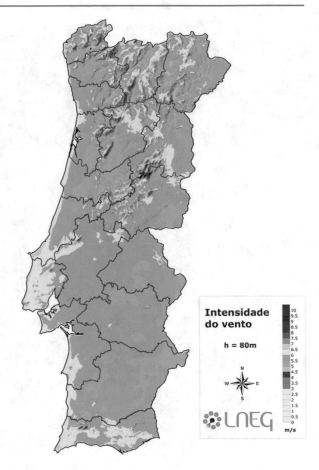

Fig. 21 Mean annual wind speed at 80 m height in onshore Portugal; *Source* LNEG, www.lneg.pt

WTG should be placed so that the prevalent wind stream is perpendicular to the rotor blades rotational plane. To cope with different wind directions, WTGs are equipped with a yaw mechanism that automatically turns the nacelle to face the wind flow in the correct direction.

Wind speed is measured using anemometers and wind direction using wind vanes. Cup anemometers, as illustrated in Fig. 27, are the most used.

6.3 Wind Forecast

Wind power predictions directly impact the activities of the different agents in the electricity sector. Transmission Systems Operators (TSOs) and Distribution System Operators (DSOs) oversee the reliability of their networks, ensuring the supply of electricity to all the customers, complying with legal requirements and high standards of quality. Therefore, the safe operation of a network with high penetration of wind energy requires appropriate wind power forecasts to avoid negative impacts

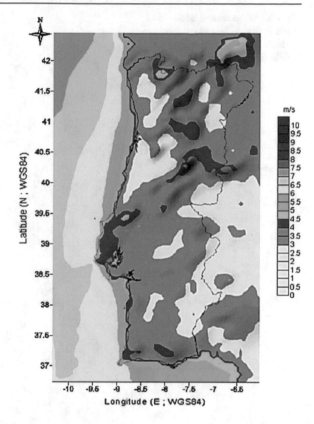

Fig. 22 Mean annual wind speed at 10 m height in Portugal; *Source* LNEG, www.lneg.pt

on its reliability. Furthermore, in the market, wind power producers must make injected power predictions for a specific period. They must provide a generation schedule for a considered time horizon where any deviation from this schedule imposes some undesirable penalties for them. To minimize those penalties, an accurate wind power forecast is required.

Another reason for accurate wind power predictions is to minimize the operation costs of the power plants, as thermal, hydro, and gas power plants. In these power plants, start-up and shut-down costs of a unit are considerable and the ramp times for some could be large, in order of hours. Thus, knowing the power produced by other sources of energy, like wind and Photovoltaic (PV), is crucial to allow an optimal unit commitment and economic dispatch.

Roughly, we can divide forecast methods into two categories. One is called physical models, which use some physical considerations (terrain features, altitude, roughness, obstacles, etc.) to reach the best estimate and make use of Numerical Weather Prediction (NWP) models. These can approximately estimate the evolution of some variables of interest—temperature, wind speed, humidity, and pressure, in the points of a mesh, by solving complex equations that govern the motions in the atmosphere.

Fig. 23 Mean annual wind
speed at 100 m height in
offshore Portugal; *Source*
LNEG, www.lneg.pt

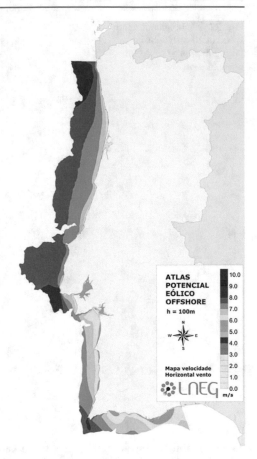

Another category is statistical models. It comprises purely statistical methods, like Auto-Regressive Moving Average (ARMA), Auto-Regressive Integrated Moving Average (ARIMA), and Artificial Intelligence (AI) methods, such as Artificial Neural Networks (ANNs), Radial Basis Function Network (RBFN), Elman recurrent network (ELM), Adaptive Neuro-Fuzzy Inference System (ANFIS), and Neural Logic Network (NLN), among others.

The objective of all methods is to perform better than the benchmark model, called the persistence model. This forecasting model states that the value for the next time period is equal to the last known value of the time series.

We will now briefly describe two of the most used wind forecasting methods: the ARMA and the ANN models. In general, the predictor aims to build relationships between the inputs and the output variables which are to be forecasted. As in all forecasting models, the future is predicted based on the past behaviour of the

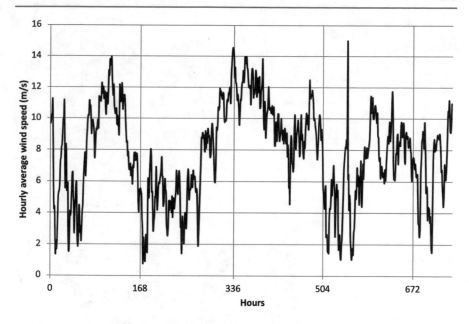

Fig. 24 Hourly average wind speed variation with time, during a period of 1 month

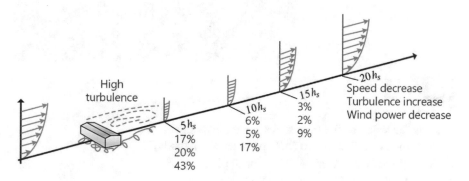

Fig. 25 Effect of obstacles in wind speed and wind power decrease; h_s: obstacle height

variable. Therefore, a historic time series is needed in order to train the model and adapt it to what really happened in the past. This being done, it is expected that the future behaves similarly to the past.

The ARMA models can be characterized by the equation:

$$X_t = \sum_{j=1}^{p} \phi_j X_{t-j} + \sum_{k=1}^{q} \theta_k e_{t-k} + C + \varepsilon_t \qquad (6.1)$$

Fig. 26 Recommended
practice to space WTG apart
inside a wind park; scale:
rotor diameters

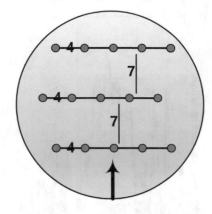

Fig. 27 Cup anemometer
and wind vane

where

- X_t is the value of the time series for the instant t.
- ϕ_j is the Auto-Regressive (AR) parameter for the lag j.
- θ_k is the Moving Average (MA) for the lag j.
- e_t is the value of the error for the instant t.
- C is a constant.
- p is the AR order (number of AR parameters).
- q is the MA order (number of MA parameters).
- ε_t is the white noise in ideal conditions.

The first step of the ARMA model is to choose the number of model parameters, i.e., the order of the model (p, q). Next, the estimation of the value of the parameters is carried out through an iterative process that adjusts the model parameters to the training historic time series. The least-square method is used to minimize the errors between the real-time series values and the predicted output of the model. This enables the computation of future predictions.

ANN is the technique that has shown itself to be the reference in the wind forecasting methods which use AI. ANN are structures that simulate human thinking and its capability to adapt to complex problems, as well as learning by experience. ANN consist of a set of interconnected processing units, the neurons, which can produce outputs from inputs with each neuron connection having an associated weight. The weights are adjusted through a supervised learning process until the produced output corresponds to the desired output. To create an ANN, a historical dataset (observed inputs and the desired output) is required. These samples are divided into three subsets: training, validation, and test.

In the first step, the network is faced with the training historical subset (observed inputs and the desired output). The inputs are processed, and an output is obtained. The process is repeated, adjusting the weights through an iterative process that minimizes an error function between the produced and the desired output. The training phase continues as the performance is improved in the validation subset. Finally, the test subset is used to obtain an independent assessment of the network's ability to solve the problem.

The processing units are divided into layers. There may be several hidden layers in which the data processing is carried out. The inputs of each layer are the outputs of the previous layer, affected by a weight. The output a of each unit is computed by applying an activation function f to the sum of all the inputs p, already multiplied by the respective weights w, with a constant bias b, as shown in

$$a = f\left(\sum wp + b\right) \tag{6.2}$$

In summary, given a dataset with N samples, (p_i, a_i), where p_i are the inputs and a_i are the outputs, the ANN with K hidden layers and activation function f, is given by

$$\psi_K(p_j) = \sum_{i=1}^{K} \beta_i f(w_i p_j + b_i), j = 1, ..., N \tag{6.3}$$

where w_i is the weight vector between the hidden layer i and the input layer, β_i is the weight vector between the hidden layer i and the output layer, b_i is a constant bias related to the hidden layer i and $f(w_i p_j + b_i)$ is the output of the hidden layer i in relation to the input p_j.

The number of neurons is very important. If it is too small, the ANN does not achieve the required level of detail but if the number of neurons is too big, the ANN learns a level of detail that exceeds what is required.

During the training phase, all the weights and biases are adjusted so that an error surface is minimized, i.e., the surface that is obtained by plotting the error between the produced and the desired outputs, for all possible combinations of weights and bias.

The most used error minimization algorithm is one described by Levenberg–Marquardt, whose objective function is defined by

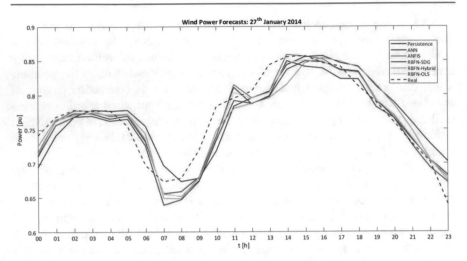

Fig. 28 Comparison of several wind power forecasting methods; one hour ahead—Example 1

$$C = \sum_{j=1}^{N} \left(\sum_{i=1}^{K} \beta_i f(w_i p_j + b_i) - a_i \right)^2 \qquad (6.4)$$

This method tries to find the minimum of a function defined as the sum of the squares of non-linear functions, by using an interpolation of Gauss–Newton and gradient descent methods.

In Figs. 28 and 29, two examples of a comparison of several wind power forecasting methods are offered. The forecasts are made one hour ahead, ANN, ANFIS, and RBFN methods are considered and the persistence benchmark method is also shown, as well as the real wind power.

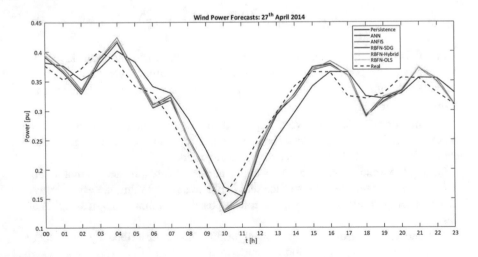

Fig. 29 Comparison of several wind power forecasting methods; one hour ahead—Example 2

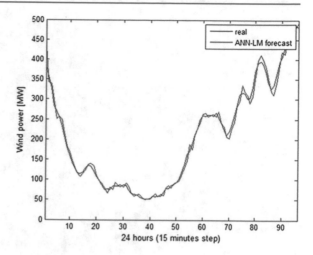

Fig. 30 15 min ahead wind power forecast

The forecasting results improve when the forecasting horizon decreases. Figure 30 shows the results for 15 min ahead forecasts.

7   Some Features of Modern WTG

7.1   Wind Turbine Generator Components

Most of the manufacturers offer horizontal axis wind turbines, in which the axis of rotation is parallel to the ground. Such WTG may be divided into three main parts: rotor, nacelle, and tower.

The rotor is composed of the blades which are fixed to the nacelle in a hub. The rotor may be located upwind, if the wind turns the blades and then impacts the tower, i.e., the rotor faces the wind, or downwind, if the wind comes from the back, i.e., impacts the tower and then turns the blades. Nowadays, all WTG are upwind, as the tower introduces turbulence to the wind stream, which is much undesirable. Also, most of the WTG offered in the market are three-bladed, as this was found to be the number of blades that maximizes efficiency. The blades are flexible and are made of Fibre-Reinforced Polymers (FRPs), such as glass and carbon fibre-reinforced plastics (GFRP and CFRP).

The nacelle is the place where the WTG components are installed. The main components inside the nacelle are shown in Fig. 31. We can mention the low-speed shaft (about 15 rpm) connected to the rotor blades and the high-speed shaft (about 1500 rpm) connected to the electrical generator. In between, there is a gearbox (there are WTG models that do not have a gearbox). Also, we can see the yaw motor that drives the nacelle to be always facing the wind stream, when the

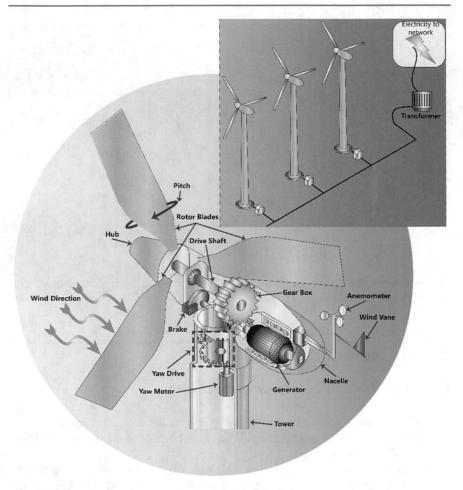

Fig. 31 Main components inside the nacelle of a WTG

direction changes, as given by the wind vane. The pitch system is used to limit the
output power to the rated power for high wind speeds (as given by the anemometer)
and will be discussed later.

Currently, the majority of wind turbines are supported by conical tubular steel
towers.

Taller towers for wind turbines make sense. For instance, an 80 m tower can let
2 to 3 MW WTG produce more power, and enough to justify the additional cost of
20-m more, than if installed at 60 m.

A very popular WTG model is the Vestas V90. The rated power is 3 MW; it has
a rotor diameter of 90 m and a hub height that can vary between 80 and 105 m,
depending on the specifics of the installation location. The nacelle alone weighs
more than 75 tons, the blade assembly weighs more than 40 tons, and the tower
itself weighs about 152 tons, for a total weight of 267 tons.

7.2 The Need for Variable Speed Operation

Let us define the quantity Tip Speed Ratio as

$$TSR = \lambda = \frac{\omega_T R}{u} \tag{7.1}$$

where ω_T is the angular speed of the rotor blades in rad/s, R is the radius of the circle defined by the rotation of the blades (equal to the blade's length), in m, and u is the wind speed in m/s.

A typical graphic of the variation of the power coefficient (WTG efficiency) with TSR is illustrated in Fig. 32.

An analytic equation for the curve in Fig. 32 is

$$C_P = 0.22 \left[116 \left(\frac{1}{\lambda} - 0.035 \right) - 5 \right] exp \left[-12.5 \left(\frac{1}{\lambda} - 0.035 \right) \right] \tag{7.2}$$

As can be seen from Fig. 32, if the angular speed of the turbine is kept constant, or nearly constant, either the wind speed increases (TSR decreases) or decreases (TSR increases), the WTG efficiency, C_P, is not at its maximum value, and therefore there is a particular TSR that maximizes C_P.

This operation mode is undesirable, because, in the power curve zone where the output power increases with the wind speed, we want maximum efficiency, so that to extract the maximum possible power from the wind.

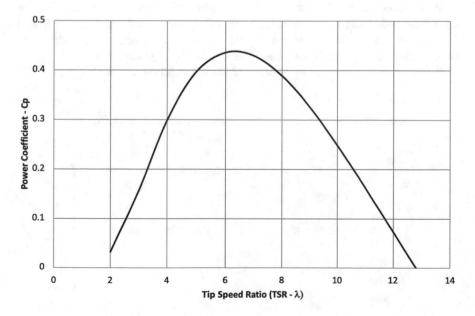

Fig. 32 Typical variation of the power coefficient (C_P) with the TSR

The way to overcome this situation is to let the angular speed change in pace with the wind speed. For instance, when wind speed increases, we want the turbine to rotate faster so that TSR is at its optimal value and the maximum efficiency is obtained.

Let us remember that the turbine angular speed is an image of the generator angular speed, if a gearbox exists, or it is equal to the generator angular speed in gearless WTG. In one way or another, electrical generators that support variable speed are required.

Conventional synchronous generators operate at a fixed speed and as so are not adequate; conventional asynchronous generators operate at a nearly fixed speed and are not adequate either. Modern WTG electrical generators use power electronics to achieve variable speed, as it is the case of Double Fed Induction Generators (DFIG), in WTG with gearbox, or Direct Driven DC-Link Synchronous Generator, in gearless WTG.

7.3   Power Control

Let us recall a typical WTG power curve that we repeat hereafter in Fig. 33 for convenience.

For wind speeds higher than the rated wind speed, the output power is regulated to the WTG rated power, because rated power cannot be exceeded, or the WTG would be damaged.

How is this achieved? Power control is achieved by pitching the blades, i.e., by rotating them over a longitudinal axis.

When the wind speed, as measured by a dedicated anemometer installed on the top of the nacelle, is higher than the rated wind speed, as defined by the manufacturer, an order is sent to the blade pitch mechanism, which pitches (turns) the rotor blades slightly out of the wind. The system is in general either made up of electric motors and gears or hydraulic cylinders and a power supply system.

Fig. 33 Typical WTG power curve

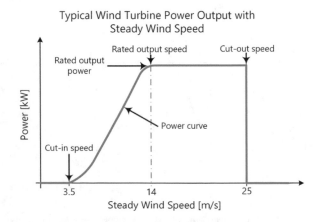

By pitching the blades (increase pitch angle), the C_P is decreased, as seen in Fig. 34, where the C_P as a function of the TSR is depicted for different pitch angles (β). Also, recall Fig. 13.

We can see that the WTG efficiency decreases as the β angle increases, as desired; maximum efficiency is obtained for $\beta = 0$ (no pitch). When the efficiency decreases, the output power also decreases. In this case, the pitch angle is controlled, so that constant output power is obtained, through proper variation of the efficiency.

An analytic equation for obtaining the curves of Fig. 34 is

$$C_P = 0.22 \left(\frac{116}{\lambda_i} - 0.4\beta - 5 \right) exp\left(-\frac{12.5}{\lambda_i} \right) \tag{7.3}$$

where

$$\lambda_i = \frac{1}{\frac{1}{\lambda + 0.08\beta} - \frac{0.035}{\beta^3 + 1}} \tag{7.4}$$

The typical variation of the pitch angle with the wind speed is shown in Fig. 35. For wind speeds lower than the rated wind speed, the pitch angle is zero, because,

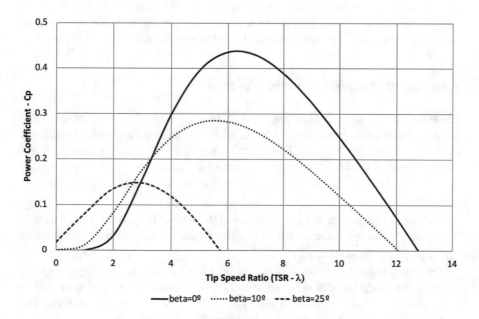

Fig. 34 Typical variation of the power coefficient (C_P) with the TSR, for different pitch angles (beta)

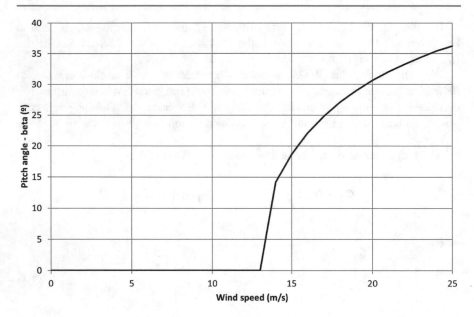

Fig. 35 Typical variation of the pitch angle with the wind speed

in this power curve zone, the objective is to maximize the efficiency. For wind speeds higher than the rated wind speed, the pitch angle continuously increases, so that to keep the output power constant, by constantly decreasing the efficiency.

7.4 Simple Analysis of Wind Turbines

There are two initial velocities active on a blade of a WTG. The incoming wind velocity V, which is perpendicular to the blade and the rotating motion, which is approximated by the tangential velocity of the blade $U = \omega r$, where ω is the angular speed of the blade and r is the radius of the circle described by the rotating blades (actually, the length of the blade). Figure 36 shows the velocities just introduced.

As seen in Fig. 36, the blade experiences a velocity of magnitude $\overrightarrow{W}$ and angle γ which will produce a force on the blade. This is the relative velocity of the blade and is given by $\overrightarrow{W} = \overrightarrow{V} - \overrightarrow{U}$.

The operating principle of a WTG blade is very similar to a plane wing. The shape of the blade induces the generation of low-pressure air on the more curved side, whereas high-pressure air is generated on the other side of the aerofoil. The result of this is a force that produces torque and ultimately power.

It is convenient to represent the force applied to the blade by a component parallel to the relative velocity (Drag—D) and another one perpendicular to it (Lift

Fig. 36 Velocities in a WTG blade

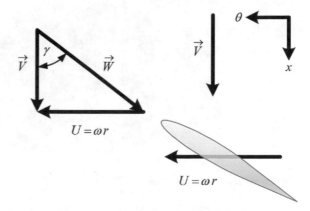

—L). To obtain useful output power, a force in the tangential direction must be produced: L and D forces must be resolved in the tangential direction to give a force F_θ, which contributes to the torque. We recall that the torque is $T = F_\theta \times r$ and the power is $P = T \times \omega$. On the other side, the force perpendicular to the tangential velocity direction is denoted by F_x and contributes to the movement (thrust). The situation is further explained in Fig. 37.

The relevant forces are given by

$$F_x = L\sin\gamma + D\cos\gamma \tag{7.5}$$

$$F_\theta = L\cos\gamma - D\sin\gamma \tag{7.6}$$

Fig. 37 Velocities and forces in a WTG blade

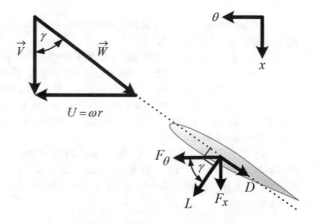

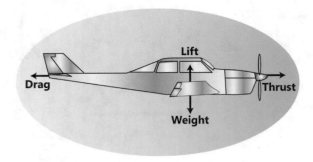

Fig. 38 Forces acting in a plane

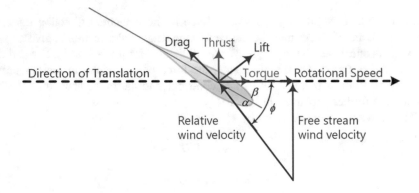

Fig. 39 Relevant angles

To better understand the forces involved, an example with the forces acting in a plane is shown in Fig. 38.

It is important to define some relevant angles that are represented in Fig. 39.

The relevant angles are α, the angle of attack, β, the pitch angle, and ϕ, the inflow angle. We define the chord line as the line that connects the leading edge and the trailing edge of the blade profile. α is defined as the angle between the chord line and the relative velocity; β is the angle between the blade motion plane and the chord line; and $\phi = \alpha + \beta$.

From Fig. 39, it is apparent that when the wind speed increases, the angle of attack also increases. The lift force increases with the angle of attack to a certain angle where lift reaches the maximum value. From this angle on, the lift decreases,

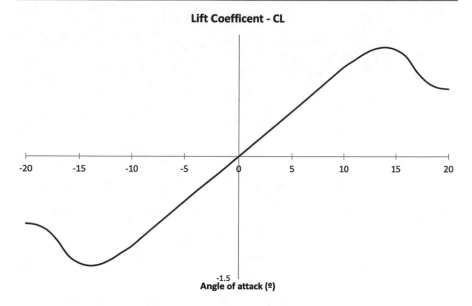

Fig. 40 Lift coefficient as a function of the angle of attack

and the blade is said to stall. The retarding drag force also increases with the angle of attack. For angles of attack higher than the angle that produces maximum lift, the drag increases dramatically.

The said phenomena can be seen with the aid of the lift coefficient:

$$C_L = \frac{L}{\frac{1}{2}\rho A W^2} \tag{7.7}$$

and the drag coefficient:

$$C_D = \frac{D}{\frac{1}{2}\rho A W^2} \tag{7.8}$$

The representation of the two coefficients as a function of the angle of attack is illustrated in Fig. 40 and Fig. 41, respectively.

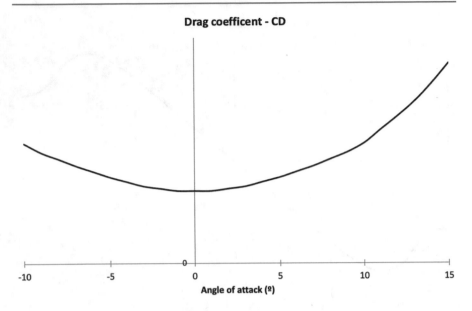

Fig. 41 Drag coefficient as a function of the angle of attack

8 Conclusions

Winds are caused by differences in pressure along the earth's surface since the solar radiation received on earth is greater in the equatorial zones than in the polar zones. The origin of wind is, therefore, solar radiation.

This chapter has been mainly dedicated to electricity production from Wind Turbine Generators (WTGs). The characterization of the wind profile at a site is obtained through wind speed measurements taken over a sufficiently long period of time. The result of the measurement campaign allows the construction of a frequency table depicting the occurrence of each wind speed class, i.e., the number of yearly hours the wind speed is between two values. This is called a wind speed histogram.

The measurements performed at a site also allow obtaining the statistical distributions that best fit the characteristics of the wind resource. It was observed that the Weibull distribution is the one that leads to the closest representations of the real wind speed at a given location. When the available information refers only to the mean annual wind speed, the Rayleigh distribution is used.

Measurements are not normally taken at the height at which the WTG rotor is to be installed. As the parameter of interest is the wind speed at the hub height, the measured values need to be corrected for the rotor height. Prandtl law is often used for this purpose.

In addition to the representation of the wind profile, the process of computing the electricity WTG requires the knowledge of the WTG power curve, i.e., the variation of the electrical power output as a function of the input wind speed. Manufacturers provide the WTG power curves in their catalogues.

To broaden the scope of the studies, it is useful to have analytical expressions for the input data. We have seen that the WTG power curves are well approximated by sigmoid-type functions, while Weibull probabilistic distributions are the most adequate to describe the wind profile. The produced annual electrical energy is obtained based on the integration, for the wind speeds of interest, of the product of the said two characteristics. When the available data does not allow for this detailed model to be used, simplified models to obtain rough estimates of the electricity production have been offered. It was stressed that the model to be used depends upon the quantity and quality of the available data.

It was shown that the wind speed is very sensitive to the existence of obstacles in its path, so this aspect must be considered when locating WTG on the ground to maximize the wind available for conversion into electricity. Attention was drawn to the need for a careful location of the WTG within a wind farm, to minimize the wake effect.

An aspect of utmost importance is wind forecasting, which has obvious implications for the operation of the power system. Two forecasting techniques were presented, the Auto-Regressive Moving Average method (ARMA) and the Artificial Neural Networks (ANNs), which belong to the group of the so-called statistical models. Both methods, using however different algorithms and techniques, start from the same point, i.e., they predict future values from the analysis of correlations found in historical data. The application to case studies has shown that both methods can make reasonable predictions in the short term. For long-term forecasts it advisable to use the so-called physical models, which incorporate the mass movements in the atmosphere.

This chapter also includes a section on the technical aspects of WTG. The advantages of operating WTG with variable speed have been demonstrated since this corresponds to allow the maximum conversion efficiency to be obtained. For high wind speeds, the output power of the WTG must be limited not to exceed their rated power. The blade's pitch control strategy that forces the output power limitation was discussed. It has been shown that the output power can be fine-tuned using the power electronics capabilities of the electrical generators used in modern WTG.

9 Proposed Exercises

Problem WE1
Consider a 3 MW Wind Turbine Generator (WTG) with a rotor diameter equal to 112 m. The rated power is obtained at a rated (nominal) wind speed of 12 m/s. The cut-in wind speed is 3 m/s, and the cut-out wind speed is 25 m/s. Assume that the power curve can be simulated by a sigmoid function with parameters

$c_1 = 8.0686; c_2 = 1.2623$, between the cut-in wind speed and the rated wind speed. At the WTG location, the wind speed pdf is described through a Weibull distribution with parameters $k = 2.4800; c = 6.0683 \, \text{m/s}$.

Compute

(1) The power coefficient (C_P) at rated wind speed.
(2) An estimate of the yearly produced electrical energy. Hint: use the trapezoidal integration method.
(3) An estimate of the yearly non-produced electrical energy because the wind generator is stopped for wind speeds greater than the cut-out wind speed.

Solution

(1)

$$
C_p(u_N) = \frac{P_e(u_N)}{P_{avail}(u_N)} = \frac{P_e(u_N)}{\frac{1}{2}\rho A u_N^3} = \frac{P_N}{\frac{1}{2}\rho_{air} \frac{\pi (RD)^2}{4} u_N^3} = 28.65\%
$$

(2)

$$
E_a = 8760 \int_{u_0}^{u_{max}} P_e(u) f(u) du = 8760 \left[\int_{u_0}^{u_N} P_e(u) f(u) du + P_N \int_{u_N}^{u_{max}} f(u) du \right]
$$
$$
= A + B
$$

Let us look at integral A.

$$
A = 8760 \int_{u_0}^{u_N} P_e(u) f(u) du = 8760 \int_{u_0}^{u_N} \frac{P_N}{1 + \exp\left[-\frac{u - c_1}{c_2}\right]} \frac{k}{c} \left(\frac{u}{c}\right)^{k-1} \exp\left(-\left(\frac{u}{c}\right)^k\right) du
$$

Using integration formulas, we obtain $A = 5097.4736 \, \text{MWh}$, which seems to be very complicated!!!

The smart alternative is to use the trapezoidal integration method (Fig. 42).

$$
\int_a^b f(x) dx \approx [f(x_0) + f(x_1)] \frac{(x_1 - x_0)}{2} + [f(x_1) + f(x_2)] \frac{(x_2 - x_1)}{2} + \ldots
$$
$$
+ [f(x_{n-1}) + f(x_n)] \frac{(x_n - x_{n-1})}{2}
$$

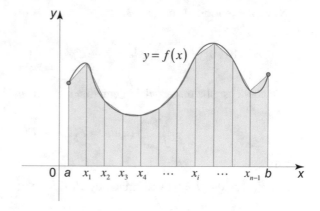

Fig. 42 Trapezoidal integration method (Problem WE1)

$$\int_a^b f(x)dx \approx \frac{\Delta x}{2}[f(x_0) + 2f(x_1) + 2f(x_2) + \ldots + 2f(x_{n-1}) + f(x_n)] =$$
$$= \Delta x\left[\frac{f(x_0)}{2} + f(x_1) + f(x_2) + \ldots + f(x_{n-1}) + \frac{f(x_n)}{2}\right]$$

where $\Delta x = \frac{b-a}{n}$ and $x_i = a + i\Delta x$.

As $n \to \infty$ the right-hand side of the expression approaches the definite integral $\int_a^b f(x)dx$.

Application: $g(u) = 8760\frac{3}{1+\exp\left[-\frac{u-8.0686}{1.2623}\right]}\frac{2.4800}{6.0683}\left(\frac{u}{6.0683}\right)^{1.4800}\exp\left(-\left(\frac{u}{6.0683}\right)^{2.4800}\right)$

Using the integration formula, we obtain $A = \int_3^{12} g(u)du = 5085.0754\,\text{MWh}$. Note that $\Delta x = 1$ (Table 8).

The error is -0.24%, which is a very good approximation bearing in mind that we are using a unitary integration step (1 m/s).

A better approximation would be obtained if we decrease the integration step, but this would come with a working burden, that in this type of application is not necessary.

Table 8 Application of the trapezoidal integration method (Problem WE1)

u	g(u)
3	56.3462
4	155.5398
5	351.1513
6	649.6334
7	957.7802
8	1080.9140
9	912.8394
10	586.1901
11	297.9340
12	129.8401

Now, let us look at integral B.

$$B = 8760 P_N \int_{u_N}^{u_{max}} f(u) du = 8760 P_N \int_{12}^{25} \frac{k}{c} \left(\frac{u}{c}\right)^{k-1} \exp\left(-\left(\frac{u}{c}\right)^k\right) du$$

We could use the trapezoidal integration method again, but this time we note that

$$\int_{u_N}^{u_{max}} f(u) du = F(u_N) - F(u_{max})$$

where $F(u)$ is the Weibull cdf given by $F(u) = \exp\left(-\left(\frac{u}{c}\right)^k\right)$

$$B = 8760 P_N [F(12) - F(25)] = 115.8177 \, \text{MWh}$$

And finally,

$$E_a = A + B = 5200.8931 \, \text{MWh}$$

$$h_a = \frac{E_a}{P_N} = 1733.63 \, \text{h}$$

(3)

$$E_{np} = 8760 \times P_N \int_{25}^{\infty} f(u) du = 8760 P_N [F(25) - F(\infty)] = 7.59 \times 10^{-8} \, \text{kWh}$$

The "lost" energy is negligible. In fact, it is not lost, because it could never have been produced, even though we have an ideal turbine.

Problem WE2

A 1 MW Wind Turbine Generator (WTG) has 3 blades with a rotor diameter of 54 m and a hub height of 45 m. Table 9 depicts the WTG power curve. The wind speed pdf, at rotor height, can be approximated by a Weibull distribution, in which the scale parameter is $c = 7.45 \, \text{m/s}$ and the shape parameter is $k = 1.545$. The soil where the WTG will be installed has a roughness length $z_0 = 10^{-2}$ m. Calculate

(1) The average annual wind speed at rotor height and at 10 m.
(2) The number of hours per year in which the WTG is regulated to operate at rated power and the energy it produces during that period.

Table 9 Power curve for Problem WE2

u (m/s)	P (kW)
0	0
1	0
2	0
3	0
4	13
5	55
6	116.1
7	204
8	317.4
9	444.7
10	583.1
11	715.6
12	822.1
13	906.8
14	963.7
15	991
16	1000
17	1000
18	1000
19	1000
20	1000
21	1000
22	1000
23	1000
24	1000
25	1000
26	0

Solution

(1)

$$u_{ma} = c\Gamma\left(1 + \frac{1}{k}\right) \approx 0.9c$$

$$u_{ma}(45) = 0.9c = 6.70\,\text{m/s}$$

$$u(z_1) = u(z_2)\frac{\ln\left(\frac{z_1}{z_0}\right)}{\ln\left(\frac{z_2}{z_0}\right)}$$

$$u_{ma}(10) = u_{ma}(45)\frac{\ln\left(\frac{10}{z_0}\right)}{\ln\left(\frac{45}{z_0}\right)} = 5.50\,\text{m/s}$$

(2)

$$F(u) = \exp\left(-\left(\frac{u}{c}\right)^k\right)$$

$$n_{(P=P_N)} = 8760\int_{16}^{25} f(u)du = 8760[F(16) - F(25)] = 323.84\,\text{h}$$

$$E_{aP=P_N} = 323.84\,\text{MWh}$$

Problem WE3

In a place where the roughness length of the ground is 0.03 m, it is known that the annual average wind speed at 10 m height is 5 m/s. At this site, we intend to install a standard 2 MW WTG, 3-bladed, 80 m rotor diameter, mounted on an 80 m tower height, with cut-in, rated and cut-out wind speeds equal to 4, 15, and 25 m/s, respectively. Calculate

(1) The efficiency of the wind-electrical conversion at rated wind speed.
(2) The average annual wind speed at 80 m rotor height.
(3) The designer of the wind energy conversion system says that, in this place, you can get about 1,750 rated power equivalent hours (utilization factor). Do you agree with the statement of the designer? Justify your answer. Hint: remember $\Gamma(3/2) \approx 0.9$.

Solution

(1)

$$C_p(u_N) = \frac{P_e(u_N)}{P_{avail}(u_N)} = \frac{P_e(u_N)}{\frac{1}{2}\rho A u_N^3} = \frac{P_N}{\frac{1}{2}\rho_{air}\frac{\pi(RD)^2}{4}u_N^3} = 19.17\%$$

(2)

$$u_{ma}(80) = u_{ma}(10)\frac{\ln\left(\frac{80}{z_0}\right)}{\ln\left(\frac{10}{z_0}\right)} = 6.79\,\text{m/s}$$

(3)

Approximation: $k = 2; c = \frac{u_{ma}}{0.9} = 7.54 \, \text{m/s}$

$$P_{avg} = P_N \left(\frac{e^{-\left[\left(\frac{u_0}{c}\right)^k\right]} - e^{-\left[\left(\frac{u_N}{c}\right)^k\right]}}{\left(\frac{u_N}{c}\right)^k - \left(\frac{u_0}{c}\right)^k} - e^{-\left[\left(\frac{u_{max}}{c}\right)^k\right]} \right) = 0.40 \, \text{MW}$$

$$E_a = 8760 P_{avg} = 3510 \, \text{MWh}$$

$$h_a = \frac{E_a}{P_N} = 1755 \, \text{h}$$

Problem WE4

A WTG with a nominal capacity of 810 kW has three blades, a rotor diameter equal to 48 m and hub height that equals 40 m. The cut-in, nominal, and cut-off wind speeds equal 2, 14, and 25 m/s, respectively. The wind speed cdf, at rotor height, in the place where the turbine is to be installed is given by (u in m/s): $F(u) = \exp(-0.0175u^2)$. In a specific hour, at the height of 20 m, the following 15 min average wind speeds were measured: 11.38 m/s; 11.59 m/s; 13.77 m/s; 14.38 m/s. Assume that the mean power coefficient (C_P) for wind speeds comprised between 12 m/s and the nominal wind speed equals 35% and that soil roughness length is $z_0 = 0.01$ m. Compute

(1) The number of annual hours in which the WTG operates at nominal power.
(2) The shape and scale parameters of the wind speed Weibull pdf, using the cdf linearization method.
(3) The energy produced in this particular hour by the WTG.

Solution

(1)

$$n_{(P=P_N)} = 8760 \int_{14}^{25} f(u)du = 8760[F(14) - F(25)] = 283.55 \, \text{h}$$

(2)

$$F(u) = \exp(-0.0175u^2) = \exp\left[-\left(\frac{u}{\sqrt{\frac{1}{0.0175}}} \right)^2 \right]$$

$$k = 2; c = \sqrt{\frac{1}{0.0175}} = 7.56 \, \text{m/s}$$

But we do not want to solve the problem this way. We want to use the cdf linearization method.

$$F(u) = \exp\left(-\left(\frac{u}{c}\right)^k\right) \leftrightarrow \ln F(u) = -\left(\frac{u}{c}\right)^k \leftrightarrow \ln[-\ln F(u)] = \ln\left[\left(\frac{u}{c}\right)^k\right]$$

$$\ln[-\ln F(u)] = k\ln u - k\ln c \leftrightarrow y = mx + b$$

$$k = m; c = \exp\left(-\frac{b}{m}\right)$$

$$y_1 = \ln\left[-\ln F(14)\right]; \; x_1 = \ln 14$$

$$y_2 = \ln\left[-\ln F(25)\right]; \; x_2 = \ln 25$$

$$m = k = 2; c = \exp\left(-\frac{b}{m}\right) = 7.56 \, \text{m/s}$$

(3)

We will exemplify for the first 15 min. interval.

$$u_1(20) = 11.38 \, \text{m/s} \leftrightarrow u_1(40) = 12.42 \, \text{m/s}$$

$$P_1(12.42) = C_p 0.5 \rho A u^3 = 745.84 \, \text{kW}$$

$$E_{1h} = 0.25 \sum_{i=1}^{4} P_i = 788.43 \, \text{kWh}$$

Problem WE5

In a particular location, a wind mast is placed. The registers indicate that the average annual wind speed equals 5.5 m/s and 6.4 m/s, at 10 m and 30 m heights, respectively. It foreseen the installation of a 2 MW WTG. The WTG has 3 blades, and the rotor diameter equals 80 m, the hub height being 70 m. The cut-in, nominal, and cut-out wind speeds equal 4 m/s, 15 m/s, and 25 m/s, respectively. From the WTG power curve, the following pairs of values $(u; P_e)$ are known: (10; 1,279), (11; 1,590), (12; 1,823), (13; 1,945), (14; 1,988), for u in m/s and P_e in kW. Compute

(1) The roughness length of the soil.
(2) The annual average wind speed at rotor height.
(3) The annual energy produced by each WTG, when it is regulated to operate at nominal power.
(4) The annual energy produced by the WTG, when operating with wind speeds comprised between 13 and 15 m/s.

Solution

(1)

$$u(30) = u(10)\frac{\ln\left(\frac{30}{z_0}\right)}{\ln\left(\frac{10}{z_0}\right)} \xrightarrow{yields} \ln z_0 = \frac{u(10)\ln 30 - u(30)\ln 10}{u(10) - u(30)} \xrightarrow{yields} z_0 = 0.0121 \text{ m}$$

(2)

$$u(70) = u(10)\frac{\ln\left(\frac{70}{z_0}\right)}{\ln\left(\frac{10}{z_0}\right)} = 7.09\frac{\text{m}}{\text{s}} = u_{ma}$$

(3)

$$E_{a(P=P_N)} = 8760 P_N (F(15) - F(25)) = 522.06 \text{ MWh}$$

$$F(u) = \exp\left[-\frac{\pi}{4}\left(\frac{u}{u_{ma}}\right)^2\right]$$

(4)

$$E_{a(13<u<15)} = 8760\left[(F(13) - F(14))\frac{P_e(13) + P_e(14)}{2} + (F(14) - F(15))\frac{P_e(14) + P_e(15)}{2}\right] = 722.28 \text{ MWh}$$

Wind Energy Conversion Equipment

6

Abstract

To optimize the efficiency of the conversion of the available power in the wind into electrical output power, the electrical generators that equip the Wind Turbine Generators (WTGs) must be operated at a variable speed. The connection of a variable-speed electrical generator to a constant frequency grid requires the use of a power electronics interface. In this chapter, we address the two main types of electrical generators that equip the modern WTG—the Double-Fed Induction Generators (DFIGs) and the Direct-Driven DC-Link Synchronous Generators (DDSGs). Both generators are based on classical and widely used induction and synchronous (alternator) conventional generators, but they add a power electronics interface, basically a frequency converter, to allow the connection to a constant frequency grid. The operating principle and main features of both the DFIG and the DDSG are introduced. Steady-state models to describe the behaviour of the said electrical generators are presented. These models are based on equivalent circuits, whose equations are written using the basics of electrical circuits presented in Chap. 2.

1 Introduction

When wind power has begun to develop back in the '80s, the objective was to make the electrical conversion equipment as simple as possible, so that Wind Turbine Generators (WTGs) could be cheap and reliable. The obvious choice of the electrical generator was the well-known and robust squirrel-cage induction (or asynchronous) machine. This type of machine was and keeps on being widely used in the industry as a motor. However, like any electrical machine, it is reversible, meaning that it can also be used as a generator of real power, providing the turbine shaft rotates faster than the electrical grid frequency. To keep costs down and reliability high, the first WTGs were equipped with Squirrel-Cage Induction Generators (SCIGs). SCIG portray the Type#1 wind generators.

The rotor speed of asynchronous generators is variable but in a very narrow range. This speed change as a percentage of the synchronous speed is known as the slip. Normally, the slip of squirrel-cage induction machines is about 2–3%. As seen before in Chap. 5, this is a serious drawback because it does not allow for maximum efficiency in the conversion of the available power in the wind into electrical output power. Maximum efficiency is obtained if the rotor speed varies over a wider range, following wind speed changes. A classical solution to accomplish this was to use a wound-rotor induction generator connected through slip rings to thyristor-controlled variable resistances. This is a Type#2 wind generator. Alternatively, the resistors and electronics can be mounted on the rotor, eliminating the slip rings—this is the Weier design.

As wind power developed more and more, the issue of costs became less important, because cost savings were achieved in other ways, namely due to the increasing demand for WTG. On the other hand, power electronics showed a huge development allowing for its implementation in electrical generators. This gave rise to new electrical generators connected to the grid through power electronics devices, therefore allowing for a wider variation of the rotor speed with the consequent efficiency increase.

Type#1 and Type#2 wind generators are no more used. Instead, the electrical generators that equip modern WTG are of two types: Double-Fed Induction Generators (DFIGs)[1] (Type#3) and Direct-Driven DC-Link Synchronous Generators (DDSGs) (Type#4).

Type#3 generators take the Type#2 design to the next level. The DFIG is a wound-rotor asynchronous generator whose rotor is connected to the grid through a power electronics interface. The rotor excitation is supplied via slip rings by a current regulated voltage source converter, which can adjust the rotor currents' magnitude and phase. This rotor-side converter is connected to another converter, the grid-side converter, which controls the power factor at the interconnection point.

The converters are typically sized to about 30% of the rated power of the generator. Nevertheless, a great deal of control is achieved in the output with a small amount of power injected into the rotor circuit. Power can be injected into the grid both in the stator and in the rotor when the generator is moving faster than synchronous speed. When the generator is moving slower than synchronous speed, real power flows from the grid to the rotor, and real power flows from the stator to the grid, an operation mode that is not possible in a conventional SCIG. These two modes, made possible by the four-quadrant nature of the two converters, allow a much wider speed range, both above and below synchronous speed. Typically, there is a gearbox connecting the turbine's low-speed shaft to the high-speed

---

[1]The DFIG is also known as Double-Output Induction Generator (DOIG). Given that in wind power applications this machine operates as a generator, the designation of DOIG seemed more appropriate. However, in the literature, this machine is mostly known as DFIG.

generator shaft, the increase ratio being about 100 times. Type#3 generator's technology is the most popular around the world.

Type#4 DDSG is a synchronous generator (alternator), with a power electronics interface connection to the grid in the stator. The generator, which is directly connected to the turbine's shaft, without a gearbox, sends power to the grid through a full-scale back-to-back frequency converter. The frequency converter feeds power to the grid with the appropriate voltage and frequency characteristics.

The rotating machines of this type have been constructed as wound-rotor synchronous machines, similar to conventional generators found in hydroelectric plants, with separate field (excitation) current and high pole numbers, known as Electrically Excited Synchronous Generators (EESGs), or as self-excited Permanent Magnet Synchronous Generators (PMSGs).

Type#1 and Type#2 WTGs can typically not control voltage. Instead, these WTGs typically use power factor correction capacitors to maintain the power factor or reactive power output to a setpoint. On the other hand, Type#3 and Type#4 WTGs can control voltage. The four-quadrant power control capability of the voltage-source converters used in Type#3 and Type#4 wind turbines allows them to either generate or consume reactive power, therefore providing voltage control.

Taking advantage of the power electronics, WTGs are being called to assist during faults in the grid.[2] Low-Voltage Ride-Through (LVRT) capability refers to the ability of the wind turbines to withstand credible fault conditions and support network voltage recovery by injecting reactive current. Depending on the wind turbine design, a LVRT threshold is defined at around 0.8–0.9 pu. When the WTG voltage drops below this threshold, the WTG suspends normal operation and starts injecting reactive current, while control of active power is given lower priority. As the voltage returns to normal, active and reactive currents start to ramp back to their pre-fault values.

As mentioned earlier, Type#1 and Type#2 generators are no longer offered by the manufacturers. Therefore, they will not be dealt with in this chapter. In this text, we will introduce the basic concepts associated with Type#3 (DFIG) and Type#4 (DDSG) electrical generators used in modern WTG. The main objective of the chapter is to propose steady-state models able to describe the electrical behaviour of the said generator's types.

We will begin with the DFIG by presenting its operating principle and the main components and associated connections. This allows for an equivalent circuit to be established. The steady-state model is grounded on the equations governing this equivalent circuit. For this purpose, the electrical circuits concepts presented in Chapter 2 will be used. The different operating modes—above and below the synchronous speed—will be presented and discussed.

---

[2]For further details, see AEMO, Australian Energy Market Operator, "Wind turbine plant capabilities report", available in https://www.aemo.com.au/-/media/Files/PDF/Wind_Turbine_Plant_Capabilities_Report.pdf/.

As far as the DDSG is concerned, we will begin by recalling the basic working principle of the alternator, as the DDSG is built based on an alternator. The main components of the DDSG, namely the power electronics ones, and the connections between them are also presented. We will then introduce the steady-state model, differentiating the round rotor from the salient pole types of generators. For the latter, it is not possible to set an equivalent circuit, so the model is based on phasor diagrams. Overall, the steady-state models are based on the well-known EMF behind a synchronous impedance model. The Permanent Magnet Synchronous Generator is gaining momentum in recent years, hence a model applied to this type of DDSG is presented at the end.

2 Double-Fed Induction Generator

In classical induction machines, there are two types of rotors: squirrel-cage and wound. The squirrel-cage rotor is composed of a cylinder of steel with aluminium or copper conductors embedded in its surface and short-circuited in the tops. Figure 1 shows an image of a squirrel-cage rotor of an induction machine.

As far as the wound rotor is concerned, the rotor windings are connected through slip rings to external resistances, as seen in Fig. 2.

The typical Power versus Speed curve of a squirrel-cage induction machine is depicted in Fig. 3. The rotor speed is measured in pu of the synchronous speed. For instance, rotor speed equal to 1 pu means the rotor is running at synchronous speed. Rotor speeds higher than 1 pu corresponds to generator operating mode and negative slips. Rotor speeds lower than 1 pu corresponds to the motor operating mode and positive slips. Of course, for wind power applications, only the generator mode is relevant.

The inversion of the electrical power is apparent: in the generator convention, the electrical power is positive when electrical power is supplied at the machine terminals, therefore running as a generator. The normal operation is in the upward zone, corresponding to a slip of 2–3% as already mentioned. The squirrel-cage induction machine is said to operate at a quasi-constant speed.

2.1 Operating Principle

If the rotor is wounded, the picture completely changes. The Power versus Speed curve changes as one changes the resistance in series with the rotor winding, as can be seen in Fig. 4.

We observe that, for a given electrical power, it is possible to change the rotor speed by changing the rotor resistance. In a modern DFIG, an AC/DC/AC power electronics-based conversion system is used instead of rotor resistances.

Fig. 1 Squirrel-cage rotor of an induction machine

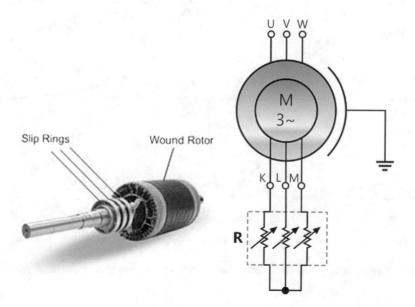

Fig. 2 Wound rotor of an induction machine

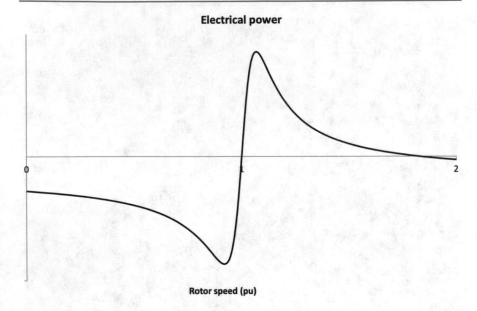

Fig. 3 Power versus Speed curve of a squirrel-cage induction machine

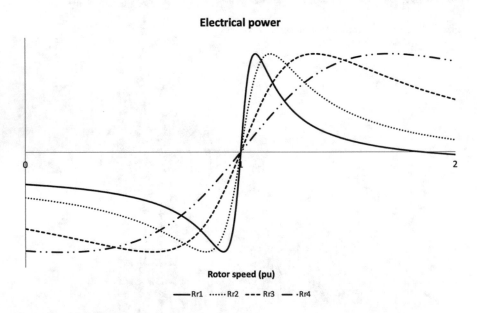

Fig. 4 Power versus Speed curves of a wound-rotor induction machine; $R_{r4} > R_{r3} > R_{r2} > R_{r1}$

The generator's rotor is connected to the turbine's rotor (blades) through a gearbox to adapt the two speeds. Therefore, the two speeds are related through a constant. When the wind speed changes, it changes the turbine speed and consequently the generator rotor speed. It is not possible to connect a variable-speed rotor (which is an image of the frequency) to a constant frequency (50 Hz) grid. Hence, rotor's variable frequency quantities are transformed into zero-frequency (DC) quantities using an AC/DC converter (rectifier). Then, the DC quantities are inverted to the 50 Hz constant frequency of the grid using a DC/AC converter (inverter). We note that electrical power is transferred in the rotor of a DFIG, which is not possible in a Squirrel-Cage Induction Generator (SCIG). Moreover, this machine allows for generator mode with positive slip, as we will see later in this chapter. This implies the power to be supplied in the rotor.

This is the operating principle of a DFIG: until the output rated power is reached, power is controlled so as to optimize the tip speed ratio and therefore maximize the C_p (see Chap. 5 for further details); when the rated power is reached, the output power is regulated to the constant rated power.

We note that all of the above happens in the DFIG rotor. The DFIG stator is equal to a SCIG stator: it is directly connected to the grid, without any converters.

2.2   Main Equipment

The typical connections scheme of a DFIG is shown in Fig. 5.

In the WTG equipped with a DFIG, there is a gearbox to adapt the low-speed shaft of the turbine to the high-speed shaft of the generator. Typically, the turbine blades rotate in an interval between 9 and 18 rpm and the generator between 1050 and 1950 rpm ($\pm 30\%$ slip), for a typical 2 pairs of poles generator. Both stator and rotor are grid connected. The stator is directly connected to the grid. The rotor is connected to the grid through an AC/DC/AC converter.

There are indeed two converters: a rotor-side converter, which is an AC/DC converter, and a grid-side converter, which is a DC/AC converter. Both converters are six pulses bridges equipped with Insulated Gate Bipolar Transistors (IGBTs) with a control system by Pulse Width Modulation (PWM). The DC/AC converter connected to the step-up transformer controls the capacitor DC voltage of the DC-Link and controls the power factor at the Point of Common Coupling (PCC), typically between 0.9 inductive and 0.9 capacitive. This avoids the need for an external capacitor bank for power factor compensation. The AC/DC converter directly connected to the rotor controls the amplitude and phase of the current injected/extracted by the rotor circuit. The converters' PWM control system can impose an AC sinusoidal waveform with adjustable frequency, amplitude, and phase.

The overall objective of the control system is threefold: (i) to maximize the C_P when power control is not active, (ii) to uphold a given power factor at the PCC, and (iii) to keep the total power drawn by both the stator and rotor constant when

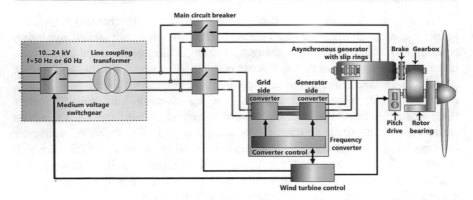

Fig. 5 Typical connections scheme of a DFIG

power control is active. The latter action is complemented by the blades' pitch angle control system.

2.3 Equivalent Diagram

Figure 6 presents the equivalent diagram of a DFIG. The power electronics converters connected to the rotor are not represented. Therefore, it is indeed the equivalent diagram of a wound-rotor induction machine. The rotor parameters are referred to the stator. All quantities are in pu. Bold denotes phasor quantities.

The diagram is quite similar to the SCIG diagram, except that the DFIG is a wound-rotor machine, therefore the rotor is accessible by two external terminals. That is why we find a rotor voltage V_r.

In Fig. 6, R_s, $X_{\ell s}$ and R_r, $X_{\ell r}$ are the resistances and leakage reactances of the stator and rotor windings, respectively, X_m is the magnetization reactance, and s is the slip. P_s is the power transferred in the stator: it is positive if it is being delivered in the stator, towards the grid (generator operating mode) and it is negative (motor operating mode) contrarywise. P_{ag} is the power transferred in the airgap; it is positive if it is flowing from the rotor to the stator (generator operating mode) and it is negative (motor operating mode) contrarywise. P_{mec} is the mechanical (motion) power supplied to the rotor: it is positive if it is being supplied to the rotor (generator operating mode) and it is negative (motor operating mode) contrarywise. P_r is the power transferred in the rotor: it is positive if it is flowing from the grid towards the rotor and it is negative contrarywise, i.e., from the rotor to the grid. We note that, in general, P_r can be positive or negative, independently of the operating mode of the machine. We will come back later to this issue.

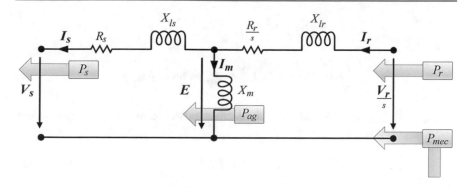

Fig. 6 Equivalent diagram of a DFIG, without the power electronics converters

2.4 Steady-State Model

Based on the equivalent diagram, we can write the following equations that describe the DFIG in steady state[3]:

$$V_s = -(R_S + jX_{\ell s})I_s + E \tag{2.1}$$

$$\frac{V_r}{s} = \left(\frac{R_r}{s} + jX_{\ell r}\right)I_r + E \tag{2.2}$$

in which the Electromotive Force (EMF) and the slip are given by (ω_r is the rotor angular speed and ω_s is the synchronous speed)

$$E = jX_m I_m \tag{2.3}$$

$$s = \frac{\omega_s - \omega_r}{\omega_s} \tag{2.4}$$

We note that by making $V_r = 0$, we obtain the steady-state equations of the SCIG.

The active and reactive power in the stator and in the rotor is given by

$$P_s = -R_s I_s^2 + P_{ag} = \text{Re}\{V_s I_s^*\} \tag{2.5}$$

$$Q_s = -X_{\ell s} I_s^2 + Q_{ag} = \text{Im}\{V_s I_s^*\} \tag{2.6}$$

---

[3]As usually, we use bold notation to represent complex quantities and regular notation to represent real quantities.

$$\frac{P_r}{s} = \frac{R_r}{s}I_r^{\,2} + P_{ag} = \text{Re}\left\{\frac{V_r}{s}I_r^{\,*}\right\} \tag{2.7}$$

$$\frac{Q_r}{s} = X_{\ell r}I_r^{\,2} + X_m I_m^{\,2} + Q_{ag} = \text{Im}\left\{\frac{V_r}{s}I_r^{\,*}\right\} \tag{2.8}$$

P_{ag} and Q_{ag} are the active and reactive power transferred in the airgap, respectively, given by

$$P_{ag} = \text{Re}\{\boldsymbol{E}\boldsymbol{I_s}^*\} \tag{2.9}$$

$$Q_{ag} = \text{Im}\{\boldsymbol{E}\boldsymbol{I_s}^*\} \tag{2.10}$$

We recall that the rotor is rotating, the slip being s. The rotor frequency is therefore $\omega_s - \omega_r = s\omega_s$. The rotor leakage and magnetization reactances are $sX_{\ell r}$ and sX_m, respectively, and the air-gap reactive power is sQ_{ag}. Hence, the reactive power effectively transferred in the rotor is $\frac{Q_r}{s}$ (Eq. 2.8).

As for the active power, the reactances-related issue is not relevant, the active power being only concerned with the resistances. Therefore, the active power transferred in the rotor is P_r.

$$P_r = R_r I_r^{\,2} + sP_{ag} = \text{Re}\{V_r I_r^{\,*}\} \tag{2.11}$$

The losses are

$$P_{Loss} = R_s I_s^2 + R_r I_r^2 \tag{2.12}$$

$$Q_{Loss} = X_{\ell r}I_r^2 + X_m I_m^2 + X_{\ell s}I_s^2 \tag{2.13}$$

The power balance equation is

$$P_{ag} = P_r - R_r I_r^{\,2} + P_{mec} \tag{2.14}$$

It is common sense that the active power transferred in the airgap is equal to the rotor power plus the mechanical power (input power) minus the rotor losses.

From Eq. 2.11, it is possible to write

$$sP_{ag} = P_r - R_r I_r^{\,2} = P_{sl} \tag{2.15}$$

P_{sl} is often called "slip power". This allows us to conclude that from the active power transferred in the airgap, P_{ag}, the parcel sP_{ag} is equal to the rotor power minus the rotor losses. The other part is the mechanical power. Therefore,

$$P_{mec} = (1 - s)P_{ag} \tag{2.16}$$

Regarding a SCIG, it is $V_r = 0$ and therefore $P_r = Q_r = 0$. So, in Eq. 2.15, if the slip is negative, the machine operates as a generator ($P_{ag} > 0$); if the slip is positive, the machine operates as a motor ($P_{ag} < 0$). In the case of the DFIG, the situation is a bit more complicated: the DFIG can operate as a generator with positive slip, providing power is supplied in the rotor in a quantity larger than the rotor losses. We will come back shortly to this issue.

Finally, two more equations can be written from the above:

$$P_s = P_r + P_{mec} - P_{Loss} \tag{2.17}$$

$$Q_s = \frac{Q_r}{s} - Q_{Loss} \tag{2.18}$$

The reader is invited to demonstrate Eqs. 2.17 and 2.18 and to show their physical meaning.

2.5   Operating Modes

Tables 1 and 2 show the operating modes of the SCIG and DFIG, respectively. The basis for the construction of these tables is Eq. 2.15.

It is important to highlight that the DFIG can operate as a generator with positive slip, providing active power is supplied in the rotor, from the grid, in a quantity larger than the rotor losses. Another noted situation occurs when the DFIG supplies active power both by the stator ($P_s > 0$) and the rotor ($P_r < 0$).

2.6   Introduction to Control

The slip power increases with $|s|$. Typically, the DFIG operates within a $\pm 30\%$ slip range. It is important to highlight that the power electronics that are connected to the DFIG's rotor need to be rated for only the slip power. This derate leads to significant cost savings and simplifies the design of the power electronics.

Let us retake Eqs. 2.1 and 2.2 and bear in mind that from the equivalent diagram of Fig. 6 we can write

$$I_r = I_s + I_m \tag{2.19}$$

Hence, we come to

$$V_s = -(R_s + jX_s)I_s + jX_m I_r \tag{2.20}$$

$$\frac{V_r}{s} = \left(\frac{R_r}{s} + jX_r\right)I_r - jX_m I_s \tag{2.21}$$

Table 1 SCIG operating modes

	SCIG			
	P_r	P_{ag}	P_{mec}	Mode
$s > 0$	0	<0	<0	Motor
$s < 0$	0	>0	>0	Generator

Table 2 DFIG operating modes

	DFIG			
	P_r	P_{ag}	P_{mec}	Mode
	>0	<0; $P_r < R_r I_r^2$	<0	Motor
$s > 0$	>0	>0; $P_r > R_r I_r^2$	>0	Generator
	<0	<0	<0	Motor
	>0	<0; $P_r > R_r I_r^2$	<0	Motor
$s < 0$	>0	>0; $P_r < R_r I_r^2$	>0	Generator
	<0	>0	>0	Generator

where

$$X_s = X_{\ell s} + X_m \tag{2.22}$$

$$X_r = X_{\ell r} + X_m \tag{2.23}$$

Writing Eq. 2.20 in order to I_r and replacing in Eq. 2.21, we obtain an equation relating $\frac{V_r}{s}$ with V_s and I_s, which is equivalent to have $\frac{V_r}{s}$ related to the stator power, P_s and Q_s.

$$\frac{V_r}{s} = f(I_s, V_s) = g(P_s, Q_s) \tag{2.24}$$

Just for curiosity, the formal relationships are

$$\frac{V_r}{s} = Z_{Is}I_s + K_{Vs}V_s \tag{2.25}$$

$$Z_{Is} = \left(\frac{X_r R_s + \frac{X_s R_r}{s}}{X_m}\right) + j\left(\frac{X_r X_s - X_m^2 - \frac{R_r R_s}{s}}{X_m}\right) \tag{2.26}$$

$$K_{Vs} = \frac{X_r}{X_m} + j\frac{-\frac{R_r}{s}}{X_m} \tag{2.27}$$

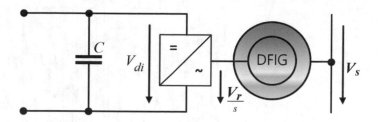

Fig. 7 Rotor voltage control system diagram

But what really matters at this stage is that it is possible to control both active and reactive power in the stator through the rotor voltage.

In Fig. 7, we show the relevant part of the DFIG's rotor and respective electronic converters. To control the rotor voltage, it is necessary that the inverter DC input voltage, V_{di}, remains constant. Condenser C plays this role.

A power electronics converter is composed of many electronic switches. These switches are nothing but power electronic devices, like IGBT, for instance. In our case, the rotor-side AC/DC converter is equipped with IGBT and the control system is PWM type.

PWM is a technique used to control the power electronic switches, by applying a PWM signal. The PWM signal is the signal that makes the IGBT being in "on state" (closed) or "off state" (open). The average voltage fed to the rotor is controlled by turning the switch between supply and rotor on and off at a fast rate. The longer the switch is on, the higher the total power supplied.

To obtain a PWM pulse train corresponding to a given signal, the intersective method is used (Fig. 8): one signal (here the sine wave) is compared with a saw-tooth waveform. When the latter is less than the former, the PWM signal is in a high state (1), and the switch is on. Otherwise, it is in the low state (0) and the switch is off. The duty cycle is the ratio of "on" time to the regular interval or period of time, and it is expressed in percentage.

Two modulation indexes are generally defined: the amplitude modulation index —ratio of the sine to triangular waves amplitudes, and the frequency modulation index—ratio of the frequencies of the sine to triangular waves.

$$m_a = \frac{V_{\sin}}{V_{triang}} \tag{2.28}$$

$$m_f = \frac{f_{\sin}}{f_{triang}} \tag{2.29}$$

In practical applications, the amplitude of the triangular wave is fixed, and the frequency can be fixed or variable. Both the amplitude and frequency of the sine wave are variable and are the variables that allow the rotor voltage control to be carried on. The frequency of the sine wave is tuned to the frequency of the rotor's

Fig. 8 PWM control system

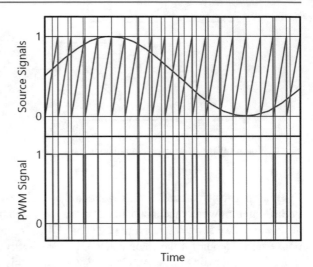

electrical quantities ($\omega_s - \omega_r$). The sine wave amplitude is adjusted so that the rotor voltage amplitude has the desired value. Moreover, the control of the sine wave phase allows for the rotor voltage phase control. Hence, the amplitude modulation index is the tool to control the rotor voltage.

In steady state, the relationship between the AC rotor voltage (amplitude), $\frac{V_r}{s}$, and the inverter DC input voltage, V_{di}, is as follows:

$$\frac{V_r}{s} = \frac{\sqrt{3}}{\sqrt{2}} \frac{m_a}{2} V_{di} = 0.612 m_a V_{di} \tag{2.30}$$

If the phase of $\frac{V_r}{s}$ is θ_r, one can successively write

$$\frac{\boldsymbol{V_r}}{s} = \frac{V_r}{s}(cos\theta_r + jsin\theta_r) \tag{2.31}$$

$$\frac{\boldsymbol{V_r}}{s} = 0.612 V_{di}(m_a cos\theta_r + jm_a sin\theta_r) \tag{2.32}$$

$$\frac{\boldsymbol{V_r}}{s} = 0.612 V_{di}(m_{ar} + jm_{ai}) \tag{2.33}$$

We note that

$$\boldsymbol{m_a} = m_{ar} + jm_{ai} \tag{2.34}$$

is the complex amplitude modulation index or, in a simplified form, the modulation index.

It is apparent from Eq. 2.33 that it is possible to control the rotor voltage from the modulation index (in the complex form). Given the DFIG equivalent circuit parameters, slip, stator voltage (supposed fixed), and stator complex power, it is possible to compute the modulation index that imposes the given stator complex power, $S_s = P_s + jQ_s$. The computation process is as follows.

1. Stator current

$$I_s = \left(\frac{S_s}{V_s}\right)^*$$

(2.35)

2. EMF

$$E = V_s + (R_S + jX_{\ell s})I_s$$

(2.36)

3. Rotor current

$$I_r = I_s + \frac{E}{jX_m}$$

(2.37)

4. Rotor voltage

$$\frac{V_r}{s} = \left(\frac{R_r}{s} + jX_{\ell r}\right)I_r + E$$

(2.38)

5. Modulation index

$$\frac{V_r}{s}\frac{1}{0.612V_{dI}} = m_{ar} + jm_{ai}$$

(2.39)

Note that the reactive power supply in the stator can be set positive (supplying) or negative (absorbing) by properly adjusting the modulation index. This capability is very important because the DFIG can operate with a controllable power factor, so it is not necessary to install power factor correcting capacitor banks, which were necessary for SCIG.

Let us look at an example that will help the understanding of the computation process.

Example 2—1

A DFIG is operating as a generator with a slip equal to –15%. The amplitude modulation index is 0.5804 + j0.0874 and the inverter DC voltage is 1.8/0.612 pu. The electrical complex power in the rotor is $P_r + j\frac{Q_r}{s} = -0.1059 + j0.5718$ pu. The stator voltage is 1 pu. DFIG equivalent circuit parameters (pu): R_s=0.0055; $X_{\ell s}$=0.055; R_r=0.0055; $X_{\ell r}$=0.1375; X_m=3.4925. Compute the stator complex power.

Solution (in pu):

$$\frac{V_r}{s} = 0.612 V_{dl}(m_{ar} + jm_{ai}) = 1.0447 + j0.1573 = 1.0565 e^{j8.5636}$$

$$S_r = P_r + jQ_r = V_r I_r^*;$$

$$I_r = \left(\frac{S_r}{V_r}\right)^* = 0.7414 - j0.4357 = 0.8599 e^{-j30.4409}$$

$$E = \frac{V_r}{s} - \left(\frac{R_r}{s} + jX_{\ell r}\right) I_r = 1.0120 + j0.0394 = 1.0128 e^{j2.2298}$$

$$I_s = \frac{E - V_s}{R_S + jX_{\ell s}} = 0.7309 - j0.1451 = 0.7452 e^{-j11.2240}$$

$$S_s = V_s I_s^* = 0.7309 + j0.1451.$$

3   Direct-Driven DC-Link Synchronous Generators

The basic element of a DDSG is the well-known alternator, which is widely used in conventional thermal and large hydropower plants. So, we will begin by reviewing the basic working principle of the alternator, and then we will introduce the necessary changes to allow its use in wind energy conversion systems.

3.1   Review on the Basic Working Principle of the Alternator

We have seen in Chap. 2 that the rotation of a winding inside a uniform magnetic field produces a sinusoidal EMF. This is based upon Faraday's law of electromagnetic induction which says an EMF is induced in the conductor inside a magnetic field when there is a relative motion between the conductor and the magnetic field. This is the actual working principle of an alternator.

In practice, the construction of a real alternator is a little bit different. For the establishment of the basic working principle of the alternator, we have considered that the magnetic field is stationary, and the conductors are rotating. In the practical construction of the alternator, the stator (also known as armature) conductors are steady and field magnets (rotor) rotate between them. The rotor of the alternator is mechanically coupled to the shaft of the turbine blades, which is made to rotate at synchronous speed (that is why the alternator is also known as a synchronous generator). The rotor houses an excitation (or field) winding, responsible for the creation of the magnetic flux, in which a DC current flows.[4] The flux varies because the rotor is rotating. The magnetic flux cuts off the stationary armature conductors housed on the stator. As a direct consequence of this flux cutting, an induced EMF is produced in the armature conductors. Figure 9 shows the construction of an alternator, with separate DC field excitation.

To produce the magnetic field, a permanent magnet could be used instead of a separate DC field excitation mounted in the rotor. The operating principle is the same, except that the magnetic field is created by the permanent magnet (Fig. 10).

3.2   Equipment General Description

A schematic of a DDSG is displayed in Fig. 11.

Contrary to the WTGs equipped with DFIG, the WTGs equipped with DDSG have no gearbox.[5] The rotor of the DDSG is directly connected to the turbine shaft, therefore the frequency of the generator's electrical quantities is proportional to the variable turbine's frequency, typically changing between 17 and 36 rpm, depending on the wind speed. To allow the generator to rotate at such low speeds, it has a high number of pairs of poles, typically more than 30 pairs of poles.[6] This means that this is a synchronous generator that does not rotate at synchronous speed, instead it rotates at a variable low speed.

It is not possible to directly connect a variable-speed generator to a constant frequency grid. To allow for this connection to be made, the generator must be decoupled from the grid, and a power electronics interface system is needed. This AC/DC/AC interface is composed of a rectifier (AC/DC conversion), a DC-link (DC/DC conversion), and an inverter (DC/AC conversion). High-frequency harmonics are produced which may be mitigated by filters. The variable-frequency and magnitude AC electrical quantities of the generator are rectified (DC) and then inverted to the appropriate AC grid fixed frequency (50 Hz) and magnitude. We note that these converters must be sized to handle all the power to be supplied to the grid (full-rated converters), contrary to what happens in the DFIG's case (converters sized to the slip power).

In practical applications, the AC/DC converter is a six-pulse bridge equipped with thyristors with a constant fire angle, meaning they have no control function assigned. The DC/DC converter ensures that the DC voltage is kept constant as

---

[4]The DC current is obtained from the alternator output by using a rectifier (AC/DC converter).
[5]Cost, vibration, and noise associated with gearbox are therefore eliminated.
[6]This means the generator is very large, contrarywise to the DFIG which is compact.

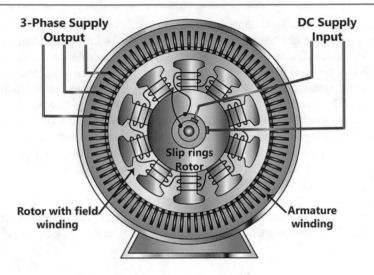

Fig. 9 Construction of an alternator with separate DC field excitation

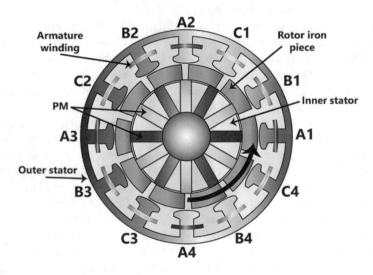

Fig. 10 Construction of an alternator with permanent magnets

required for control purposes. In principle, the DC voltage control is handled by the excitation system of the alternator for high speeds. For low speeds, the excitation system is not able to provide such control, as so a DC/DC converter is required. The DC/AC converter is a six-pulse bridge equipped with IGBTs and a PWM control system. It controls both active power and power factor at the PCC. Also, it is possible to control the speed, so that the Tip Speed Ratio (TSR) is optimal, and the efficiency is maximized.

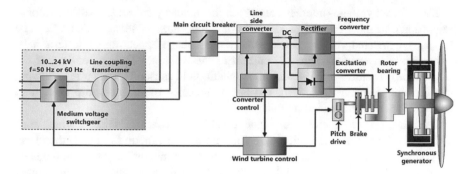

Fig. 11 Schematic of a DDSG

3.3 Operating Principle

The inverter control system finds the AC reference current based on the grid AC voltage measuring and active power available for the wind speed measuring at a particular time. The AC reference current is imposed in the AC system by controlling the inverter output voltage. This requires the DC voltage to be constant, which is achieved with help of the DC-Link, as seen before. The inverter AC voltage imposes both AC and DC currents. If the rotor speed is such that the alternator's excitation system is capable of providing the necessary DC voltage, then the DC/DC converter is switched off and the voltage and current are equal both in rectifier and inverter. On the opposite, the DC/DC converter is switched on and the computation of the rectifier voltage and current is required, using the duty ratio. Please remember that the alternator's EMF is proportional to the rotor speed.

Electromagnetic torque is imposed by the inverter, through proper stator current control, based on the available active power measuring for that particular wind speed. If the turbine's mechanical torque is greater than the electromagnetic torque, then the machine increases speed; on the opposite, the machine decreases speed.

3.4 Steady-State Model

Synchronous generators produce electricity whose frequency is synchronized with the mechanical rotational speed by the following relationship:

$$n_r = \frac{60f}{p} \tag{3.1}$$

where n_r is the rotor speed in rpm, f is the grid frequency (50 Hz), and p is the number of pairs of poles.

The induced voltage (internal voltage) in the stator by the rotation of the rotor is proportional to the frequency and to the magnetic flux, as seen in Chap. 2.[7] The internally generated voltage is not usually the voltage appearing at its terminals, except when there is no armature current in the machine. This is because of the following: (i) Distortion of the air-gap magnetic field caused by the current flowing in the stator (armature reaction), (ii) Self-inductance of the armature coils, and (iii) Resistance of the armature coil.

The armature current creates a magnetic field in the machine's stator, which adds to the rotor (main) magnetic field affecting the total magnetic field and, therefore, the voltage. Hence, the voltage at the generator's terminals has two components: one derived from the internal voltage due to the main magnetic flux and another related to the armature reaction voltage.

This leads to the well-known synchronous generator model of an EMF behind an impedance.

$$V_s = E - (R_s + j\omega_r L_s)I_s \qquad (3.2)$$

V_s is the terminal voltage, E is the internal voltage, I_s is the stator current, R_s is the stator winding resistance per phase, and $X_s = \omega_r L_s$ is the synchronous reactance, which includes the armature reactance and self-inductance. The steady-state equivalent diagram of Fig. 12 can be drawn.

The phasor diagram of a synchronous generator with a lagging power factor is shown in Fig. 13.

The angle between the voltage and the current is the well-known power factor angle,[8] ϕ; the angle between the voltage and the EMF is called the power angle (δ).

The model just reviewed is valid for machines with a round rotor. If, as usual, the rotor is salient, we must take into account the existence of two axes: the direct axis, d, and the quadrature axis, q. The rotor flux is aligned with the d-axis and the armature reaction flux has components along with both the d- and q-axis. Therefore, we need to define two synchronous reactances, $X_d = \omega_r L_d$ and $X_q = \omega_r L_q$, along d- and q-axis, respectively. Furthermore, the stator current is decomposed along d- and q-axis as

$$I_s = I_{ds} + I_{qs} \qquad (3.3)$$

We can write similarly as we did in Eq. 3.2:

---

[7]Further reading: EE340 Synchronous Generators I, available in http://www.ee.unlv.edu/~eebag/EE%20340%20Sync%20Generators%20I.pdf.

Lecture Notes on Electrical Machines II, available in https://www.cet.edu.in/noticefiles/226_ELECTRICAL_MACHINE-II.pdf.

Paul Krause, Oleg Wasynczuk, Scott Sudhoff, Steven Pekarek, "Analysis of electric machinery and drive systems", IEEE Press, 3rd edition, 2013.

Mohamed E. El-Hawary "Electrical Energy Systems", CRC Press, 2000.

A.E. Fitzgerald, Charles Kingsley JR., Stephen D. Umans, "Electric Machinery", 6th edition, McGraw Hill, 2003.

[8]Recall that angle Ø (power factor angle) is positive for a lagging power factor.

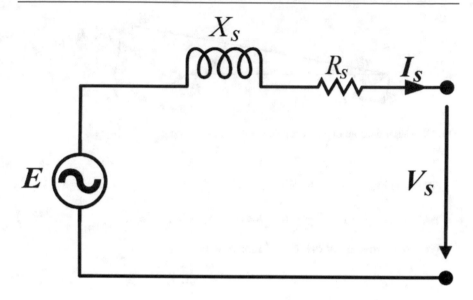

Fig. 12 Equivalent steady-state diagram

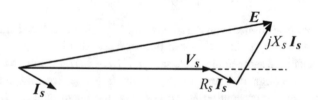

Fig. 13 Phasor diagram; lagging power factor

$$V_s = E - R_s I_s - j\omega_r L_d I_{ds} - j\omega_r L_q I_{qs} \tag{3.4}$$

This equation is represented in the phasor diagram of Fig. 14. We note that the existence of two-axis makes it impossible to establish an equivalent diagram as it was the case of round rotor machines.

In using the phasor diagram of Fig. 14, the armature current must be resolved into its d- and q-axis components. This resolution assumes that the phase angle $(\phi + \delta)$ of the armature current, I_s, with respect to the internally generated voltage, E, is known, i.e., we know the location of the d- and q-axis. This is not the case in a practical situation: we often know the power factor angle, but the power angle is unknown. To overcome this difficulty, we will proceed as follows.

Replacing Eq. 3.3, Eq. 3.4 becomes

$$V_s = E - R_s I_s - j\omega_r L_d I_{ds} - j\omega_r L_q (I_s - I_{ds}) \tag{3.5}$$

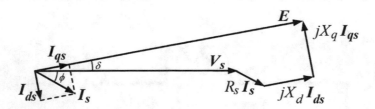

Fig. 14 Phasor diagram of a salient-pole synchronous generator

After manipulation, we obtain

$$V_s = E - R_s I_s - j\omega_r L_q I_s - j\omega_r (L_d - L_q) I_{ds} \tag{3.6}$$

We can rearrange, by defining phasor E_{qd} as

$$E_{qd} = V_s + R_s I_s + j\omega_r L_q I_s \tag{3.7}$$

The EMF can be written as

$$E = E_{qd} + j\omega_r (L_d - L_q) I_{ds} \tag{3.8}$$

We note that E is aligned with the q-axis (see Fig. 14) because the flux is aligned with the d-axis. The important feature of phasor E_{qd} is that it is aligned with phasor E, i.e., they share the same angle. Note that $j\omega_r(L_d - L_q)I_{ds}$ is aligned with q-axis, so E_{qd} and E are aligned. Therefore, by computing E_{qd}, we find out the power angle, before actually computing E. To compute the magnitude of E, we use the following equation which holds for the magnitudes:

$$E = E_{qd} + \omega_r (L_d - L_q) I_{ds} \tag{3.9}$$

The phasor diagram of Fig. 15 shows how it is possible to go forward without the advanced knowledge of δ angle.

From the phasor diagram in Fig. 15, the following useful relationships may be extracted[9]:

$$I_{ds} = I_s sin(\delta + \phi) \tag{3.10}$$

$$I_{qs} = I_s cos(\delta + \phi) \tag{3.11}$$

$$I_s cos\phi = I_{qs} cos\delta + I_{ds} sin\delta \tag{3.12}$$

---

[9]The reader is invited to explain these equations.

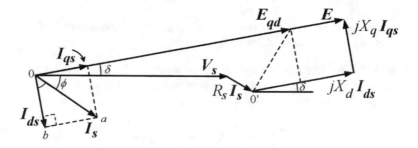

Fig. 15 Constructing the phasor diagram with no advanced knowledge of δ

$$I_s \sin\phi = I_{ds} \cos\delta - I_{qs} \sin\delta \qquad (3.13)$$

and for the phasors:

$$I_{ds} = I_{ds} e^{-j(90-\delta)} \qquad (3.14)$$

$$I_{qs} = I_{qs} e^{j\delta} \qquad (3.15)$$

$$I_s = (I_{qs} - jI_{ds}) e^{j\delta} \qquad (3.16)$$

Note that the phase angle between E and I_s is $(\delta + \phi)$. The angle ϕ is positive for a lagging power factor, the corresponding current's angle being negative. In general, the desired angle is equal to the sum between the power angle and the power factor angle.

Example 3—1[10]
A 100 kW, 450 V, 50 Hz, Δ-connected, three-phase synchronous generator has a direct axis reactance of 0.15 Ω and quadrature axis reactance of 0.07 Ω. Its armature resistance may be neglected. It is connected to a power electronics converter that acts as a current source and generates 150 A at a power factor of 0.8 lagging. The machine is rotating at a rated speed.

(a) *Find the internally generated voltage E of this generator at full load, assuming that it has cylindrical rotor reactance X_d.*
(b) *Find the generated internal voltage E of this generator at full load assuming that it has a salient-pole rotor.*

---

[10]Taken from Lecture Notes on Electrical Machines II, available in https://www.cet.edu.in/noticefiles/226_ELECTRICAL_MACHINE-II.pdf.

Solution:

The machine is running at a rated speed, so the rotor speed is the synchronous speed and $X_d = \omega_r L_d$ *and* $X_q = \omega_r L_q$.

(a) *The EMF (or internal voltage) is given by Eq. 3.2. To use a single-phase equivalent diagram, we need to consider phase quantities. The machine is Δ connected, so the line current is 150 A and the phase current is*

$$I_s = \frac{150}{\sqrt{3}} e^{-jacos(0.8)} = 86.6025 e^{-j36.8699} A$$

The power factor angle ϕ *is 36.8699°. We note that* ϕ *is symmetrical of the current's angle.*

The voltage applied to the Δ is the phase-to-phase voltage. Therefore (note the reference angle, 0°, is located in the terminal voltage, as usual),

$$V_s = 415 e^{j0} V$$

The EMF is

$$E = V_s + jX_s I_s = 422.9219 e^{j1.4080} V$$

(b) *The EMF is given by Eq. 3.6, but first, we need to determine the location of the q-axis by computing:*

$$E_{qd} = V_s + jX_q I_s = 418.6654 e^{j0.6637} V$$

The power angle δ *is 0.6637° and this determines where the q-axis is located (and the angle of the EMF,* E). *Now, we can compute the magnitude of the d-component of the stator current,* I_{ds}:

$$I_{ds} = I_s sin(\delta + \phi) = 52.7606 A$$

and the magnitude of the EMF:

$$E = E_{qd} + (X_d - X_q) I_{ds} = 422.8862 V$$

The EMF phasor is

$$E = 422.8862 e^{j0.6637} V$$

The same result would have been obtained if we used the phasor equation:

$$E = E_{qd} + j(X_d - X_q)I_{ds} \text{ where } I_{ds} = I_{ds}e^{-j(90-\delta)}$$

We conclude that the magnitude of the EMF is not much affected by making the simplification of considering a salient rotor as cylindrical. However, the power angle is very different, and this error is of significance especially when studying the transient behaviour of a system including many synchronous machines.

The output power is

$$P = \sqrt{3}V_s\sqrt{3}I_s cos\phi = 86.256\,\text{kW}$$

$$Q = \sqrt{3}V_s\sqrt{3}I_s sin\phi = 64.692\,\text{kvar}$$

If the converters are each 97% efficient, the real power that reaches the power system is $86.256(0.97)^2 = 81.158\text{kW}$*. The reactive power, however, does not have to flow through. The grid-side converter can generate (or absorb) an amount of reactive power that is independent of Q.*

This problem can be solved in the pu system. The reader is invited to go through the following MATLAB® code for a better understanding.

```
% code to solve example 3-1 in pu
% data
Sn=100e3; Vn=415; f=50; Zb=3*Vn^2/Sn; Xd=0.15/Zb;
Xq=0.07/Zb;
pf=0.8; V=1*exp(1j*0); Ib=Sn/(sqrt(3)*Vn); I=150/Ib;
% results
Is=I*exp(-1j*acos(pf));
Is_m=abs(Is);
fi=-angle(Is); fi_g=fi*180/pi
% cylindrical rotor
E1=V+1j*Xd*Is;
E1_m=abs(E1)
delta1=angle(E1)*180/pi
% salient pole
Eqd=V+1j*Xq*Is;
Eqd_m=abs(Eqd)
delta=angle(Eqd); delta_g=delta*180/pi
deltafi=delta+fi;
Ids_m=Is_m*sin(deltafi)
Iqs_m=Is_m*cos(deltafi)
Ids=Ids_m*exp(-1j*(pi/2-delta));
Iqs=Iqs_m*exp(1j*delta);
E_m=Eqd_m+(Xd-Xq)*Ids_m
E=Eqd+1j*(Xd-Xq)*Ids
% power output
P=V*Is_m*pf
Q=V*Is_m*sin(fi)
Pgrid=P*0.97^2
```

3.5   Introduction to Control

The idea is to control the power output. In pu, the power output is given by

$$S_s = P_s + jQ_s = V_s I_s^* \tag{3.17}$$

The voltage is given by Eq. 3.6, one can write, neglecting the stator resistance,

$$S_s = \left[E - j\omega_r(L_d - L_q)\boldsymbol{I}_{ds}\right]\boldsymbol{I}_s^* - j\omega_r L_q I_s^2 \tag{3.18}$$

Now, we will take into account Eq. 3.16 for the stator current and take note that $\boldsymbol{I}_{ds}$ and $\boldsymbol{I}_{qs}$ are given by Eqs. 3.14 and 3.15, respectively, and further that $\boldsymbol{E} = E e^{j\delta}$. After some manipulation, we finally obtain

$$P_s = E I_{qs} + \omega_r(L_q - L_d) I_{ds} I_{qs} \tag{3.19}$$

$$Q_s = E I_{ds} - \omega_r L_d I_{ds}^2 - \omega_r L_q I_{qs}^2 \tag{3.20}$$

It can be seen that the power output control can be realized by controlling the stator current qd-axes components. If the cylindrical rotor approximation holds ($L_d = L_q$), then the real power is proportional to I_{qs}, and I_{ds} is kept equal to zero, so as to minimize the ohmic losses. For salient-pole generators, there is an infinite number of I_{qs}, I_{ds} combinations that produce the desired power output, conditional on the rotor speed.

3.6   The Permanent Magnet Synchronous Generator (PMSG)

A schematic of a PMSG[11] is shown in Fig. 16. The connection to the rotor shaft can be made through a gearbox, or alternatively, it can be directly connected (direct link). The two options are possible.

Usually, the power electronics topology employs two back-to-back IGBT-based converters with a DC-link. Another common topology is similar to the one seen for the DDSG with a field excited alternator: an uncontrolled diode rectifier is connected at the stator side instead of a fully controllable converter. This helps to reduce the complexity and cost of the power electronics but does not provide as much flexibility to control the stator currents. In the sequence, we assume the fully controllable topology, therefore considering we have full control over the stator current magnitude and phase angle.

---

[11]Further reading: Dionysios C. Aliprantis, Fundamentals of Wind Energy Conversion for Electrical Engineers, available in: https://engineering.purdue.edu/~dionysis/EE452/Lab9/Wind_Energy_Conversion.pdf.

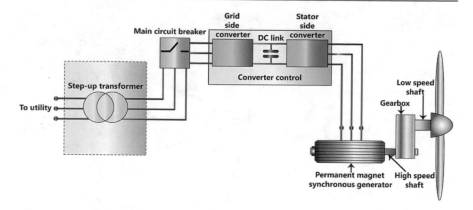

Fig. 16 Typical connections scheme of a PMSG

As for separated excitation generators, the EMF, $\boldsymbol{E_{pm}}$, depends both on the flux linkage generated by the permanent magnets in the rotor, λ_{pm}, and, due to the armature reaction, on a component of the stator current, denoted by I_{ds}.

$$\boldsymbol{E_{pm}} = \frac{1}{\sqrt{2}}\omega_r\left[\lambda_{pm} - \left(L_d - L_q\right)I_{ds}\right]e^{j\delta} \qquad (3.21)$$

In Eq. 3.21, we have not used the pu formulation and most of the variables are already known: L_q and L_d are the qd-axes inductances, these inductances being defined due to the rotor saliency; ω_r is the rotor speed; λ_{pm} is the constant flux linkage generated by the permanent magnets; δ measures the rotor's angle relative to the phase voltage (in steady state, the rotor and the voltage have the same frequency, so δ is constant); and finally, I_{ds} is the stator current component along the d-axis.

The phasor diagram is shown in Fig. 17, where the stator voltage is chosen as the angle reference, as usual. The phasor diagram is like the separated excitation machine one. However, this figure is important because it highlights two sets of orthogonal axes: (i) The real and imaginary axes are stationary, and they represent the coordinate system for phasors. (ii) The q- and d-axes are positioned on the rotor, and in particular the d-axis is aligned with the magnetic flux generated by the permanent magnets. The figure shows how the stator current phasor can be decomposed into two other components I_{qs} and I_{ds} (instead of $Re\{I_s\}$ and $Im\{I_s\}$) along these axes.

Fig. 17 Phasor diagram of a
PMSG in SI units

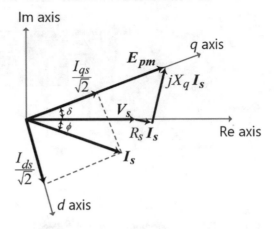

4 Conclusions

Recent developments in power electronics made it possible to connect variable
speed generators to a constant frequency grid. The need for variable-speed generators is justified by the significant increase in the efficiency of the conversion of the
power available in the wind into electrical power output. In modern Wind Turbine
Generators (WTGs), two types of variable-speed electrical generators are used:
Double-Fed Induction Generator (DFIG) and Direct-Driven DC-Link Synchronous
Generator (DDSG).

The DFIG is a classical induction generator whose rotor is connected to the grid
through an AC/DC/AC power electronics converter. This assembly allows for the
speed to vary over a wide range, therefore increasing the conversion's efficiency.
Electrical power can be supplied to the grid both in the stator and in the rotor when
the speed is above the synchronous speed. When the speed is below the synchronous speed, the operation as a generator is still possible (this is not possible in
conventional induction generators), providing a small amount of power is delivered
to the rotor through the bi-directional converters. There is a gearbox to adapt the
speeds of the turbine and the generator, therefore making the size of the generator
very compact. The power electronics converters are located only in the rotor, so
they are sized to a fraction of the machine's rated power.

The operating principle and building of the DDSG are completely different. In
this case, we have a synchronous generator (alternator) that rotates at a variable
speed. This is made possible because the alternator is decoupled from the grid
where it is connected. The connection to the grid is made through a full-scale
AC/DC/AC converter by which all the converted electrical power is transferred to
the grid. In this assembly, there is no gearbox, the rotor blades rotating at the same
speed as the generator. This is possible because the alternator is built with a high
number of poles, therefore making the generator look very big. The DC excitation
of the alternator was initially provided by a separate excitation system, but more

recently configurations with self-excited permanent magnet excitation are being proposed by the manufacturers.

In this chapter, we have provided steady-state models for both types of generators that allow for a description of their electrical behaviour. The DFIG model is based on the equivalent circuit of a wound-rotor induction (also known as asynchronous) machine, where the rotor terminals are accessible. The possibility of this machine to operate as a generator for a wide range of rotor speeds, even for rotor speeds below the synchronous speed, has been explored and discussed. The basics of the operating principle of the power electronics converters that made possible this versatile operation have been presented, namely the main functions associated with each one of the two converters.

The DDSG model is based on the alternator's usual modelling, i.e., an EMF behind a synchronous impedance, valid for round rotor machines. This model was extended to incorporate the case of salient-pole synchronous generators, introducing the theory based on the definition of the direct and quadrature axis. The working principle of the AC/DC/AC electronic system that performs frequency conversion has been addressed, as well as the role of each one of the three converters (AC/DC, DC/DC, and DC/AC). Also, the basics of the control system have been provided. Finally, the case of the permanent magnet synchronous generator has been briefly investigated.

5 Proposed Exercises

Problem WECE1

In a particular operating point, measurements on a DFIG with a rotor resistance equal to 0.0085 pu showed the following values:

$$P_s - P_r = 0.374\,\text{pu}; \quad V_r = 0.00515e^{j217.9°}\,\text{pu};$$
$$I_r = 0.51e^{-j41.3°}\,\text{pu}; \quad P_{mec} = 0.379\,\text{pu}$$

Compute the following:

(1) The slip.
(2) The active power losses.

Solution

Given: $P_s - P_r; V_r; I_r; P_{mec}$

(1)

$$sP_{ag} = P_r - R_r I_r^2$$

$$P_r = Re\{V_r I_r^*\} = -4.9216 \times 10^{-4}\,\text{pu}$$

$$P_{ag} = P_r - R_r I_r^2 + P_{mec} = 0.3763\,\text{pu}$$

$$s = -0.0072$$

(2)

$$P_s = P_r + P_{mec} - P_{Loss}$$

$$P_s = P_r + 0.374 = 0.3735\,\text{pu}$$

$$P_{Loss} = 0.0050\,\text{pu}$$

$$P_{Ls} = P_{ag} - P_s = 0.0028\,\text{pu}$$

$$P_{Lr} = P_r + P_{mec} - P_{ag} = 0.0022\,\text{pu}$$

$$P_{Loss} = P_{Ls} + P_{Lr}$$

Problem WECE2

A wind turbine generator is equipped with a DFIG. The induction machine has the following characteristics:

$$V_n = 1000\,\text{V};\, S_n = 2750\text{kVA};\, n_{pp} = 2$$

$$R_s = 0.0055\,\text{pu};\, X_{ls} = 0.055\,\text{pu};\, X_m = 3.4925\,\text{pu};\, R_r = 0.0055\,\text{pu};\, X_{lr} = 0.1375\,\text{pu}$$

The gearbox ratio is 104.53. In a particular operating point, the speed of the turbine equals 14.43 rpm and the complex power in the stator of the machine equals $0.73 + j0.146$ pu. Assume that the voltage at the stator of the machine equals $1 + j0$ pu.

Compute the following:

(1) The air-gap active and reactive power.
(2) The slip of the machine.
(3) The active power in the rotor.

Solution

Given: equivalent circuit parameters; n_{pp}; GB; N_T; S_s; V_s

(1)

$$P_{ag} = P_s + R_s I_s^2$$

$$Q_{ag} = Q_s + X_{ls} I_s^2$$

$$\mathbf{I_s} = \left(\frac{S_s}{V_s}\right)^* = 0.7300 - j0.1460 \, \text{pu}$$

$$P_{ag} = 0.7330 \, \text{pu}$$

$$Q_{ag} = 0.1765 \, \text{pu}$$

(2)

$$s = \frac{\omega_s - \omega_r}{\omega_s} = \frac{N_s - N_r}{N_s}$$

$$N_s = \frac{60f}{n_{pp}} = 1500 \, \text{rpm}$$

$$N_r = N_T GB = 1508.4 \, \text{rpm}$$

$$s = -0.0056$$

(3)

$$sP_{ag} = P_r - R_r I_r^2$$

$$V_s = -(R_S + jX_{\ell s})I_s + E$$

$$E = 1.0120 + j0.0393 \, \text{pu}$$

$$E = jX_m I_m$$

$$I_m = 0.0113 - j0.2898 \, \text{pu}$$

$$I_r = I_m + I_s = 0.7413 - j0.4358 \, \text{pu}$$

$$P_r = -2.2810 \times 10^{-5} \, \text{pu}$$

Problem WECE3

The machine of problem WECE2 is now in an operation point characterized by

$$\mathbf{V_r} = 0.0228e^{j1.5074°} \text{ pu}; \ \mathbf{I_r} = 0.9283e^{-j27.476°} \text{ pu}; \ P_{mec} = 0.72 \text{ pu}$$

Compute the following:

(1) The active power in the airgap.
(2) The slip.
(3) The complex power in the stator.
(4) The stator voltage.

Solution

Given: equivalent circuit parameters; $V_r; I_r; P_{mec}$

(1)

$$P_{ag} = P_r - R_r I_r^2 + P_{mec}$$

$$P_r = Re\{V_r I_r^*\} = 0.0185 \text{ pu}$$

$$P_{ag} = 0.7338 \text{ pu}$$

(2)

$$sP_{ag} = P_r - R_r I_r^2$$

$$s = 0.0188$$

(3)

$$\frac{V_r}{s} = \left(\frac{R_r}{s} + jX_{\ell r}\right)I_r + E$$

$$E = 0.9139 + j0.0442 \text{ pu}$$

$$E = jX_m I_m$$

$$I_m = 0.0127 - j0.2617 \text{ pu}$$

$$I_s = I_r - I_m = 0.8109 - j0.1666 \text{ pu}$$

$$S_s = P_s + jQ_s$$

$$P_s = P_{ag} - R_s I_s^2$$

$$P_s = 0.7300 \, \text{pu}$$

$$Q_s = Q_{ag} - X_{ls} I_s^2$$

$$Q_{ag} = Im\{EI_s^*\} = 0.1881 \, \text{pu}$$

$$Q_s = 0.1504 \, \text{pu}$$

(4)

$$S_s = V_s I_s^*$$

$$V_s = 0.9 \, \text{pu}$$

Problem WECE4

The parameters of the equivalent electrical circuit of a DFIG are (in pu)

$$R\_s = 0.0097; X_{ls} = 0.042; R_r = 0.0085; X_{lr} = 0.0714; X_m = 2.9981$$

The machine is running at the following operating point (in pu):

$$V_s = 1; \ s = 1\%; \ S_s = 0.1164 - j0.0602$$

Compute the following:

(1) The stator current.
(2) The rotor current.
(3) The air-gap active power.
(4) The rotor active power.
(5) The modulation index, given that the DC voltage is $V_{dl} = 1.8/0.612 \, \text{pu}$.

Solution

Given: equivalent circuit parameters; $s; S_s; V_s; V_{dl}$

(1)

$$I_s = \left(\frac{S_s}{V_s}\right)^* = 0.1164 + j0.0602\,\text{pu}$$

(2)

$$V_s = -(R_S + jX_{\ell s})I_s + E$$

$$E = jX_m I_m$$

$$I_r = I_m + I_s = 0.1182 - j0.2729\,\text{pu}$$

(3)

$$P_{ag} = P_s + R_s I_s^2 = 0.1166\,\text{pu}$$

(4)

$$sP_{ag} = P_r - R_r I_r{}^2$$

$$P_r = 0.0019\,\text{pu}$$

(5)

$$\frac{V_r}{s}\frac{1}{0.612V_{dI}} = m_{ar} + jm_{ai} = \boldsymbol{m_a}$$

$$\frac{V_r}{s} = \left(\frac{R_r}{s} + jX_{\ell r}\right)I_r + E = 1.1186 - j0.2180\,\text{pu}$$

$$m_a = 0.6214;\, m_{ai} = -0.1211$$

Problem WECE5

A DFIG is operating as a generator with slip equal to –15% and stator voltage equal to 1.0 pu. The amplitude modulation index is $0.5804 + j0.0874$. The electrical active power in the rotor is –0.1059 pu, the RMS rotor current is 0.8601 pu, and the inverter DC voltage is 1.8/0.612 pu.

DFIG equivalent circuit parameters (pu):

$$R_s = 0.0055; X_{ls} = 0.055; R_r = 0.0055; X_{lr} = 0.1375; X_m = 3.4925.$$

(1) Compute the rotor voltage.
(2) Compute the rotor current angle and show that its value is −30.45°.
(3) Compute the power factor in the stator.

Solution

(1) All results in pu

$$\frac{V_r}{s} = 0.612 V_{di}(m_{ar} + jm_{ai}) = 1.0477 + j0.1573$$

$$V_r = -0.1567 - j0.0236 \,\text{pu} = 0.1585 e^{-j171.44°}$$

(2)

$$\cos\phi_r = \frac{P_r}{V_r I_r} = -0.7769$$

ϕ_r is the angle between the rotor voltage ($\alpha = -171.44°$) and the rotor current (β).

$$\phi_r = \text{acos}(\cos\phi_r) \xrightarrow{yields} \phi_{r1} = 140.98° \bigvee \phi_{r2} = 219.02°$$

The rotor current must be lagging the rotor voltage.

$$\beta = (\alpha - \phi_{r2}) + 2\pi = -30.45°$$

(3) All results in pu

$$E = \frac{V_r}{s} - \left(\frac{R_r}{s} + jX_{\ell r}\right)I_r = 1.0120 + j0.0394$$

$$I_s = \frac{V_s - E}{-(R_s + jX_{\ell s})} = 0.7306 - j0.1444$$

$$S_s = V_s I_s^* = 0.7447 e^{j11.1823°}$$

$$pf = \cos 11.1823° = 0.9810 \,\text{ind.(generator convention)}$$

Problem WECE6[12]

The reactances X_d and X_q of a salient-pole synchronous generator are 1.00 and 0.60 per unit, respectively. The armature resistance may be considered negligible.

(1) Compute the internal generated voltage when the generator is loaded by the power electronics converter to 0.73 per-unit kVA, unity power factor at a terminal voltage of 0.98 per unit.
(2) Repeat considering the rotor is cylindrical with synchronous reactance equal to X_d.

Solution

Given: $X_d; X_q; S_s; pf; V_s$

(1)

$$I_s = \left(\frac{S_s}{V_s}\right) e^{-j a \cos(pf)} = 0.7449\,\text{pu}$$

$$E_{qd} = V_s + jX_q I_s = 1.0771 e^{j24.5160°}\,\text{pu}$$

$$E = E_{qd} + (X_d - X_q)I_{ds}$$

$$I_{ds} = I_s \sin(\delta + \phi) = 0.3091\,\text{pu}$$

$$E = 1.2007\,\text{pu}; \ \delta = 24.5160°$$

(2)

$$E = V_s + jX_s I_s = 1.2310 e^{j37.2380°}\,\text{pu}$$

---

[12]Taken from A.E. Fitzgerald, Charles Kingsley JR., Stephen D. Umans, "Electric Machinery", 6th edition, McGraw Hill, 2003.

Offshore Wind Electrical Systems

7

Abstract

To take advantage of higher and steadier winds, producing electricity from wind farms located offshore is becoming progressively interesting as significant reductions in investment costs are being witnessed in recent years. While the background picture for this chapter is offshore wind power, it specifically addresses the electrical system that makes possible electricity to be collected and transported to shore. The topological options of the collector system are reviewed by presenting the most used designs and proposing some innovative ones. The sizing of the interconnection cables is assessed by computing the adequate cross-section using information available in the manufacturer's datasheets. Models are introduced to assess the electrical losses in both the collector and the interconnection systems. As far as the interconnection transmission system is concerned, the two common options—High-Voltage Alternate Current (HVAC) and High-Voltage Direct Current (HVDC) are introduced and discussed. Finally, a brief overview of the hydrogen production from offshore wind is offered. A list of application problems is proposed, together with the guidelines of the solution.

1 Introduction

Offshore wind power is developing all over the world, particularly in Europe. The pace is not as high as other Renewable Energy Sources (RES), namely onshore wind and solar PV, the reason being the usual one: it is not yet economically competitive. Offshore wind power is significantly more expensive than onshore, the Levelized Cost Of Energy (LCOE) being almost three times higher, for the time being.

© The Author(s), under exclusive license to Springer Nature Switzerland AG 2022
R. Castro, *Electricity Production from Renewables*,
https://doi.org/10.1007/978-3-030-82416-7_7

In Europe, five countries (UK, Germany, Denmark, Belgium, and Netherlands) account for 99% of the total installed offshore wind power capacity, which was around 25 GW, as of 2020.[1]

The wind in the sea is more intense (higher wind speeds) and steadier because there are no obstacles. 2019 reports from Wind Europe indicate a global capacity factor of 38% (around 3330 h of utilization factor) for offshore wind installations in Europe.

This led the manufacturers to propose offshore Wind Turbine Generators (WTGs) with increasingly higher installed capacity per unit and the investors to bet in wind farms with higher installed capacity. According to Wind Europe, the average unit capacity was 8 MW per offshore WTG, and the average offshore wind farm capacity was 800 MW, both figures as of 2020.

Moreover, the current trend is to go further away from shore and into deeper waters. As of 2020, there are offshore wind farms under construction located 100 km from the coast and in 40 m deep waters.

In this chapter, we will focus on the electrical aspects of the wind farms located offshore, namely the attention will be directed to the collector system—the system that gathers the electricity produced by each Wind Turbine Generator (WTG) and delivers it to the offshore substation—and the interconnection transmission system, which connects the offshore substation to the onshore substation.

The most used topologies of the collector system, i.e., the different options to connect each WTG to the neighbours and the wind farm to the offshore substation, are reviewed, from the straightforward radial design to the more complex star design. Moreover, innovative designs that are recently being proposed are also analysed.

The aim of Sect. 3 of this chapter is to compute the adequate cross-section of the interconnection submarine cables. An engineering method based on rating factors is offered based on the information available in the manufacturers' datasheets.

The computation of the electrical losses in both the collector and interconnection systems is an important topic as far as offshore wind parks are concerned. Two models are discussed to aid in this task. A temperature-dependent model was proposed by H. Brakelmann, where a temperature-dependent correction factor is used and the classical load flow model, where the bus voltages are computed, and the losses are determined as a function of the current flowing in the cable.

Depending on the power to be transmitted and on the distance to shore, two options are available for the interconnection transmission system: High-Voltage Alternate Current (HVAC) and High-Voltage Direct Current (HVDC). When the distance overcomes a critical length, HVDC must be used because the huge amounts of reactive power produced by AC submarine cables prevent AC to be used. The main components of both HVAC and HVDC systems are presented and discussed. For the HVAC systems, the reactive power compensation devices, namely the ones based on (Flexible AC Transmission System (FACTS) technology,

---

[1]Wind Europe, www.windeurope.org.

are discussed. When HVDC is used, the systems based on modern Voltage Source Converters (VSCs) are preferred.

Finally, a brief overview of the hydrogen production from offshore wind farms is offered. The components of the system—electrolyser and fuel cell—are described and the two possible configurations for the location of the electrolyser—offshore and onshore—are explained.

2 Collector System

The purpose of the collector system is to connect all WTG, collect the energy produced by them, and deliver it to the offshore substation. It consists of a series of submarine conductors that are buried on the ocean floor, typically 1–2 m, as seen in Fig. 1. Therefore, it is one of the components of offshore wind farms on which improvements can mostly increase reliability.

In the past, the collector system only represented a minor part of the total investment. This was because most offshore wind farms were small and had radial collectors. However, current large offshore wind farms with several hundreds of MW come to change that.

The purpose of using redundancy in a wind farm collector system is to keep as many wind turbines as possible connected during an equipment fault. Offshore wind farms have demonstrated that the required repair times are much higher than on onshore sites. A cable fault shows estimated repair times between 720 h (during summer) and 2160 h (during winter). So, the existence of redundancies in offshore wind farms is important and needs to be economically assessed. If the cost of the

Fig. 1 Offshore wind farm collector system

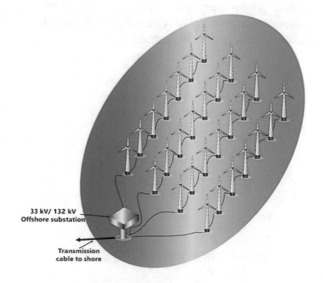

33 kV/ 132 kV
Offshore substation

Transmission
cable to shore

redundancy is less than the cost of the amount of power that is not transmitted due to the fault, it is demonstrated that the redundancy is profitable.

Referring to the different layouts existing in offshore wind farms, four basic designs can be identified: radial design, single-sided ring design, double-sided ring design, and star design. Additionally, two innovative designs may be economically appealing and reliability increasing: the single-return design and the double-sided half-ring design. Next, we will discuss these topological options.

2.1 Topology Options

2.1.1 Radial Design

The most common and straightforward arrangement is that in which several WTG are connected to a single cable feeder within a string, as seen in Fig. 2.

The number of WTG connected to each string feeder is dependent on the rating of the sea cables and on the capacity of the WTG. The biggest advantages of this design are the low cost and how simple it is to control. The low cost derives from the reduced cable length (no redundancies) and the possibility of narrowing the ratings of the cables away from the hub (offshore platform). The last cable's branch is sized to carry the rated current of one WTG, the former is sized to carry twice the rated current, and so on. The major disadvantage is its poor reliability, since any type of fault (cable or switchgear) that happens near the hub prevents all WTG in that array to export their energy. In general, a fault in a particular WTG prevents all the WTG till the end of the string to deliver energy.

2.1.2 Single-Sided Ring Design

To increase the reliability of the radial design, a possible solution is to connect the last WTG of the string to the hub with a long cable as seen in Fig. 3, in an assembly called single-sided ring design. Therefore, this design requires an additional redundant cable sized to handle the full power flow of the string in the event of a fault in the primary link near the hub.

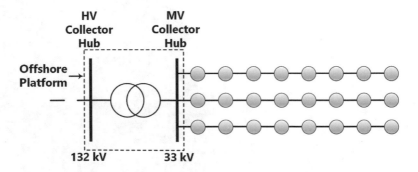

Fig. 2 Wind farm radial design

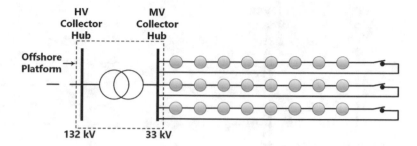

Fig. 3 Wind farm single-sided ring design

This solution allows all WTG to keep operating, except the faulty WTG. This additional security comes with an extra cost due to the new longer cable and the higher cable rating requirements throughout the string circuit. In this case, the narrowing of the cables' ratings away from the hub is not viable.

2.1.3 Double-Sided Ring Design

To get the same reliability as in the single-sided ring design but decrease its cost, the last turbine of one string is connected to the last turbine of another string, as presented in Fig. 4, giving rise to the double-sided ring design.

When compared to the previous single-sided ring design, there is a decrease in the total cable length required, while maintaining the same level of redundancy. The downside of this design is that the connection between the hub and the first turbine of the string needs to handle the output of twice as many WTG as those on its string.

Regardless of the increase in reliability provided by both the single-sided and double-sided ring designs, small offshore wind farms do not use them. In small wind farms, the fault probability is lower, and the unit cost associated with having additional equipment is higher, therefore the redundancy is not economically profitable. In contrast, for large offshore wind farms, the situation changes, because repair downtimes are significantly longer, and the redundancy becomes economically profitable. Large offshore wind farms show layouts with redundancy implementation.

2.1.4 Star Design

To reduce cable ratings and to provide a high level of reliability, one possible solution is the star design, since one cable outage only affects one WTG, except for the cable from the central WTG to the hub (Fig. 5).

This design provides a general reduction of cable ratings. Only the cable from the centre turbine to the hub needs to handle all power from all turbines in the string. On the disadvantages side, one mentions the longer diagonal cables and the more complex switchgear arrangement required for the WTG at the centre of the star.

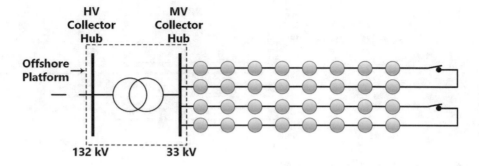

Fig. 4 Wind farm double-sided ring design

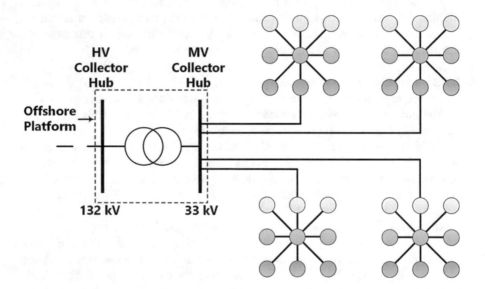

Fig. 5 Wind farm star design

2.1.5 Single Return Design

New designs are being proposed to minimize the overall cost of the ringed collector together with a good steady-state performance. These designs are derived from single-sided and double-sided ring designs. One option consists of having all the end WTG of all strings connected between them and one redundancy cable returning to the main hub. This solution is called a single-return design and is presented in Fig. 6.

The redundancy circuit is sized to support the full output power of a single string. This design does not support the failure of two string simultaneously since the probability of this occurrence is low. The cost of this design solution is

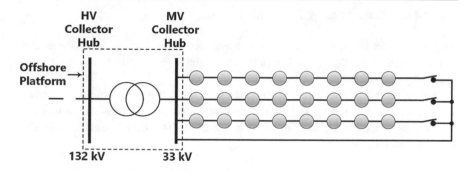

Fig. 6 Wind farm single-return design

significantly lower than the single-sided ring design and is competitive when compared to the radial design.

2.1.6 Double-Sided Half Ring Design

This new design is a variant of the double-sided ring. The redundancy is a connection between the middle turbine of one string to the middle turbine of another, as seen in Fig. 7.

In this layout, only half of the WTG in the string are shut down in case of a fault near the hub, instead of all as in the radial design. The design should include a remote-controlled load switch in each string and in the redundancy. In this way, it is possible to remotely isolate half WTG at the time. The layout only adds little modifications to the radial design, while having a significantly lower cost than the double-sided ring design.

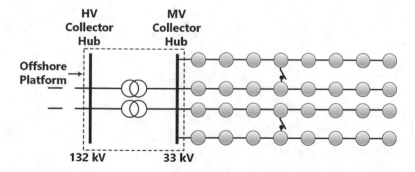

Fig. 7 Wind farm double-sided half-ring design

2.2   Basic Concepts of Reliability

Reliability is an important issue in power systems in general and in offshore wind farms in particular. The special importance of the latest is due to the high repair times required offshore.

We define the failure rate of a component as the probabilistic failure frequency and denote it as λ (year^{-1}), and the outage time in failure state of a component by μ (h). We note that the Mean Time Between Failures ($MTBF$) is measured in time units and is given by:

$$MTBF = \frac{1}{\lambda} \tag{2.1}$$

The unavailability of a component (h/year) is given by

$$U = \lambda\mu \tag{2.2}$$

or in percentage:

$$U\% = \frac{\lambda\mu}{8760} 100\% \tag{2.3}$$

These indexes apply to the reliability of a single component. However, they do not reflect the problem from the customers' side, i.e., the capability of the power system to supply the customers with an acceptable level of power quality. Therefore, the following customer-related indexes are defined:

$$SAIFI = \frac{\sum \lambda_i N_i}{\sum N_i} \tag{2.4}$$

$$SAIDI = \frac{\sum U_i N_i}{\sum N_i} \tag{2.5}$$

$$CAIDI = \frac{\sum U_i N_i}{\sum \lambda_i N_i} \tag{2.6}$$

$$ASAI = \frac{\sum N_i \cdot 8760 - \sum U_i N_i}{\sum N_i \cdot 8760} \tag{2.7}$$

where

- $SAIFI$(interruption $\cdot$ year^{-1} $\cdot$ customer^{-1}): System Average Interruption Frequency Index, defined as the total number of customer interruptions divided by the total number of customers served; this index is designed to give information about the average frequency of sustained interruptions per customer.

- $SAIDI$(hour · year^{-1} · customer^{-1}): System Average Interruption Duration Index, defined as the sum of all customer interruption durations divided by the total number of customers served; the index is designed to provide information about the average time the customers are interrupted.
- $CAIDI$(hour/interruption): Customer Average Interruption Duration Index, defined as the sum of all customer interruption durations divided by the total number of customer interruptions; $CAIDI$ gives the average outage duration that any given customer would experience; it is the ratio of $SAIDI$ to $SAIFI$.
- $ASAI$(%): Average Service Availability Index, defined as the customer hours of available service divided by the customer hours of service demand; it gives the fraction of time the customer has power during the reporting time, usually one year.

and λ_i is the failure rate at load point i, U_i is the unavailability (outage time) at load point i, and N_i is the number of customers at load point i.

The above-mentioned indexes normally apply to measure the power quality experienced by the consumers. However, the indexes can also be applied to a wind farm if we use a reverse power flow approach, in which each WTG is considered a load, and the delivery point is considered a generator.

Nevertheless, for the assessment of wind farms' reliability, the index that is most used is the ENS(MWh/year), Energy Not Supplied. ENS is affected by both failure rate and outage time which reflects the reliability of a wind farm. Therefore, it is appropriate to calculate how much of the expected generated energy will not be produced due to reliability issues. Moreover, it is useful to address the benefits of reliability improvements in different wind farm topologies. ENS is defined as

$$ENS = \sum U_i L_{avg_i} \qquad (2.8)$$

where L_{avg_i} is the average load (MW) at load point i, and U_i is the unavailability at load point i. We recall that the average load (or generation) is related to the peak load (or installed capacity) by means of the load factor (or capacity factor):

$$a = \frac{L_{avg}}{L_{peak}} \qquad (2.9)$$

Wind farms are often radial systems in which the components are series-connected. For the series-connected system to be in operating mode, it is required that all the components operate simultaneously. For a radial system, comparable to the collection system of an offshore wind farm, the following indexes are defined (the subscript s denotes a radial system composed by i series-connected components):

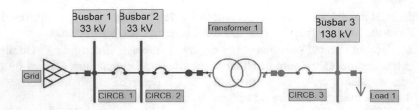

Fig. 8 Grid of Example 2—1

Table 1 Reliability data of Example 2—1

Element	Failure rate (year^{-1})	Outage time (h)
Busbar	0.001	2
Circuit breaker	0.02	24
Transformer	0.015	15

$$\lambda_s = \sum \lambda_i \tag{2.10}$$

$$U_s = \sum \lambda_i \mu_i \tag{2.11}$$

$$\mu_s = \frac{U_s}{\lambda_s} \tag{2.12}$$

where λ_s is the average failure rate (year^{-1}), U_s is the average annual unavailability (h/year), and μ_s is the total average outage time (h).

Example 2—1
Consider the grid depicted in Fig. 8[2]:
The reliability data is given in Table 1.
The load is a constant 400 MW load, supplying 80 customers.
Compute the reliability indexes: SAIFI, SAIDI, CAIDI, and ASAI.

Solution:

The grid failure rate and MTBF are

$$\lambda_s = \sum \lambda_i = 3\lambda_{BB} + 3\lambda_{CB} + \lambda_T$$
$$= 0.001 + 0.02 + 0.001 + 0.02 + 0.015 + 0.02 + 0.001 = 0.078 \, \text{year}^{-1}$$

---

[2]This example is adapted from: Tomas Winter. Reliability and economic analysis of offshore wind power systems-A comparison of internal grid topologies. MSc Thesis. Chalmers University of Technology. 2011.
 https://odr.chalmers.se/bitstream/20.500.12380/173938/1/173938.pdf.

$$MTBF = \frac{1}{\lambda_s} = 12.820 \, \text{year}$$

The grid unavailability per year is given by

$$
\begin{aligned}
U_s = \sum \lambda_i \mu_i &= 3\lambda_{BB}\mu_{BB} + 3\lambda_{CB}\mu_{CB} + \lambda_T \mu_T \\
&= 0.001 \times 2 + 0.02 \times 24 + 0.001 \times 2 + 0.02 \times 24 + 0.015 \times 15 + 0.02 \\
&\quad \times 24 + 0.001 \times 2 \\
&= 1.671 \, \text{h/year}
\end{aligned}
$$

The total outage time for the load is, therefore,

$$\mu_s = \frac{U_s}{\lambda_s} = U_s MTBF = \frac{1.671}{0.078} = 21.423 \, \text{h}$$

The four reliability indexes we are looking for are

$$SAIFI = \frac{\sum \lambda_i N_i}{\sum N_i} = \frac{0.078 \times 80}{80} = 0.078 \, \text{interruption} \cdot \text{year}^{-1} \cdot \text{customer}^{-1}$$

$$SAIDI = \frac{\sum U_i N_i}{\sum N_i} = \frac{1.671 \times 80}{80} = 1.671 \, \text{h} \cdot \text{year}^{-1} \cdot \text{customer}^{-1}$$

$$CAIDI = \frac{\sum U_i N_i}{\sum \lambda_i N_i} = \frac{1.671 \times 80}{0.078 \times 80} = 21.423 \, \text{h} \cdot \text{interruption}^{-1}$$

$$ASAI = \frac{\sum N_i \cdot 8760 - \sum U_i N_i}{\sum N_i \cdot 8760} = \frac{80 \times 8760 - 1.671 \times 80}{80 \times 8760} = 0.99981$$

As we have a single load with 80 customers, we have a single load point and therefore,

$$SAIFI = \lambda_s; \; SAIDI = U_s; \; CAIDI = \mu_s; \; ASAI = 1 - \frac{SAIDI}{8760}$$

We can also compute the Energy Not Supplied:

$$ENS = \sum U_i L_{avg_i} = 1.671 \times 400 = 668.4 \, \text{MWh}$$

3   Sizing AC Offshore Interconnection Cables

The power produced in each WTG composing an offshore wind park must be carried to an offshore substation and then exported to the onshore substation. To this end, interconnection submarine cables are used; we will now pay attention to them.

3.1   The Reactive Power Problem

The biggest electrical difference between underground cables and overhead lines is the large capacitance of the first ones. This phenomenon increases the reactive power generated by the cables, decreasing its capacity to transmit active power, especially over long distances. This implies the need to provide reactive compensation at the cables' ends. Contrary to what happens when we compensate the power factor at industrial facilities, by installing capacitors, in this case, reactors (inductances) are installed to absorb the excess capacitive reactive power.

We are going to look now at an example using a very rough approximation model just to give an overview of the nature of the problem.

Example 3—1

An offshore wind park has an installed capacity of 40 MW and is interconnected to shore through a 500 mm² aluminium submarine cable with 50 km length. The relevant parameters of the cable are $L = 0.33 \, \text{mH/km}$ and $C = 0.32 \, \mu\text{F/km}$. The rated voltage of the cable is 45 kV. The park is operating at rated capacity with a unitary power factor. Find a fast estimate of the reactive power absorbed and generated by the cable and the reactive power that is injected in the onshore bus.

Solution:

The solution we are finding is an estimate because the reactive power depends on the voltage at the offshore busbar, which is unknown. The exact methodology to solve problems of this type will be offered further in this text. For the time being, we are going to assume that the voltage in the offshore busbar is constant and equal to the rated voltage.

The rated current generated by the offshore wind park and injected in the submarine cable is (we note that the power factor is 1):

$$I_2 = \frac{P_{G2}}{\sqrt{3}V_N \times 1} = \frac{40 \times 10^6}{\sqrt{3} \times 45000} = 513.20 \, \text{A}$$

The cable is modelled by the well-known π-model, with a longitudinal impedance $Z_L = j\omega L = jX_L$ and two transversal admittances located at the ends of the cable, each one given by $\frac{Y_T}{2} = j\frac{\omega C}{2} = j\frac{B_T}{2}$.

The amplitude of the current flowing in the equivalent capacitor near the offshore busbar is (take note that the voltage is the phase-to-neutral voltage as the system is three phase, but we are looking only at one phase):

$$I_C = \frac{B_T}{2} \frac{V_N}{\sqrt{3}} = \frac{2\pi 50 \times 0.32 \times 10^{-6} \times 50 \times 45000}{2\sqrt{3}} = 65.297 \, A$$

This current is small as compared to the rated current, so, for the sake of simplification, we are going to neglect it and consider that the current across the cable is constant and equal to the rated current.

$$I_L = I_2 = 513.20 \, A$$

Now, we have determined all the relevant currents and can compute the required reactive powers.
The reactive power absorbed by the cable's equivalent inductance is (note the factor 3 because the cable is three phase):

$$Q_L = 3X_L I_L^2 = 3 \times 2\pi 50 \times 0.33 \times 10^{-3} \times 50 \times (513.20)^2 = 4.0957 \, \text{Mvar}$$

The reactive power generated by the two equivalent capacitances is (the minus sign indicates that the capacitors are generating reactive power):

$$Q_C = -3B_T V_N^2 = -3 \times 2\pi 50 \times 0.32 \times 10^{-6} \times 50 \times \left(\frac{45000}{\sqrt{3}}\right)^2 = -10.179 \, \text{Mvar}$$

The reactive power that reaches the onshore busbar is

$$Q_T = Q_L + Q_C = -6.0831 \, \text{Mvar}$$

This is the reactive power balance at the onshore bus. We note that this is a significant value, about 15% of the active power injected in the offshore bus, which limits the active power transmission capability. This problem increases with the transmission distance and with the transmitted power, therefore implying reactive power compensation by installing reactors at both ends of the submarine cable, for longer distances and larger transmitted power.

3.2 Cables' General Features

3.2.1 Layers

Figure 9 shows the different layers a cable is composed of.

A submarine cable is composed of several layers. From the inside to the outside, the cable has a conductor core, insulation, shield, sheath, armour, optic fibre, and protecting sheath. The conductor core carries the power produced by the turbines

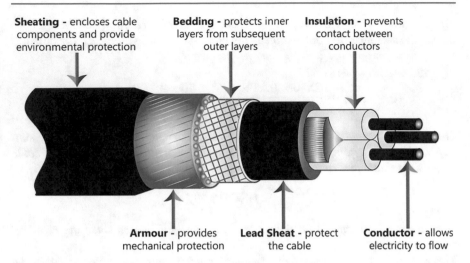

Sheating - encloses cable components and provide environmental protection

Bedding - protects inner layers from subsequent outer layers

Insulation - prevents contact between conductors

Armour - provides mechanical protection

Lead Sheat - protect the cable

Conductor - allows electricity to flow

Fig. 9 Cable layers

and is formed by threaded wires joined in a circular section. The insulation retains the propagation of the electrical current and isolates the conductor core. To smooth out the electrical field and avoid field concentration zones, the cables have a shield. The sheath is made from metallic materials and is connected to the earth, it also serves as a barrier for water. To provide mechanical strength and anti-corrosion protection, the cable has one layer called armour. The optical fibre is used to communicate and monitor cable purposes. And, finally, there is the protecting sheath, which is the final outer protection layer.

3.2.2 Configuration

The cables shown in Fig. 9 are called three-core cables, where there are three cores bundled together in the same cable. This is the most used configuration to perform the interconnection of offshore wind parks. A less used solution is a single-core configuration, where there is a single core for each cable meaning that, for a three-phase system, three cables in flat formation are needed.

Although single-core cables have higher ratings available than three-core ones, they not only exhibit higher losses (because the armouring will carry almost the same current as the core, increasing the temperature of the cable), but also its installation is much more expensive due to the number of cables up for laying. Furthermore, for small inter-cable distance, the currents in each cable suffer the influence of the others. Since the cables are not transposed that influence may cause a difference of impedances among them, creating an unbalance in the 3-phase system. Regarding the three-core cables, despite their lower costs, losses, and easiness of installation, they have drawbacks like being heavier to handle onshore and having laying depth restrictions due to the cable size.

3.2.3   Insulation

Regarding the type of insulation, there are three options available: Low-Pressure Oil-Filled (LPOF) or Low-Pressure Fluid-Filled (LPFF), where the core is involved in a hollow shaft that contains a pump-circulated high-pressure oil; Cross-linked Polyethylene (XLPE), where the insulation is composed of a solid (extruded) dielectric; and Ethylene Propylene Rubber (EPR), which have the same building principle as XLPE cables.

LPOF cables have several drawbacks: risk of fluid leakage; maximum transmission distance of 50 km due to the difficulty of maintaining oil pressure; the need for extra-protection when buried; and auxiliary equipment's high cost. On the other hand, XLPE stands as the most cost-effective solution, not only because it can reach longer distances, but also because, due to its constitutional characteristics (better bending capability, lower weight, higher mechanical resistance, and lower capacitance), it is less expensive than LPOF cables. EPR cables have the same characteristics as the XLPE cables, but, beyond a certain voltage level in AC applications, they have a higher capacitance, which implies higher losses.

3.3   Choosing the Cross-Section

When we talk about cable sizing, it is mainly about choosing the cross-section of the cable. The cross-section must be dimensioned to withstand the current under rated operational conditions. For the case of the interconnection AC submarine cable of an offshore wind park with installed capacity, S_N, (apparent power) at first sight, this nominal current is given by

$$I_N = \frac{S_N}{\sqrt{3}V_N} \tag{3.1}$$

where V_N is the rated phase-to-phase voltage of the cable.

This current is very straightforward to compute, given the characteristics of the park. However, several factors may change its value in specific ambient conditions. The most relevant of these factors, called rating factors, are

- Laying depth: the deeper the laying depth, the lower the ampacity, ampacity being defined as the maximum current a cable can carry continuously under the conditions of use without exceeding its temperature rating; it is also called the current-carrying capacity.
- Ground temperature: the higher the ground temperature, the lower the ampacity, for reasons related to the cable's maximum temperature rating.
- Ground thermal resistivity: the larger the ground thermal resistivity, the lower the ampacity; thermal resistivity (K m/W; K stands for Kelvin) is the reciprocal of thermal conductivity, a property of a material indicating its ability to conduct heat; if the ground is a poor heat conductor, the cable's ampacity decreases.

Therefore, we must compute the adjusted current (current for specific ambient conditions) from the equation:

$$I_N = I_r \prod_{i=1}^{3} f_i \tag{3.2}$$

where I_r is the adjusted current (a sort of equivalent current to choose the cross-section) and f_i are the rating factors.

The following condition must hold:

$$I_r \leq I_z \tag{3.3}$$

where I_z is the ampacity (current-carrying capacity) of the cable, chosen from the manufacturer's datasheets.

From what has been said, it is clear that the cross-section choice must be based on the adjusted current, I_r, but actually, the cable will withstand its ampacity, I_z.

Let us analyse the two possible situations:

(1) The specific ambient conditions are favourable, meaning that $\prod f_i > 1$. This implies $I_N > I_r$. The cross-section can be decreased in relation to what it would be if the nominal current was considered. Actually, the equivalent current is lower than the nominal current and the cable only has to withstand that current.

(2) The specific ambient conditions are adverse, meaning that $\prod f_i < 1$. This implies $I_N < I_r$. The cross-section should be increased in relation to what it would be if the nominal current was considered. Actually, the equivalent current is higher than the nominal current and the cable must withstand that current.

In the datasheets of the manufacturers, it is possible to find the data needed to perform the calculations mentioned above. Let us begin by the rating factors, which are shown in Tables 2, 3, and 4 (data retrieved from the manufacturer's datasheet):

Table 2 Rating factor for laying depth f_1

Laying depth (m)	Rating factor
0.05	1.10
0.70	1.05
0.90	1.01
1.00	1.00
1.20	0.98
1.50	0.95

Table 3 Rating factor for ground temperature f_2

Conductor temperature (°C)	Ground temperature (°C)							
	10	15	20	25	30	35	40	45
90	1.07	1.04	1	0.96	0.93	0.89	0.84	0.80
65	1.11	1.05	1	0.94	0.88	0.82	0.74	0.66

Table 4 Rating factor for ground thermal resistivity f_3

Thermal resistivity (Km/W)	0.7	1.0	1.2	1.5	2.0	2.5	3.0
Rating factor	1.14	1.00	0.93	0.84	0.74	0.67	0.61

After computing the adjusted current (current rating), from the nominal current and rating factors, f_1, f_2, f_3, the cross-section of the submarine can be chosen consulting Table 5.

Table 5 refers to three-core submarine cables with steel wire armour, the most used configuration in offshore wind parks' interconnection. In the consulted publication "XLPE Submarine Cable Systems: Attachment to XLPE Land Cable Systems—User's Guide" there are more data available, namely: current rating for XLPE single-core cables, technical data for XLPE submarine cable systems (namely inductance and capacitance), including single-core cables with lead sheath, three-core cables with copper wire screen, and three-core cables with lead sheath.

Example 3—2

Consider a wind park composed of five 5 MW WTG connected to a substation by one 36 kV, XLPE three phase, three-core submarine cable with aluminium conductor. The apparent power of each WTG is 5.55 MVA. The cable is to be installed under the following conditions: Temperature in seabed: 15 °C; Laying depth in seabed: 1.0 m; Seabed thermal resistivity: 0.7 K m/W; Operating temperature of XLPE cable: 90 °C.

Compute the current-carrying capacity (ampacity) of the cable and cross-section of the cable.

Table 5 Current rating for XLPE three-core submarine cables with steel wire armour

Cross section (mm^2)	Copper conductor	Aluminium conductor
	A	A
10-90 kV XLPE 3-core cables		
95	300	235
120	340	265
150	375	300
185	420	335
240	480	385
300	530	430
400	590	485
500	655	540
630	715	600
800	775	660
1000	825	720
100-300 kV XLPE 3-core cables		
300	530	430
400	590	485
500	655	540
630	715	600
800	775	660
1000	825	720

Source https://new.abb.com/docs/default-source/ewea-doc/xlpe-submarine-cable-systems-2gm5007.
pdf

Solution:

The rated current flowing in the cable is:

$$I_N = \frac{S_N}{\sqrt{3}V_N} = 445.49 \text{ A}$$

This is an unadjusted value. We now need to adjust it to the specific ambient conditions. Taking into account the problem's data and looking at Tables 2, 3 and 4, we pick up the relevant rating factors as

$$f_1 = 1, f_2 = 1.04, f_3 = 1.14$$

The adjusted current is

$$I_r = \frac{I_N}{f_1 f_2 f_3} = 375.75 \text{ A}$$

The cross-section is chosen in Table 5, so that $I_r \leq I_z$. *We select* $s = 240\,\text{mm}^2$, *the corresponding ampacity being* $I_z = 385$ A, *for an aluminium conductor.*

3.4   A Note on DC Cables

As we will see later, the transmission to shore can be made using a DC system with HVDC cables. The elements comprising a DC cable are the same as the ones in AC cables. As so, like the HVAC cables, the HVDC cable main technologies available differ in the electrical insulation used: Oil Filled (OF) cables; Mass Impregnated (MI) cables; Cross-linked Polyethylene (XLPE) cables.

In Oil-Filled cables (also known as SCFF—Self Contained Fluid-Filled), the insulation consists of paper impregnated with low-viscosity oil. The core of the cable is covered by a hollow shaft where oil is circulated by pumps at both ends of the line. The need for pumping at each end of the system may prove difficult and may significantly limit the maximum length of the cable to less than 50 km. The danger of oil spill and the need for cable protection are also obvious disadvantages of the Oil Filled cable technology.

Mass Impregnated cables have a similar construction, but the paper insulation is impregnated in resin and high viscosity oil, and no oil circulation system is required. It was the most used cable technology in existing HVDC systems and so the track record and high reliability are some of the advantages of the MI cable.

In XLPE cables, the insulating material is made of a solid dielectric, also known as an extruded dielectric. It is a relatively new technology, developed to overcome some of the limitations of the previously referred technologies. XLPE cables have all the advantages of the MI cables and additionally can carry nominal current with a cable temperature of 90 °C. Increased bending capability, higher mechanical resistance, and lower weight are also benefitting XLPE cables, since the installation process is easier than for other cables. All this bunch of advantages made XLPE the most used technology in HVDC submarine cables for offshore wind farms.

4   Electrical Losses

The losses in a cable conductor are dependent on the operating temperature. Since in field installations this temperature is unknown, the estimation of the losses may prove a difficult task.

4.1 Temperature-Dependent Losses Model

We will present the basic steps of a model developed by H. Brakelmann[3] that allows us to determine the losses of long three-phase AC cables, taking into account the longitudinal distributions of current and temperature.

The steady-state model of an arbitrary long transmission line (or cable) can be described by the well-known line equations:

$$V_x = V_e \cosh(\gamma x) - Z_0 I_e \sinh(\gamma x) \tag{4.1}$$

$$I_x = -\frac{V_e}{Z_0} \sinh(\gamma x) + I_e \cosh(\gamma x) \tag{4.2}$$

V_x and I_x are the voltage and current at distance x from the emission, respectively; V_e and I_e are the voltage and current at the emission, respectively; and the surge impedance and the propagation constant are, respectively, given by

$$Z_0 = \sqrt{\frac{R + j\omega L}{G + j\omega C}} \tag{4.3}$$

$$\gamma = \sqrt{(R + j\omega L)(G + j\omega C)} \tag{4.4}$$

where R, L, G, and C are the line parameters, resistance, inductance, conductance, and capacitance in per unit of length (p.u.l.).

The resistance p.u.l. R represents all current-dependent active losses of the cable and is temperature-dependent. Hence, for a conductor temperature θ_L, the ohmic losses, P_L, in one conductor are changed according to

$$P_L = R I^2 = R_{20} I^2 [1 + \alpha_T (\theta_L - 20)] \tag{4.5}$$

where R_{20} is the AC conductor resistance p.u.l. for 20 °C and α_T is the temperature coefficient of the conductor. We note that the value of R_{20} is usually know from the manufacturer's datasheets.

Furthermore, the equivalent resistance p.u.l. of the conductor must take into account loss factors for sheath and screen losses (λ_1) and for armour losses (λ_2). Therefore,

$$R' = R_{20}[1 + \alpha_T (\theta_{Lmax} - 20)](1 + \lambda_1 + \lambda_2) \tag{4.6}$$

R' is usually determined for the maximum permissible conductor temperature θ_{Lmax}, thus being inserted into the line equations as a constant parameter. The losses

---

[3]H. Brakelmann. Loss Determination for Long Three-Phase High-Voltage Submarine Cables. *European Transactions on Electrical Power.* Vol.13, No3, May/June 2003.

are determined as the difference of the active powers at the entry (emission) and at the end (reception) of the line. This neglect of the temperature-dependence is permissible as long as the surge impedance and the propagation constant are dominated by their inductance p.u.l. and their capacitance p.u.l. However, non-negligible errors arise for long cables, depending on the load situation, and for growing differences of the currents at the emission and at the reception of the cable, as well as for their difference to the nominal current I_N.

It is possible to demonstrate[4] that the temperature change, $\Delta\theta_L$, of the conductor for a given current, I, can be related to the maximum admissible temperature change, $\Delta\theta_{Lmax}$, for the rated current, I_N, through:

$$\frac{\Delta\theta_L}{\Delta\theta_{Lmax}} = \frac{c_\alpha\left(\frac{I}{I_N}\right)^2}{c_m - \Delta\theta_{Lmax}\alpha_T\left(\frac{I}{I_N}\right)^2} = \frac{\alpha_T\Delta\theta_L + c_\alpha}{c_m}\left(\frac{I}{I_N}\right)^2 \tag{4.7}$$

where θ_U is the ambient temperature:

$$c_\alpha = 1 + \alpha_T(\theta_U - 20) \tag{4.8}$$

$$c_m = 1 + \alpha_T(\Delta\theta_{Lmax} + \theta_U - 20) = c_\alpha + \alpha_T\Delta\theta_{Lmax} \tag{4.9}$$

Using the same reasoning, the temperature-dependent ohmic losses, P'_L, can now be expressed by means of the nominal values, P_{LN} and I_N, as

$$P'_L = P_{LN}\left(\frac{I}{I_N}\right)^2 \frac{\alpha_T\Delta\theta_L + c_\alpha}{c_m} \tag{4.10}$$

The ohmic losses p.u.l. of the cable can be calculated from the squared current ratio if an additional correction term for the temperature-dependence is introduced:

$$P'_L = P_{LN}\left(\frac{I}{I_N}\right)^2 v_\theta \tag{4.11}$$

where the correction factor is

$$v_\theta = \frac{c_\alpha}{c_\alpha + \alpha_T\Delta\theta_{Lmax}\left[1 - \left(\frac{I}{I_N}\right)^2\right]} = \frac{\alpha_T\Delta\theta_L + c_\alpha}{c_m} \tag{4.12}$$

and, if one neglects the loss factors for sheath and screen losses and for armour losses:

---

[4]Please refer to the paper we are following, for further details.

$$P_{LN} = 3R_{20}I_N^2[1 + \alpha_T(\theta_{Lmax} - 20)] \tag{4.13}$$

It is important to highlight that, with the proposed model, it is possible to determine the p.u.l. losses based solely on the nominal losses, operating current, maximum admissible temperature (for the nominal current), and ambient temperature: the knowledge of the temperature for the operating current is not necessary.

In addition to ohmic losses, there are dielectric losses (W/m) to be considered given by

$$P_d = 3V_{pn}^2 \omega C \tan\delta \tag{4.14}$$

V_{pn}^2 is the phase-to-neutral nominal voltage, C is the p.u.l. capacitance, ω is the angular frequency and $tan\delta$ is the loss tangent, which is dielectric dependent.

These losses are due because the dielectric between the conductors has a finite conductivity, therefore an energy loss will occur appearing in the form of heat in the dielectric material and increasing with increased frequency. The dielectric losses are much smaller than the ohmic losses and are often neglected.

The method we just presented allows for the computation of the temperature-dependent power losses given the current flowing in a branch, which depends on the output power of each WTG. To compute the annual energy losses, we must take into account the probability of occurrence of each output power, which depends on the input wind speed. This information is given by the wind speed probability density function, $f(u)$, u being the wind speed. Therefore, the annual energy losses are given by

$$W_L = 8760 \sum_{i=branch\#1}^{branch\#n} \sum_{j=u_{cut-in}}^{u_{cut-off}} P'_{Li} f(u_j) \tag{4.15}$$

Example 4—1

An offshore wind park is composed of 4×2.22 MW WTG. The distances are indicated in Fig. 10:

The length of the cables between the turbines and between the turbine that is closer to the substation and the substation is twice the distances specified above because the cables are buried in the seabed. A three-phase 95 mm² XLPE submarine cable with aluminium conductor, rated voltage 20 kV, rated current 265 A, frequency 50 Hz, is used. The parameters (@90 °C) are: $R = 0.325\,\Omega/km, L = 0.0004\,H/km, C = 2.5 \times 10^{-7}\,F/km$. Also, it is known that the loss tangent is 0.00055 and the temperature coefficient is 0.00403 K^{-1}. Consider that the ambient temperature is 20 °C and the maximum admissible temperature is 90 °C.

Compute the total losses when the wind park is operating at rated power.

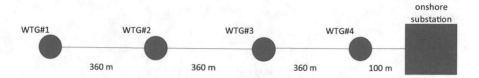

Fig. 10 Grid of Example 4—1

Solution:

We have to compute the losses in each branch because the current is different. We will exemplify with the first branch.
The current is (V_{pp} is the phase-to-phase voltage and we consider $cos\phi = 1$):

$$I_1 = \frac{1 \times P_N}{\sqrt{3}V_{pp} \times 1} = 64.0859\,\text{A}$$

The temperature-dependent p.u.l. ohmic losses are

$$P'_{L1} = P_{LN}\left(\frac{I_1}{I_N}\right)^2 \frac{c_\alpha}{c_\alpha + \alpha_T\Delta\theta_{Lmax}\left[1 - \left(\frac{I_1}{I_N}\right)^2\right]} = 4.0043 \times 0.7901$$

$$= 3.1640\,\text{kW/km}$$

where

$$P_{LN} = 3RI_N^2 = 68.4694\,\text{kW/km}$$

$$c_\alpha = 1 + \alpha_T(\theta_U - 20) = 1$$

$$\Delta\theta_{Lmax} = 90 - 20 = 70\,°\text{C}$$

The total ohmic losses in the first branch are

$$P'_{L1total} = P'_{L1} \times 0.720 = 2.2781\,\text{kW}$$

Here are the results for the other branches:

$$I_2 = 128.1718\,\text{A}$$

$$P'_{L2} = 16.0173 \times 0.8223 = 13.1710\text{kW/km}$$

$$P'_{L2total} = 9.4831\,\text{kW}$$

$$I_3 = 192.2576\,\text{A}$$

$$P'_{L3} = 36.0389 \times 0.8821 = 31.7911\,\text{kW/km}$$

$$P'_{L3total} = 22.8896\,\text{kW}$$

$$I_4 = 256.3435\,\text{A}$$

$$P'_{L4} = 64.0692 \times 0.9822 = 62.9284\,\text{kW/km}$$

$$P'_{L4total} = 12.5857\,\text{kW}$$

The total ohmic losses, considering the temperature-dependence are given by
$P'_{Ltotal} = \sum P'_{Litotal} = 47.2364\,\text{kW} = 0.5\%$ *of the total installed capacity.*
Finally, we have to add the dielectric losses, which are

$$P_d = 3V^2_{pn}\omega C \tan\delta = 3\left(\frac{20 \times 10^3}{\sqrt{3}}\right)^2 2\pi 50 \times 2.5 \times 10^{-7} \times 0.00055 \times 2.36$$
$$= 40.7779\,\text{W}$$

As it can be seen the dielectric losses are very small.

4.2   The Classical Load Flow Model

In the previous section, the losses were calculated assuming that the voltage in each WTG busbar is constant. Actually, it is not. The losses depend on the voltage in each busbar, which must be determined beforehand. This is the issue addressed by the classical load flow problem. The load flow (or power flow) problem aims at finding the voltages in every busbar except one, called the slack bus, from the knowledge of the injected powers (the difference between the generated and load powers) in every bus, except in the slack bus. In the slack bus, the voltage is known (specified) and the injected power is to be determined.

In the general case, this is a very complex problem, due to the dimension of the grids composed of a huge number of busbars. The problem is not manually feasible, requiring the use of dedicated software. However, it can be solved by hand if the number of busbars is restricted to two. Even though the offshore grid is small as compared to the dimension of a country's transmission grid, for instance, still it is composed of dozens of busbars which renders the problem unfeasible by hand.

The load flow problem formulation and solution are outside the scope of this introductory textbook. We will address only a simplified version of the problem, in which a test grid is composed of two busbars: one represents the onshore substation

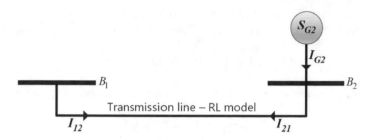

Fig. 11 Test offshore grid

and the other represents the aggregated wind park concentrated in a single offshore busbar. The two busbars are interconnected by a transmission cable. Figure 11 depicts a diagram of our test grid.

Let us analyse this system.

The equivalent generator, located at bus 2, generates given complex power, $S_{G2} = P_{G2} + jQ_{G2}$, in which P_{G2} is the real power and Q_{G2} is the reactive power. The voltage at busbar 1 (slack bus) is fixed (known), $V_1 = V_1 e^{j0} = V_1$, and the voltage at busbar 2, $V_2 = V_2 e^{j\theta} = V_2\cos\theta + jV_2\sin\theta$, is unknown. The transmission cable is modelled through the simplified R-L model, represented by a series impedance, $Z = R + j\omega L = R + jX$, in which the line parameters R (resistance[5]) and X (reactance) are given. We should highlight that this model is not entirely appropriate for representing submarine cables, because we are neglecting the capacitive effects that play an important role in cables. However, the model to include the capacitive effects is much more complex. We will later return back to this.

The objective of this exercise is to calculate the transmission losses. As we know, the losses depend on the current (in fact, on the square of the current), but the issue is that the current depends on the voltage at busbar 2, which is unknown. So, our first task will be to compute voltage V_2.

Let us write the equations of the circuit, using the pu (per unit) system. The current flowing in the transmission line is (it is now apparent that we cannot calculate the current, because it depends on the voltage):

$$I_{G2} = I_{21} = \left(\frac{S_{g2}}{V_2}\right)^* = \frac{P_{G2} - jQ_{G2}}{V_2^*} = I \tag{4.16}$$

Moreover, we can write

$$V_2 = V_1 + (R + jX)\left(\frac{P_{G2} - jQ_{G2}}{V_2^*}\right) \tag{4.17}$$

---

[5]Resistance of the conductor at the maximum admissible temperature.

This is the equation we must solve to determine V_2.

If we multiply both sides of the equation by V_2^* (conjugate of V_2), we will obtain

$$V_2^2 = V_1 V_2^* + (R + jX)(P_{G2} - jQ_{G2}) \tag{4.18}$$

Developing, manipulating, and splitting into real and imaginary part leads to

$$V_1 V_2 \cos\theta = V_2^2 - RP_G - XQ_G \tag{4.19}$$

$$V_1 V_2 \sin\theta = XP_G - RQ_G \tag{4.20}$$

The technique to solve this system of equations, with two unknowns (V_2, θ), is to square and to sum to get rid of the angle θ ($\sin^2\theta + \cos^2\theta = 1$). If we do this and after some manipulation, we obtain the following bi-quadratic equation:

$$V_2^4 - \left(2RP_G + 2XQ_G + V_1^2\right)V_2^2 + R^2\left(P_G^2 + Q_G^2\right) + X^2\left(P_G^2 + Q_G^2\right) = 0 \tag{4.21}$$

This is a bi-quadratic equation that can be solved by performing the change of variable $Y = V_2^2$. There are four solutions for this equation: two are negative and do not have physical meaning and should be disregarded; from the two solutions that are positive, one is far from 1 pu and should be disregarded too; the other is close to 1 pu and is the one that should be kept.

To compute the angle θ, one picks up Eq. 4.20 and solve:

$$\theta = \mathrm{asin}\left(\frac{XP_G - RQ_G}{V_1 V_2}\right) \tag{4.22}$$

At this time, we have computed $V_2 = V_2 e^{j\theta}$. Now, we can calculate the current using Eq. 4.16. The active and reactive losses are, respectively, given by

$$P_L = RI^2 \tag{4.23}$$

$$Q_L = XI^2 \tag{4.24}$$

The consideration of the capacitive effects in the line model, the so-called π-model, would considerably increase the complexity of the problem. Just for curiosity, we leave below the equations considering a π-model for the line.

Let us consider a transmission line represented by the π-model, with a longitudinal impedance $Z_L = R + j\omega L = R + jX$[6] and a transversal admittance $Y_T = G + j\omega C = G_T + jB_T$ (see Fig. 12).

---

[6] We recall that to be on the safe side, the resistance is taken at the maximum admissible temperature.

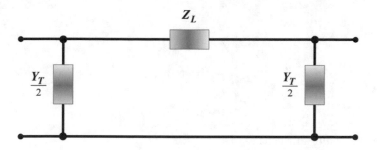

Fig. 12 Transmission line π-model

We begin by defining:

$$G_{22} = Re\left\{\frac{Y_T}{2} + \frac{1}{Z_L}\right\} \qquad (4.25)$$

$$B_{22} = Im\left\{\frac{Y_T}{2} + \frac{1}{Z_L}\right\} \qquad (4.26)$$

$$G_{21} = Re\left\{-\frac{1}{Z_L}\right\} \qquad (4.27)$$

$$B_{21} = Im\left\{-\frac{1}{Z_L}\right\} \qquad (4.28)$$

Following the same reasoning as above, we come to the following equations:

$$P_{G2} - G_{22}V_2^2 = V_2 V_1 (G_{21}\cos\theta + B_{21}\sin\theta) \qquad (4.29)$$

$$Q_{G2} + B_{22}V_2^2 = V_2 V_1 (G_{21}\sin\theta - B_{21}\cos\theta) \qquad (4.30)$$

Squaring and adding, we obtain

$$\left(G_{22}^2 + B_{22}^2\right)V_2^4 + 2\left[(B_{22}Q_{G2} - G_{22}P_{G2}) - V_1^2\left(G_{21}^2 + B_{21}^2\right)\right]V_2^2 + \\ + P_{G2}^2 + Q_{G2}^2 = 0 \qquad (4.31)$$

This is a bi-quadratic equation that allows V_2 to be obtained. The choice of the voltage to be kept follows the same rules as before.

To determine the voltage angle, we pick up Eq. 4.29. This is an equation that can be written in the form:

$$A\cos\theta + B\sin\theta = C \qquad (4.32)$$

where

$$A = V_2 V_1 G_{21} \qquad (4.33)$$

$$B = V_2 V_1 B_{21} \qquad (4.34)$$

$$C = P_{G2} - G_{22} V_2^2 \qquad (4.35)$$

The solution of this equation is

$$\theta = \phi_1 \pm \mathrm{acos}\left(\frac{C}{A}\cos\phi_1\right) \qquad (4.36)$$

where

$$\phi_1 = \mathrm{atan}\left(\frac{B}{A}\right) \qquad (4.37)$$

There are two solutions. The solution that should be retained is the closest to 0, as the voltage angle is always close to 0. Also, the power flow is from busbar 2 to busbar 1. Hence, the voltage angle in 2 is positive: power flows from the higher angles to the lower (the voltage angle in 1 is 0).

Now, it is possible to compute the losses. We note that it is necessary to calculate the current that flows through both the longitudinal and transversal branches. The current that flows in the cable near busbar 2 is

$$I_{21} = \left(\frac{S_{G2}}{V_2}\right)^* \qquad (4.38)$$

The current flowing in the longitudinal branch is

$$I_L = I_{21} - \left(\frac{Y_T}{2}\right)V_2 \qquad (4.39)$$

The active losses are, therefore,

$$P_L = RI_L^2 \qquad (4.40)$$

To compute the reactive losses, we must take into account the reactive power consumed in the inductances (taken as positive) and the reactive power generated by the capacitors (taken as negative). Therefore,

$$Q_L = XI_L^2 - \left(\frac{B_T}{2}\right)V_2^2 - \left(\frac{B_T}{2}\right)V_1^2 \qquad (4.41)$$

Example 4—2

A 10 MW offshore wind park is connected to shore through an AC submarine cable, 240 mm², aluminium conductor ($\alpha_T = 0.00403$), 45 kV rated voltage, 30 km length 90 °C maximum admissible temperature. Consider the wind park represented by an equivalent aggregate. The parameters of the cable are $R_{20} = 0.125 \ \Omega/km$, $L = 0.61$ mH/km, $C = 0.24 \ \mu F/km$. The park is generating 9 MW and 0 Mvar. The voltage at the onshore slack bus is 1.0 pu. Compute the active and reactive losses: (a) neglecting the transversal admittance; (b) considering the π-model of the cable.

Solution (in pu):

(a) Please see the following MATLAB® code.

```
% this program solves the power flow problem through the
bi-quadratic equation
% exemple#1: C=0
clear all; clc;
% data
Sb=10; Vb=45; Zb=Vb^2/Sb; V1=1.0;
alfaT=0.00403; Teta_max=90;
l=30; RL=0.125*(1+alfaT*(Teta_max-20))*l/Zb; L=0.61e-3;
XL=2*pi*50*L*l/Zb;
Pg2=9/Sb; Qg2=0/Sb;
% computation of V2
a=1; b=-2*RL*Pg2-2*XL*Qg2-V1^2;
c=RL^2*(Pg2^2+Qg2^2)+XL^2*(Pg2^2+Qg2^2);
X1=(-b+sqrt(b^2-4*a*c))/(2*a);
X2=(-b-sqrt(b^2-4*a*c))/(2*a);
V2_1=sqrt(X1); V2_2=sqrt(X2);
V2=V2_1
teta=asin((XL*Pg2-RL*Qg2)/(V1*V2));
teta_g=teta*180/pi
V2c=V2*(cos(teta)+1j*sin(teta));
% computation of losses
I21=(Pg2-1j*Qg2)/conj(V2c);
PL=RL*abs(I21)^2, QL=XL*abs(I21)^2
% check losses
I12=-I21;
SG1=V1*conj(I12), SG2=Pg2+1j*Qg2;
PL1=real(SG1+SG2), QL1=imag(SG1+SG2)
```

Here are the results:
$V2 = 1.0206$; teta_g $= 1.4346$.
$PL = 0.0185$; $QL = 0.0221$.
$SG1 = -0.8815 + 0.0221i$.
$PL1 = 0.0185$; $QL1 = 0.0221$.

(b) Please see the following MATLAB® code. It is straightforward to follow the code from the equations presented before.

```
% this program solves the power flow problem through the
bi-quadratic equation
% example#2: consider C
clear all; clc;
% data
Sb=10; Vb=45; Zb=Vb^2/Sb; V1=1.0;
alfaT=0.00403; Teta_max=90;
l=30; RL=0.125*(1+alfaT*(Teta_max-20))*l/Zb;
L=0.61e-3; XL=2*pi*50*L*l/Zb; ZL=RL+1j*XL;
C=0.24e-6; BT=2*pi*50*C*l*Zb; YT=0+1j*BT;
Pg2=9/Sb; Qg2=0/Sb;
% computation of V2
Yeq1=YT/2+1/ZL; Yeq2=-1/ZL;
G22=real(Yeq1); B22=imag(Yeq1); G21=real(Yeq2);
B21=imag(Yeq2);
a=G22^2+B22^2; b=2*(B22*Qg2-G22*Pg2)-V1^2*(G21^2+B21^2);
c=(Pg2^2+Qg2^2);
X1=(-b+sqrt(b^2-4*a*c))/(2*a); X2=(-b-sqrt(b^2-
4*a*c))/(2*a);
V2_1=sqrt(X1); V2_2=sqrt(X2); V2=V2_1
A=V2*V1*G21; B=V2*V1*B21; C=Pg2-G22*V2^2;
fi1=atan(B/A);
teta1=fi1+acos(C*cos(fi1)/A), teta2=fi1-acos(C*cos(fi1)/A);
teta=teta1; teta_g=teta*180/pi
V2c=V2*(cos(teta)+1j*sin(teta));
% computation of losses
I21=(Pg2-1j*Qg2)/conj(V2c)
I12=Yeq1*V1+Yeq2*V2c
SG1=V1*conj(I12), SG2=Pg2+1j*Qg2;
PL=real(SG1+SG2), QL=imag(SG1+SG2)
% check losses
PL1=RL*abs(I21-(YT/2)*V2c)^2
QL1=XL*abs(I21-(YT/2)*V2c)^2-(BT/2)*V2^2-(BT/2)*V1^2
```

Here are the results:
V2 = 1.0273; teta_g = 1.1051.
I21 = 0.8759 + 0.0169i; I12 = -0.8805 + 0.4474i.
SG1 = -0.8805 - 0.4474i.
PL = 0.0195; QL = -0.4474.
PL1 = 0.0195; QL1 = -0.4474.
Hereafter, some remarks concerning the results:

- *The voltage at the wind park busbar (2) is higher than at the slack bus (1) because a voltage drop exists on the cable.*
- *The voltage angle at busbar 2 is positive as power is flowing from 2 to 1.*
- *The active losses are approximately 0.2 MW, about 2% of the generated power.*

- In case (b), the currents at both ends of the cable are no longer equal, because of the capacitive effects included in the π-model.
- In case (b), the reactive losses are negative because of the excess reactive power generated by the "capacitors". This is a real problem, preventing the use of AC cables for long distances.

5   Connection to Shore: The AC Versus DC Option

One of the most important aspects of the design of offshore wind parks is the type of connection to the electrical grid located inland. The electrical energy generated in the offshore wind farm requires one or more submarine cables to transmit the power to the onshore utility grid that services the end-users.

As the receiving grid is AC, the obvious choice to transmit power to shore is the well-proven and reliable AC technology, here named High-Voltage Alternate Current (HVAC), because the transmission is made in High Voltage to decrease the losses. However, we have seen that HVAC has several limitations related to the maximum power and distance that can be transmitted. Losses rise with the voltage, and the cable length. Therefore, there is a critical length (100–150 km depending on cable type) until which the AC transmission is viable. Further increase the voltage to increase the transmission capacity does not work because reactive power does also increase with voltage, therefore reducing the available active power capacity.

This situation can be overcome by High-Voltage Direct Current (HVDC). When there is the need to transmit large amounts of power through very long distances, or if the AC grid that connects to the wind farm is weak, the choice for the transmission systems falls on the HVDC technology. In these systems, the offshore substation, before transmitting the collected energy, converts it to DC. At the onshore substation, that conversion is reversed, and the energy is delivered with the receiving grid requirements.

In a DC transmission system, there is no reactive power, therefore the complete power-carrying capability can be used in the transmission. Moreover, cable losses above a certain distance are lower in DC than in AC. The drawback of the HVDC transmission is that it is more expensive due to the AC/DC/AC conversions.

There are currently three different transmission technologies available:

- HVAC—High-Voltage Alternate Current, the conventional solution for short-distances and low power transfers.
- HVDC-LCC—High-Voltage Direct Current with Line Commutated Converters, the conventional DC solution, using old but reliable technology.
- HVDC-VSC—High-Voltage Direct Current with Voltage Source Converters, the modern DC solution, using state-of-the-art power electronics, that is becoming the new standard for long-distances and high-power transfers.

When choosing between AC or DC for connecting offshore wind farms to the grid, the main parameters to be considered are rated power, distance to shore, and the distance onshore to the nearest strong grid connection point.

5.1 High-Voltage Alternate Current (HVAC)

Connecting the wind farm to the grid by an AC cable is the most straightforward technical solution, as both the power generated by the wind farm and the onshore transmission grid are AC. The HVAC transmission offers some advantages over the DC solutions such as

- Proven and low-cost technology,
- Easy to integrate into existing power systems,
- Low losses over small distances.

On the other hand, there are some constraints of the HVAC system that significantly limit the use of this technology, namely

- There is an excessive amount of reactive power produced in AC submarines cables,
- An increase in the cable length means an increase in its capacitance which results in a reactive power increase, eventually resulting in a transmission distance limit for AC systems,
- Necessary use of reactive power compensation systems at the ends of the cable,
- Load losses are significantly higher for longer distances,
- For large wind farms, several cables may be necessary, increasing line losses.

A transmission system based on HVAC technology includes the following main components:

- AC based collector system within the wind farm,
- Three-core XLPE HVAC submarine transmission cable(s),
- Offshore substation with transformer and reactive power compensation,
- Onshore substation with transformer and reactive power compensation.

Figure 13 illustrates the layout of an HVAC based offshore wind farm, with the components mentioned above.

As already mentioned, the major problem concerning the connection of wind farms with AC submarine cables is the fact that cables generate significant amounts of reactive power. This reactive power is produced by the high shunt capacitance of cables (significantly higher than in overhead lines). In the AC system, the cable must carry both the load current and the reactive current (called charging current) generated by the cable capacitance, which reduces the effective active power rating

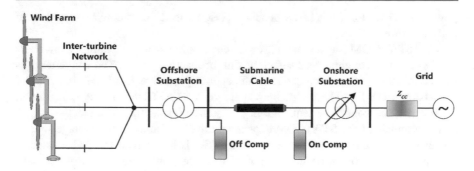

Fig. 13 Layout of an HVAC wind farm

of the cable. In other words, reactive current occupies "space" that cannot be fulfilled by the active current.

This problem is aggravated for long distances because the capacitance is distributed along the cable, and for high transmission voltages, which are necessary to reduce losses, but increase the generation of reactive power, as it depends on the square of the voltage. The maximum distances for an AC submarine cable system vary according to the rated voltage and the maximum transmitted power decreases with the length of the cable.

The solution for the large amounts of reactive power at the cable is to compensate the reactive power produced by absorbing reactive power, thus reducing the additional losses, and increasing the maximum transmitting distance. The compensation is usually realized by fixed or electronically controlled shunt reactors. The fixed shunt reactor is the simplest device but the progress in Flexible AC Transmission System (FACTS) devices, such as Static Var Compensator (SVC) or STATic Synchronous COMpensator (STATCOM), considerably extends the reactive power and voltage control possibilities offered by the switched shunt reactors.

Fixed shunt reactors have the advantage of requiring no transformer for the connection, thus having no additional power losses. One of the disadvantages is that the reactors are designed for a single operational mode, usually to compensate the cable at full load.

The SVC and the STATCOM are part of the FACTS devices family, used for voltage regulation and power system stabilization, based on power electronics. These devices are capable of both generating and absorbing reactive power. The flexibility of use is, therefore, their main advantage, since they allow the continuous variable reactive power absorption (or supply, if needed). The FACTS devices can also contribute to the improvement of the voltage stability and the recovery from network faults.

The SVC ("SVC Classic" for some manufacturers) is based on conventional capacitor and reactor banks, that are electronically controlled through thyristors, either Thyristor Controlled Reactor (TCR), used for linear absorption of reactive power or Thyristor Switched Capacitor (TSC), used for stepwise injection of reactive power.

The STATCOM device uses power electronic Voltage Source Converters (VSCs) and can act either as a source or sink of reactive power. The converter uses semiconductors with turn-on and turn-off capability, such as Insulated Gate Bipolar Transistors (IGBTs). The basic components of a STATCOM are the transformer, the inverter, the control system, and the DC source (capacitor). The reactive power at the terminals of the STATCOM depends on the amplitude of the voltage source. For example, if the terminal voltage of the VSC is lower than the AC voltage, it absorbs reactive power, as in the case of reactive power compensation for submarine cables.

An example of an AC connection with HVAC transmission is the 160 MW wind farm at Horns Rev. The wind farm is located at the Danish west coast and is sited 14–20 km offshore in the North Sea, connected to shore with AC at 150 kV. For power transportation, a single 150 kV subsea-power cable is in operation. Since the turbines are connected at 34 kV, an additional platform with the 34–150 kV transformer was necessary.[7]

5.2 High-Voltage Direct Current–Line Commutated Converters (HVDC-LCC)

Current trends in offshore wind farms point to installations rated at hundreds of MW and further from shore. These transmission requirements are not feasible using AC submarine cable transmission, given the limitations mentioned before. Connecting offshore wind farms through a DC link has then been considered since there was a significant experience in transmitting large amounts of power over long distances through the well-known HVDC links. The first technology used was a direct application of the existing onshore HVDC applications: the line-commutated converter HVDC (HVDC-LCC, also referred to as "Conventional" or "Classic" HVDC) using thyristors in the converters. The main reason for this initial choice is the HVDC-LCC proven track record since there was an accumulated experience of decades for this technology. The first commercial onshore HVDC-LCC connection was installed in 1954 and since then many other conventional HVDC links were installed all over the world.

HVDC-LCC employs line commutated, current source converters, with thyristor valves. The use of the thyristor grants this HVDC technology limited operating flexibility, since the lack of turn-off controllability of the conventional thyristor results in poor power factors and considerable waveform distortion.

However, we may list some of the most important advantages that HVDC-LCC transmission offers over AC:

---

[7]Josef Schachner. Power connections for offshore wind farms. MSc Thesis, TU Delft. https://ocw.tudelft.nl/wp-content/uploads/E_infra_master_thesis.pdf.

- Asynchronous connection, since sending and receiving end frequencies can differ,
- Transmission distance using DC is not limited by cable charging current,
- Low cable power losses,
- Higher power transmission capability per cable,
- Power flow is fully defined and controlled,
- HVDC does not transfer short circuit current.

Some of the constraints of HVDC-LCC transmission are the following:

- Production of harmonics in the converter, making the use of filters necessary,
- The converters at each end consume reactive power.

The HVDC-LCC transmission system for an offshore wind farm is represented in Fig. 14. An HVDC-LCC transmission system consists of the following main components:

- AC based collector system within the wind farm,
- Offshore three-phase two-winding converter transformers,
- Auxiliary power set (diesel generator or battery),
- Reactive power compensation (capacitors or STATCOM),
- AC and DC Filters,
- DC submarine cables,
- Onshore converter station (with transformer and LCC converter) and filters.

The LCC power converters (onshore and offshore) are the most important elements in the system, as they perform the AC/DC conversion offshore and the DC/AC conversion onshore. The LCC converters are based on thyristor valves,

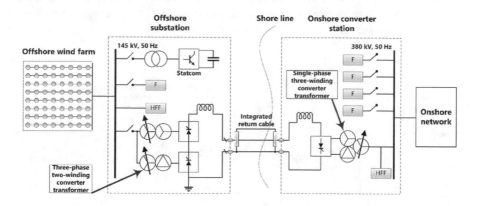

Fig. 14 Basic configuration of a wind farm using an HVDC-LCC; Legend: F = filter; HFF = High-frequency filter

capable of standing 8 kV and DC currents up to 4 kA. In the LCC converter, the current is always lagging the voltage due to the control angle of the thyristors; hence these converters consume reactive power. For this reason, reactive power compensation is necessary at both ends to provide reactive power to the system. Capacitor banks or STATCOM devices are considered for this effect.

A line-commutated converter requires an AC voltage source for its commutation, so an auxiliary power set is required at the offshore station to provide the necessary commutation voltage for the HVDC-LCC connection when there is little or no wind. This function can be provided by a STATCOM or by a diesel generator. A diesel generator or a battery is used to supply power to other devices in the wind farm when it is disconnected from the grid.

The AC filters are used to absorb harmonic currents generated by the HVDC converters, thus reducing the impact of the harmonics on the AC system. These filters also supply reactive power to the converter station.

We could not find any reference to a single operating HVDC-LCC for offshore wind farms. A huge number of HVDC-LCC projects from leading companies as ABB and Siemens are currently operating for other purposes, like for instance, connecting remote generation or loads, DC links in AC grids, interconnecting grids with different frequencies, power from shore. HVDC is currently used in projects as big as 2000 km line length, 6400 MW transmitted power, ± 800 kV DC voltage (Xiangjiaba—Shanghai, in China).

5.3 High-Voltage Direct Current–Voltage Source Converter (HVDC-VSC)

HVDC-VSC has become the new trend for long-distance offshore wind power transmission. This is a relatively recent technology (first installed in 1999) and it is known commercially as "HVDC Light", for ABB and "HVDC Plus", for Siemens. The basis of the HVDC-VSC technology is the self-commutated converters, with IGBT valves. Figure 15 depicts a general layout of an HVDC-VSC system applied to an offshore wind farm.

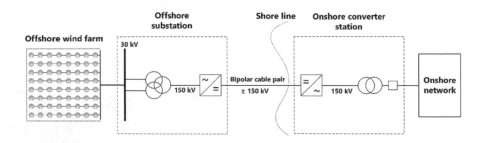

Fig. 15 Basic configuration of a wind farm using an HVDC-VSC system

VSC converters are self-commutating, not requiring an external voltage source for their operation. Also, the reactive power flow can be independently controlled at each AC network and the reactive power control is independent of active power control. These features make VSC transmission an interesting option for the connection of offshore wind farms. In addition to the referred HVDC advantages, the VSC technology also offers the following main benefits:

- Total and independent control of active and reactive power, provided by the use of PWM (Pulse Width Modulation) converters,
- Minimum risk of commutation failures, because it uses self-commutating semiconductor devices (IGBT) that no longer need a sufficiently high AC voltage to commutate,
- Smaller size of converters and filters than HVDC-LCC, due to the high switching rate of the IGBT (this however increases the losses),
- Less space occupied than HVDC-LCC, which is a benefit in offshore platforms,
- Possibility of the converters starting with a dead grid, not needing any start-up mechanism ("Black-start" capability),
- The AC voltage controller can control the voltage in the AC network and contribute to enhanced power quality.

However, there are some constraints to the use of VSC technology. The main constraint is the considerably lower experience in HVDC-VSC transmission, compared to the LCC option. HVDC-LCC has many years of service experience, with high availability rates. Also, VSC transmission has higher system power losses, when compared to an LCC system. Typically, the power loss at full load for the converters is about 4–6% for VSC and 2–3% for LCC transmission. This is due to the high switching rate of the IGBT. The IGBT valves are also much more expensive than the power thyristors used in LCC technology.

The converter is the most important module since it performs the conversion from AC to DC (rectification) at the sending end and from DC to AC (inversion) at the receiving end of the DC link. By allowing control of turn-on and turn-off operation, through the IGBT valves, the operation of the converter does not rely on synchronous machines in the connected AC system. IGBTs are switched on and off with frequencies defined by PWM algorithms. In this way, the PWM can create any desired voltage waveform. However, due to the switching process of IGBT, the AC output currents and voltages are not sinusoidal, i.e., the spectrum of their waveforms contains not only the fundamental frequency component but also higher-order harmonics. This requires the use of filters to obtain sinusoidal waveforms.[8]

According to ABB as of October 2018, 40 of the over 90 offshore wind farms in Europe have a nameplate capacity higher than 200 MW and roughly one-third of

---

[8]Asimenia Korompili, Qiuwei Wu, Haoran Zhao. Review of VSC HVDC connection for offshore wind power integration. *Renewable and Sustainable Energy Reviews*. 2016.
http://dx.doi.org/10.1016/j.rser.2016.01.064.

these are connected to the grid by HVDC transmission. By that date, there were seven HVDC-VSC offshore wind connection systems in operation and another three under construction. They are located in the area of the North Sea known as the German Bight.

In April 2015, TenneT (transmission system operator in the Netherlands and in a large part of Germany) commissioned SylWin1, currently the longest offshore grid connection in the world. SylWin1 connects the DanTysk, Butendiek, and Sandbank wind farms 70 kms off the coast of Sylt to the mainland. SylWin1, with an output of 864 MW, can feed large quantities of wind power into the German power grid via an HVDC-VSC system. To transport the energy generated by a total of 232 wind turbines to the onshore converter station in Büttel, a distance of 205 km must be bridged: 160 km of submarine cable and 45 km of land cable.[9]

6 Hydrogen Production from Offshore Wind

Hydrogen is a gas that can be easily produced using electrolysis and has several potential applications, ranging from energy source for transportation to being mixed into the natural gas grid, along with current applications in fuel refining and fertilizer production. Historically, hydrogen production is based on fossil fuels and emits a large amount of CO_2, however, in the last decades, significant advances have been made in electrolysis and renewable energy production, making the production of green hydrogen at a reasonable price point possible.

Furthermore, with governments pushing the reduction of carbon emissions and lowering the dependence on fossil fuels, the demand for green hydrogen has quickly rose and is expected to rise substantially in the coming years. With the help from incentives and policies, green hydrogen is being heavily investigated around the world with the objective of producing hydrogen without carbon emissions that with a small incentive can compete with traditional hydrogen production methods.

An electrolyser is a device that receives DC electricity and demineralized water and separates the hydrogen and oxygen atoms from the water molecule through a chemical reaction, generating high purity oxygen and hydrogen. While different technologies for electrolysers operate in slightly different ways, all have an anode and cathode that are separated by an electrolyte.

Currently, there are two technologies used in commercial applications for the production of hydrogen, Alkaline Electrolyser (AEL), and Proton Exchange Membrane Electrolyser (PEMEL). Another technology undergoing intense research and development is Solid Oxide Electrolyser (SOE), which promises high efficiencies and flexibility, at the cost of both high operating temperatures (500–1000 °C, varies according to the chemistry) and durability.

---

[9]https://www.tennet.eu/our-grid/offshore-projects-germany/sylwin1/.

Fuel cells are devices that use hydrogen to produce electricity, with the only by-products being water and heat. In recent years fuel cells have also experienced significant advancements and are starting to be used in commercial applications, like passenger cars, trucks, buses, and grid-connected dispatchable power plants. Since one of the reasons electrical grids are dependent on fossil fuels is the need to control power generation and renewable sources are intermittent (hydroelectric dams with reservoir provide some flexibility, but ultimately are dependent on rainfall upstream), the use of fuel cells can aid in reducing the use of fossil fuels in electricity generation.

Higher wind speeds and more consistent wind can be found offshore which leads to higher energy production per turbine installed, with the disadvantages being higher cost and technical challenges due to the rough sea conditions the equipment is subjected to. Considering that underwater pipeline installation is cheaper than electrical cables and that transport of a gas in a pipeline suffers smaller losses, a case can be made for the production of hydrogen offshore with pipelines to transport it to shore.

Two system configurations can be found, the first consists of an offshore wind farm, offshore electrolyser, and onshore hydrogen storage (Fig. 16), while in the second system the electrolyser is located onshore (Fig. 17). A fuel cell can be added in both systems to provide electricity in high-demand periods and act as frequency control for the grid.

For the first system, the electricity generated by the wind turbines travels a short distance to the electrolyser platform, where hydrogen is produced, compressed, and transported to shore on a pipeline. On the other hand, for the second system, the electricity is transmitted to shore by a traditional cable, where a choice can be made: sell the electricity directly to the grid or produce hydrogen. This is known as a hybrid system, where the operator can control the amount of power being sold to

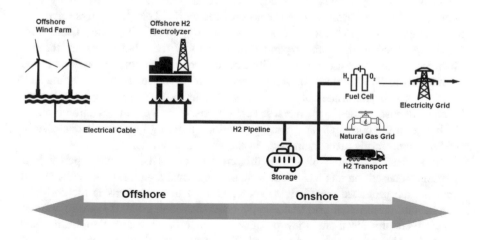

Fig. 16 Offshore electrolyser system

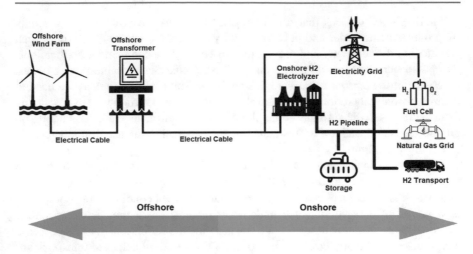

Fig. 17 Onshore electrolyser system

the grid or fed into the electrolyser, even being able to buy electricity from the grid to produce hydrogen during periods of extremely low electricity prices, which provides load flexibility to the grid operator as well. Since the source of the electricity powering the electrolyser is the wind farm, no carbon is emitted during the production of hydrogen.

7 Conclusions

Today wind power represents already a remarkable share of the electricity production, namely in Europe in other regions of the world. This situation is driven by the competitive productions costs shown by Wind Turbine Generators (WTG) grouped in the popular wind farms. Most of the wind farms are located onshore precisely because the competitive production costs apply to the wind farms located inland. The shortness of wind farms located in the sea is due not to the lack of wind, which is considerably higher and steadier offshore, but to the still high investment costs related to the harsh sea conditions. However, this situation is about to change in the coming years, many projections pointing to a visible decrease in offshore wind farms investment costs in the short/medium term.

While the background picture of this chapter was offshore wind power, our attention focused on the electrical systems that make possible to collect and transport the produced electricity to shore. One of these systems is the collector system, i.e., the internal grid that interconnects the offshore WTG one to another and delivers the electricity they produce to the offshore substation. A balance between costs and reliability is required. The current trend to increase the size of offshore wind farms (in 2019 the average size of the offshore wind farms installed in Europe was 600 MW) highlights the importance of keeping connected as many wind turbines as possible during an equipment fault. The conventional radial design shows poor reliability, so alternative designs, with redundancy, are being proposed. The available topological options have been reviewed in this chapter, from the low redundancy classical layouts to innovative designs with increased redundancy.

When designing the interconnection transmission system, one key aspect is to choose the right cross-section of the submarine cables. If the transmission is AC, one should not forget that submarine cables produce reactive power, due to increased capacitive effects characteristic of the cables (this effect is more pronounced in underground cables than in overhead lines). This feature of the cables must be considered when the sizing of the interconnection cables is to be addressed. An engineering model was proposed to guide the cables' cross-section selection. The model is based on information provided by the manufacturers in the product datasheet, namely the rating factors related to the effects of the laying depth, ground temperature and ground thermal resistivity on the adjusted nominal current carried by the cable.

Some power is lost in both the collector and the transmission systems. The determination of the losses is crucial to assess the technical and economic feasibility of the offshore wind park. For this purpose, we presented two models to compute the losses associated to an offshore wind farm. The first model is a temperature-dependent model that includes a factor to correct the nominal losses based on temperature effects. It is important to highlight that the correction factor depends on the maximum admissible temperature (for the nominal current), and ambient temperature. The second model is based on classical load flow concepts. It is assumed that the power generated by the park is known. The power flow model allows to compute the bus voltages and ultimately the current flowing in the cable; the losses depend on the square of the current.

When addressing the question of bringing to shore the electricity produced offshore, two options for the transmission system come to mind: High-Voltage Alternate Current (HVAC) and High-Voltage Direct Current (HVDC). As the receiving grid is AC, the straightforward answer is HVAC. However, the solution to this problem is not as simple as this. When the power to be transmitted and the distance to shore increase beyond a critical length, the HVAC option becomes unfeasible. This is due to the huge amounts of reactive power that are produced by the submarine cables that leave no space for the real power to be transmitted. Even when the transmitted power is smaller, and the distances are shorter, reactive power compensation systems are required. Among these compensation systems, the ones based on FACTS (Flexible AC Transmission System) devices, such as the SVC (Static Var Compensator) and the STATCOM (STATic Synchronous COMpensator), stand up. When the solution is going HVDC, the systems based on self-commutated Voltage Source Converters (VSC) are preferred, because they allow a total and independent control of real and reactive power, among other beneficial features.

Hydrogen can fulfil the role of energy storage and even as an energy carrier since it has a much higher energetic density than batteries and can be easily stored. Considering that offshore wind sector is facing significant growth and technical advances, CO^2 free hydrogen has the potential to be produced from offshore wind farms. The main components of the whole system, namely the electrolyser and the fuel cell, were presented and the possible configurations—offshore electrolyser versus onshore electrolyser—were compared and discussed.

8 Proposed Exercises

Problem OWES1

A 10 MW offshore wind park is connected to shore through an AC submarine cable, 240 mm², aluminium conductor, 45 kV rated voltage. Consider the wind park represented by an equivalent aggregate. The total length parameters of the cable, at maximum admissible temperature, are $R = 4.8090, X = 5.7491$. The park is generating 10 MW and 7.5 Mvar. The voltage at the onshore slack bus is 1.0 pu. Solve the problem in per unit (pu) in the following base: $S_b = 10\,\text{MVA}; V_b = 45\,\text{kV}$. Compute

1. The voltage magnitude (in pu) and angle (in °) at the offshore bus.
2. The active losses.
3. Repeat for $Q_G = 0\,\text{Mvar}$ and comment on the results.

Solution

(1)

$$Z_b = \frac{V_b^2}{S_b}; R = \frac{R_\Omega}{Z_b}; X = \frac{X_\Omega}{Z_b}; P_G = \frac{P_{MW}}{S_b}; Q_G = \frac{Q_{Mvar}}{S_b}; V_1 = \frac{V_{kV}}{V_b} = 1\,\text{pu}$$

$$V_2^4 - \left(2RP_G + 2XQ_G + V_1^2\right)V_2^2 + R^2\left(P_G^2 + Q_G^2\right) + X^2\left(P_G^2 + Q_G^2\right) = 0$$

$$a = 1; b = -1.0901; c = 0.0021$$

$$V_{21} = 1.0431 \bigvee V_{22} = 0.0444 \bigvee V_{23} = -1.0431 \bigvee V_{24} = -0.0444$$

The voltage is the positive solution closer to 1 pu.
Therefore, $V_2 = 1.0431\,\text{pu}$

$$\theta = \operatorname{asin}\left(\frac{XP_G - RQ_G}{V_1 V_2}\right) = 0.5811\,°$$

$$V_2 = 1.0431 e^{j0.5811°}\ \text{pu}$$

(2)

$$I = \left(\frac{S_{g2}}{V_2}\right)^* = \frac{P_{G2} - jQ_{G2}}{V_2^*} = 1.1983 e^{-j36.289°}\ \text{pu}$$

$$P_L = RI^2 = 0.0341\ \text{pu} = 0.341\ \text{MW} = 0.0341\%$$

The power injected in busbar 1 is

$$S_1 = V_1 I_{12}^* = V_1(-I)^* = -0.9659 - j0.7092\ \text{pu}$$

The active power is $-(P_G - P_L) = -0.9659\,\text{pu}.$
The reactive power is $-(Q_G - Q_L) = -(Q_G - XI^2) = -0.7092\,\text{pu}$
Note the minus sign in both the active and reactive power. This indicates that active and reactive power from the generator, discounted from the losses, is being injected into busbar 1.

(3)

$$V_2 = 1.0228 e^{j1.5906°}\ \text{pu}$$

$$P_L = 0.0227\ \text{pu} = 0.227\ \text{MW} = 0.0227\%$$

$$S_1 = S_{12} = V_1 I_{12}^* = V_1(-I)^* = -0.9773 + j0.02714\ \text{pu}$$

The losses decreased because the reactive power flowing in the line also decreased.

Note the plus sign of the reactive power indicating that busbar 1 is supplying the line reactive losses, so that 0 Mvar reach the generator.

Problem OWES2
A 10 MW offshore wind park is connected to shore through an AC submarine cable, 240 mm², aluminium conductor, 45 kV rated voltage. Consider the wind park represented by an equivalent aggregate. The total length parameters of the cable, at maximum admissible temperature, are $R = 0.023743\,\text{pu}, X = 0.028391\,\text{pu}$. The park is generating 10 MW and –7.5 Mvar. The voltage at the onshore slack bus is 1.0 pu. The voltage magnitude at the offshore busbar is 1.0014 pu. Compute the

reactive losses, in kvar, respectively. Solve the problem in per unit (pu) in the following base: $S_b = 10\,\text{MVA}; V_b = 45\,\text{kV}$.

$$\theta = \text{asin}\left(\frac{XP_G - RQ_G}{V_1V_2}\right) = 2.6443°$$

$$V_2 = 1.0014e^{j2.6443°}\,\text{pu}$$

$$I = \left(\frac{S_{g2}}{V_2}\right)^* = \frac{P_{G2} - jQ_{G2}}{V_2^*} = 1.12483e^{j39.514°}\,\text{pu}$$

Note that the current is capacitive because the generator is absorbing reactive power.

$$Q_L = XI^2 = 0.0442\,\text{pu} = 442\,\text{kvar}$$

The power injected in busbar 1 is

$$S_1 = S_{12} = V_1I_{12}^* = V_1(-I)^* = -0.9630 + j0.7942\,\text{pu}$$

The reactive power injected in busbar 1 is positive, meaning reactive power is being sent from busbar 1 to the offshore generator. Its value is

$$Q_{12} = -Q_{g2} + Q_L = 0.7942\,\text{pu}$$

In the onshore bus, reactive power must be sent so that 0.75 pu reach the offshore bus. So, the line reactive losses must be added to the 0.75 pu.

Problem OWES3

A 4 × 2.5 MW offshore wind park is connected to shore through an AC submarine cable, aluminium conductor, 20 kV rated voltage, 385 A rated current, 0.8 power factor. The length of the cable is 30 km. The cable parameters (@20 °C) are $R = 0.125\,\Omega/\text{km}, L = 0.0004\,\text{H/km}, C = 2.5 \times 10^{-7}\,\text{F/km}$, the temperature coefficient being $0.00403\,\text{K}^{-1}$. Consider that the ambient temperature is 15 °C and the maximum admissible temperature is 90 °C. Compute the temperature-dependent total losses (kW) in the interconnection cable when the wind park is operating at rated power.

Solution

$$R_{90} = R_{20}[1 + \alpha_T(\theta_{Lmax} - 20)] = 0.1603\,\Omega/\text{km}$$

$$P_{LN} = 3R_{90}I_N^2 = 71.265\,\text{kW/km}$$

$$I = \frac{P_N}{\sqrt{3}V_N pf} = 360.84\,\text{A}$$

$$c_\alpha = 1 + \alpha_T(\theta_U - 20) = 0.9798$$

$$\Delta\theta_{Lmax} = \theta_{Lmax} - \theta_U = 75\,^\circ\text{C}$$

$$v_\theta = \frac{c_\alpha}{c_\alpha + \alpha_T\Delta\theta_{Lmax}\left[1 - \left(\frac{I}{I_N}\right)^2\right]} = 0.9639$$

$$P_L = P_{LN}\left(\frac{I}{I_N}\right)^2 v_\theta = 60.340\,\text{kW/km}$$

$$P_{LT} = P_L l = 1810.2\,\text{kW}$$

Problem OWES4

A 4×2.5 MW offshore wind park is connected to shore through an AC submarine cable, aluminium conductor, 235 A rated current, 0.8 power factor. The length of the cable is 30 km. The cable resistance (@90 °C) is $R_{90} = 0.16026\,\Omega/\text{km}$, the temperature coefficient being $0.00403\,\text{K}^{-1}$. Consider that the ambient temperature is 15 °C and the maximum admissible temperature is 90 °C. When the wind park is operating at rated power, the temperature-dependent correction factor in the interconnection cable is 0.85851. Compute the rated voltage (kV) of the cable.

Solution

$$c_\alpha = 1 + \alpha_T(\theta_U - 20) = 0.9798$$

$$\Delta\theta_{Lmax} = \theta_{Lmax} - \theta_U = 75\,^\circ\text{C}$$

$$\left[1 - \left(\frac{I}{I_N}\right)^2\right] = \frac{c_\alpha(1 - v_\theta)}{\alpha_T\Delta\theta_{Lmax}v_\theta} = 0.5343 \xrightarrow{yields} I = 160.37\,\text{A}$$

$$V_N = \frac{P_N}{\sqrt{3}I pf} = 45\,\text{kV}$$

Small Hydro Plants

8

Abstract

Large hydropower plants use a renewable source and provide fast regulating capacity, which is of utmost importance for the stability of the power system. However, they have environmental impacts that should be considered. Small Hydro Plants (SHP) are hydropower plants with an installed capacity lower than 10 MW, therefore mitigating the environmental issues associated with large hydro. In this chapter, we introduce the main concepts related to the study of SHP, namely some design options—turbine and rated flow choice. The main objective of the chapter is to show how the electrical energy produced by an SHP can be computed. For this purpose, two models are presented. Firstly, a simplified model is offered, which is often used in the early stages of the SHP project design. Then, a more detailed model is introduced, providing the basis for the full project design of an SHP. Some examples of the intermediate and final required computations are given. In the final part of the chapter, a list of application problems is proposed, together with the guidelines of the solution.

1 Introduction

Hydropower plants are a valuable asset for every system operator. Since the prime mover is water, which furthermore is a storable resource, they can provide fast power regulating response, therefore being an important aid when it comes to matching the generation with the demand. Also, as this is a more and more decisive feature, hydropower uses a renewable resource. The drawback is that large hydro plants have a non-negligible environmental impact on the habitat's fauna and flora. Small Hydro Plants (SHP) use the same operating principle and show the same advantages; however, they do not have the said negative environmental issues, which makes them suitable for dispersed renewable production.

Table 1 Small Hydro Plants
classifications

	Capacity (MW)
Small hydro	<10
Mini-hydro	<2
Micro-hydro	<0.5

According to the standards, the following classifications apply (Table 1).

A drawing with the basic components of an SHP is shown in Fig. 1. Figure 2 shows the details of the powerhouse.

The water is taken at the intake and forwarded to the forebay. Then, a penstock (the hydraulic circuit) carries the water to the powerhouse where the potential and kinetic energies of the water is transformed into mechanical energy in a turbine and further converted into electrical energy using a generator. The water is finally returned to the river at the tailrace. Not all the incoming water can be used to produce electricity, a residual flow must be left in the river so that the river does not go completely dry.

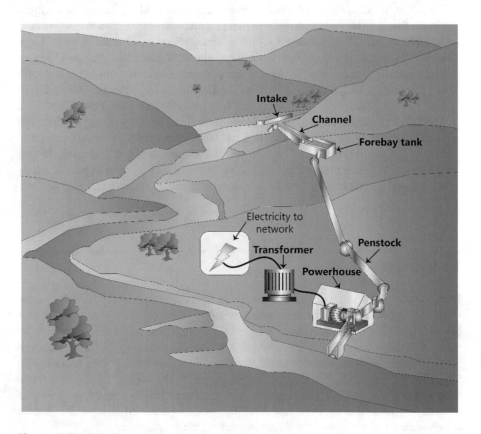

Fig. 1 A Small hydro plant

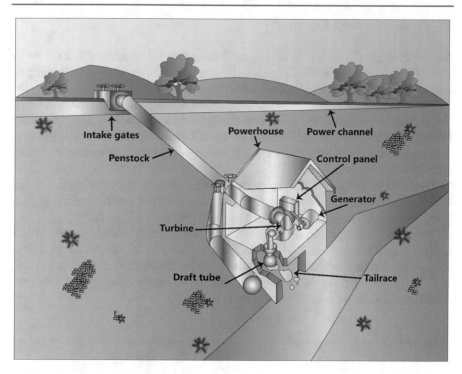

Fig. 2 Inside the powerhouse of a Small hydro plant

In this chapter, SHP are studied. The main objective of this chapter is to teach how to come up with an estimate for the electricity that can be generated by an SHP. For this purpose, two models are offered. We begin with a simplified model that makes several simplifications and assumptions to allow the electricity produced to be obtained through a straightforward equation. This model can be applied in preliminary studies to assess the potential of a particular site. Then, we propose a more detailed model that is applicable in more advanced stages of the SHP project and that involves the collaboration of several engineering branches.

The basis to apply both models is the Flow Duration Curve (a representation of the incoming flows by decreasing magnitude) and the selection of the appropriate turbine type (Pelton, Francis, Kaplan, Propeller), given the rated flow and the head. The turbine is the most relevant equipment of an SHP; therefore, its efficiency is of utmost importance. These topics are addressed beforehand.

2   Flow Duration Curve

Any assessment of a small hydro project begins with information about the water flow. This information is usually gathered in the form of a chronological flow times series, an example of which is depicted in Fig. 3.

The base time is usually a day. As so, the chronological flow time series represents the daily average incoming flow to a particular section of a river. The daily averages can be based on an average of several years. So, for instance, the time series point corresponding to January 1 is the daily average flow on that day taken as an average of several years.

Then, the flows are ordered by descending magnitude, from the highest to the lowest: the first point is the maximum flow, the second point is the second highest flow, and so on. We obtain the Flow Duration Curve (FDC) as shown in Fig. 4.

The information provided by the FDC is the number of days in the year a particular flow is exceeded. For instance, the maximum flow is exceeded for 0 days in the year; the flow equal to 2.5 m^3/s is exceeded for 50 days in the year.

If we divide all flows by the mean annual flow and the time by 365 days, we obtain the FDC in pu (per unit), as shown in Fig. 5.

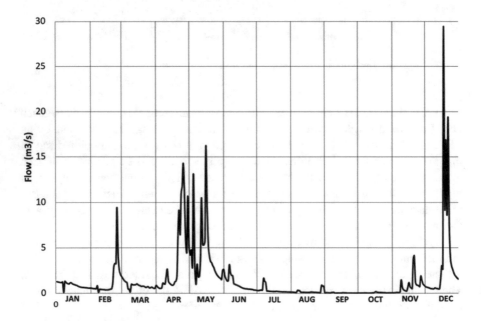

Fig. 3 Chronological flow time series

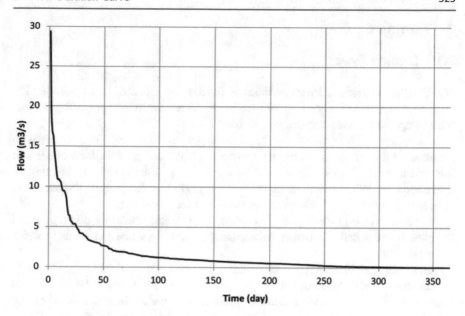

Fig. 4 Flow duration curve (FDC)

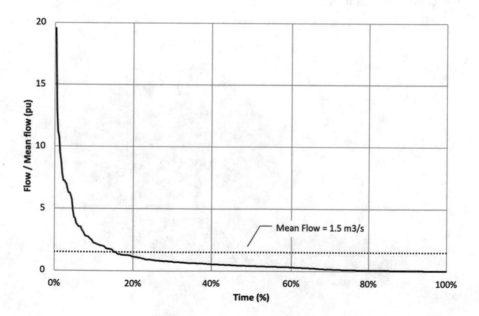

Fig. 5 Flow duration curve (FDC) in pu

3 Turbine Choice

3.1 Turbine Types

The detailed analysis of hydro turbines is outside the scope of this textbook. We offer hereafter some basics on hydro turbines. There are two types of hydro turbines: impulse turbines and reaction turbines.

In impulse turbines, the stator has nozzle jets, and the rotor is a wheel with spoon-shaped buckets, at atmospheric pressure. They are used for high heads and low flows. Examples of impulse turbines are Pelton, Turgo, and Banki-Mitchell hydro turbines, Pelton being the most known type. In Fig. 6, we show the operating principle of a Pelton turbine. Note the rotor and the nozzle jets.

As far as reaction turbines are concerned, in the stator, there is a distributor and the rotor is composed of a runner, where the pressure is not constant, and the water is accelerated.

There are mainly two types of reaction turbines: Francis, which are used in intermediate heads and flows, and Kaplan (or axial), used mainly in low heads and high flows. Kaplan turbines can have fixed runner blades, in this case, they are called propellers; on the other hand, they can have double regulation—distributor guide vanes and blades are regulated; or single regulation—only blades are regulated. A view of a Francis turbine is shown in Fig. 7, whereas in Fig. 8 we present details of Kaplan turbines.

Fig. 6 Rotor (right) and nozzle jets (left) of a Pelton turbine. *Source* Energy Education, www. energyeducation.ca

Fig. 7 View of a Francis turbine. *Source* commons. wikimedia.org

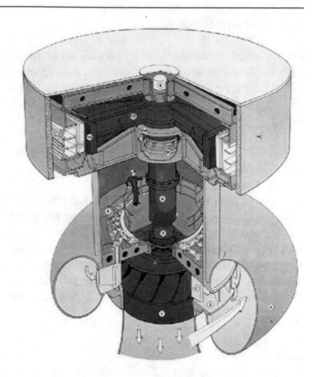

Fig. 8 Details of Kaplan turbines. Energy Education, *Source* www. energyeducation.ca

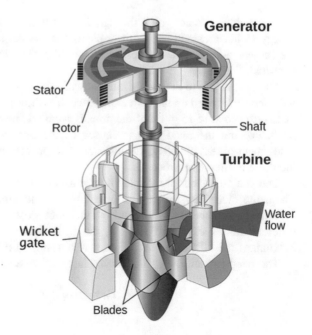

3.2 Turbine Efficiency

An important characteristic of hydro turbines is efficiency. The efficiency changes
with the flow, the maximum efficiency is usually obtained for the rated flow.[1] The
flatter the curve, the better, because it means that high efficiency is obtained for a
wide range of flow variation. Figure 9 shows typical efficiency curves for hydro
turbines.

3.3 Turbine Choice

The type of turbine depends both on the rated flow and on the head. The head is the
difference between the higher elevation and the lower elevation, i.e., the vertical
change in elevation between the head (reservoir) water level and the tailwater
(downstream) level. There are standard graphical tables that allow us to choose the
turbine type based on the knowledge of these two parameters. It goes without
saying that the rated flow and the head are the main project parameters that must be
known from the beginning. Figure 10 shows an example of a graphical table
usually used for the turbine choice.

4 Electrical Energy Yield—Simplified Model

In this chapter, we are going through a simplified model to compute the electrical
energy yield of a small hydropower plant. This model is used in the early stages of
the project and provides an estimation of the electricity production that can be
expected.

The first step is to define the rated flow. In the past, the project engineer selected
the rated flow based on his own experience, his choice usually laying in a rated flow
that is exceeded in 15–40% of the time. Nowadays, powerful computation tools are
available, and the rated flow is chosen as the one that maximizes an economic
assessment index, the NPV, for instance. The process of selecting the rated flow is
therefore an iterative process.

The first approach for the rated flow is usually to take it as the mean annual flow.
Then, the electricity production is computed and an NPV is obtained from the
investment and the operation and maintenance costs. The process is repeated for
different values of the rated flow, the final decision being the rated flow that
maximizes the NPV. In what follows, it is assumed that the rated flow is known.

The rated capacity, P_N, that can be installed in a small hydro plant is given by

$$P_N = \gamma Q_N H_b \eta_c \tag{4.1}$$

---

[1]We will discuss later how to choose the rated flow.

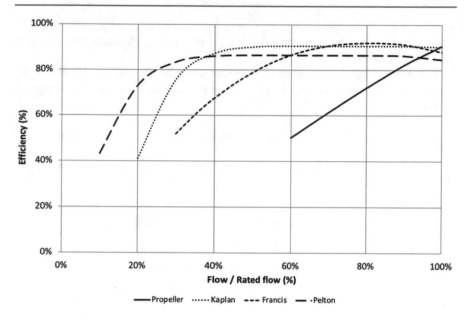

Fig. 9 Hydro turbines typical efficiency curves

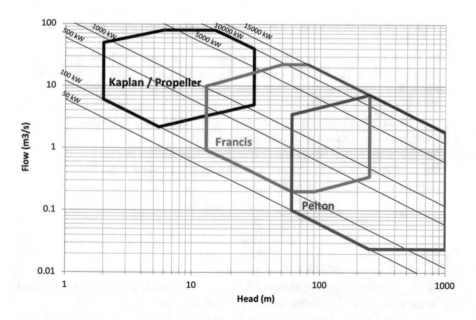

Fig. 10 Standard graphical table for turbine choice

where γ is the water-specific weight ($\gamma = \rho g = 9810\,\mathrm{N/m^3}, \rho = 1000\,\mathrm{kg/m^3}$), Q_N is the rated flow in $\mathrm{m^3/s}$, H_b is the gross head in m, and η_c is the overall efficiency of the power plant.

The global efficiency of a small hydro plant is usually around 70%. Therefore, an approximate equation for easily computing the rated capacity of a small hydro plant in kW is

$$P_N = 7Q_N H_b \tag{4.2}$$

From the relationship between power and energy, we can write that the annual electricity production is

$$E_a = \int P(t)dt = 9810 \int Q(t) h_u(t) \eta(t) dt \tag{4.3}$$

where we assume that all quantities in Eq. 4.1 can change in time. h_u is the useful head (gross head minus all the hydraulic losses) and η is the combined efficiency of the turbine, generator, transformer, and auxiliary equipment.

4.1 One Single Turbine

To begin with, we shall address the case of a small hydro plant equipped with a single turbine.

We target a simplified model. So, we assume that both the head and the efficiency are constant. The head is equal to the gross head and the overall efficiency is around 70%.

We have to take a closer look at the flow, which is impossible to assume as constant because it is apparent from the FDC that it changes. The process to take the flow variation into account will be described next.

The first step is to compute the operating area as shown in Fig. 11.

The rated flow, which is assumed to be known, is marked in the FDC. Only for the purpose of this simplified model, two factors are defined for each type of hydro turbine:

$$\alpha_1 = \frac{Q_{mT}}{Q_N}; \alpha_2 = \frac{Q_{MT}}{Q_N} \tag{4.4}$$

where Q_{MT} is the maximum turbined flow, Q_{mT} is the minimum turbined flow, and Q_N is the rated flow. The value of the two factors is defined in Table 2.

This means that for flows higher than the maximum turbined flow, the turbine is regulated to operate at maximum turbined flow, the excess water being wasted. For flows lower than the minimum turbined flow, the turbine is disconnected, because the efficiency is too low, the remaining water is not used for electricity production.

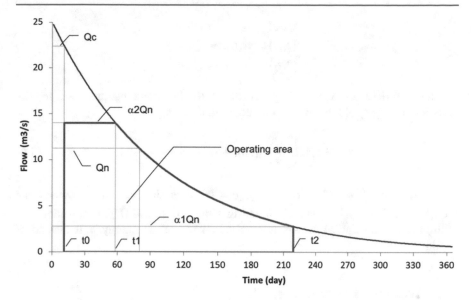

Fig. 11 Operating area computation. Q_C—flood flow; $\alpha_2 Q_n$—maximum turbined flow; Q_n—rated flow; $\alpha_1 Q_n$—minimum turbined flow; the flood flow is exceeded for t_0 days; the maximum turbined flow is exceeded for t_1 days; the minimum turbined flow is exceeded for t_2 days

Table 2 Factors α_1 and α_2 for each type of turbine; simplified model only

Turbine	α_1	α_2
Pelton	0.15	1.15
Francis	0.35	1.15
Double regulation Kaplan	0.25	1.25
Single regulation Kaplan	0.4	1.0
Propeller	0.75	1.0

In the operating area, as defined in Fig. 11, it is assumed that the turbine's efficiency is constant (please refer to Fig. 9). The operating area is further reduced by considering the flood flow, Q_c. For flows higher than the flood flow, it is considered that the turbine is disconnected due to lack of head. The flood flow is assumed to be known. The operating area is thereby the red area in Fig. 11. The available water outside this area is not used to produce electricity.

From the FDC, times t_0, t_1, and t_2 can be determined by knowing the flood flow, the maximum turbined flow, and the minimum turbined flow, respectively, as indicated in Fig. 11 and in Eq. 4.5. We recall that the rated flow is supposed to be known.

$$Q(t_0) = Q_c$$
$$Q(t_1) = \alpha_2 Q_N = Q_{MT} \tag{4.5}$$
$$Q(t_2) = \alpha_1 Q_N = Q_{mT}$$

The electricity production is proportional to the operating area. As so, the electrical energy yield in kWh can be computed through

$$E_a = 7 H_b \left(\int_{t_1}^{t_2} Q(t)dt + (t_1 - t_0)\alpha_2 Q_N \right) 24 \tag{4.6}$$

The quantity $\int_{t_1}^{t_2} Q(t)dt + (t_1 - t_0)\alpha_2 Q_N$ is the operating area. It is assumed that the global efficiency is around 70% and that the gross head, H_b, is constant. The multiplication by the 24 factors is to obtain kWh as an energy unit instead of kWday.

4.2 Two Equal Turbines

Let us now address the case of a small hydro plant equipped with two equally sized turbines. The objective is to find the differences relative to the one turbine case.

The turbines are equally sized. So, $Q_{N1} = Q_{N2} = Q_N/2$. In this situation, the clever way of operating the turbines is the following:

- For flows less than the rated flow of one turbine, let's say turbine 1, only turbine 1 is operating, with turbine 2 being disconnected.
- For flows higher than the rated flow of turbine 1, the two turbines are operating, the flow being equally split by the two turbines.

Mathematically, this translates into Table 3.

This is clever because it is better, from an efficiency point of view, to have one turbine at full capacity than two turbines at half capacity.

The maximum turbined flow is unchanged relative to the single turbine case, because

$$Q_{MT} = \sum_{i=1}^{2} \alpha_2 Q_{Ni} = \alpha_2 Q_N \tag{4.7}$$

Table 3 Two equal turbines operating mode

Q_j	Turbine 1	Turbine 2
$0 \leq Q_j < Q_N/2$	Q_j	0
$Q_N/2 < Q_j \leq \alpha_2 Q_N$	$Q_j/2$	$Q_j/2$

As for the minimum turbined flow, it is now different (actually, it is lower) from the single turbine case.

$$Q_{mT} = \alpha_1 Q_{N1} = \alpha_1 Q_{N2} < \alpha_1 Q_N \qquad (4.8)$$

In Fig. 12, this change in the minimum turbined flow is highlighted.
A new time t_3 is now defined as

$$Q(t_3) = \alpha_1 Q_{N1} = \alpha_1 \frac{Q_N}{2} \qquad (4.9)$$

The total energy produced in the case the small hydro plant is equipped with two equal turbines is

$$E_a = 7H_b \left(\int_{t_1}^{t_3} Q(t)dt + (t_1 - t_0)\alpha_2 Q_N \right) 24 \qquad (4.10)$$

Comparing Eqs. 4.6 and 4.10, it is apparent that for the same installed capacity, a small hydro plant equipped with two turbines will produce always more energy than when it is equipped with a single turbine. This is because a smaller turbine can

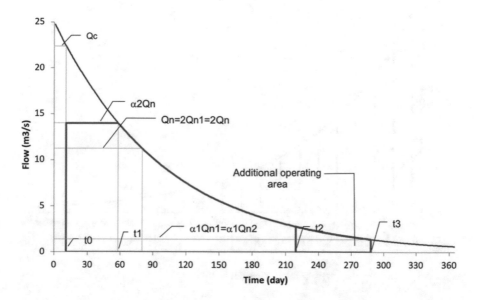

Fig. 12 Additional operating area computation for the two equal turbines case. Q_C—flood flow; $\alpha_2 Q_n$—maximum turbined flow (two turbines); Q_n—rated flow; $\alpha_1 Q_{n1} = \alpha_1 Q_{n2}$—minimum turbined flow (one turbine); the flood flow is exceeded for t_0 days; the maximum turbined flow is exceeded for t_1 days; the minimum turbined flow is exceeded for t_3 days

pick up lower flows. However, this does not mean that a two-turbine installation is better than a single turbine one. Indeed, for the same installed capacity, the cost of two equally sized turbines, with half the rated capacity each, is higher than the cost of a full-rated capacity single turbine. The same applies to the O&M costs. The best solution is provided by the NPV computation, which determines if the increased investment is compensated by increased production.

5 Electrical Energy Yield—Introduction to the Detailed Model

In a design phase, more detailed models are needed to compute the electrical energy yield of a small hydro plant. One of these models is going to be presented next.

In real life, the FDC is not described by an analytic equation. In fact, what is usually available is a curve described by a set of discrete points $(t_i, Q(t_i))$. An example is provided in Fig. 13, where 21 data points of the FDC are available.

The procedure in this method involves the computation of the power curve, i.e., the variation of the output power with time, and then compute the energy supplied as the area below the power curve.

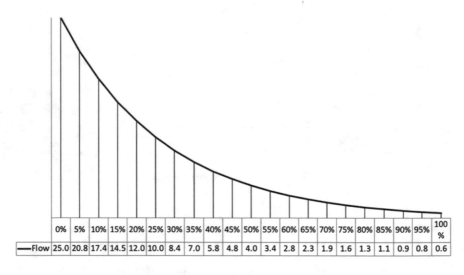

	0%	5%	10%	15%	20%	25%	30%	35%	40%	45%	50%	55%	60%	65%	70%	75%	80%	85%	90%	95%	100%
▬Flow	25.0	20.8	17.4	14.5	12.0	10.0	8.4	7.0	5.8	4.8	4.0	3.4	2.8	2.3	1.9	1.6	1.3	1.1	0.9	0.8	0.6

Time (%)

Fig. 13 Example of a usually available discrete FDC

5.1 Computing the Output Power as a Function of the Flow

The general equation to compute the output electrical power as a function of the incoming flow is

$$P(Q_i) = \gamma Q_{i\_u}\left[H_b - h_{hydr}(Q_{i\_u}) - h_{flood}(Q_i)\right]\eta_t(Q_{i\_u})\eta_g\eta_{transfo}(1 - p_{other}) \quad (5.1)$$

where

- P the electrical power output in W.
- Q_i the incoming flow at point i of the FDC in m^3/s.
- γ the water-specific weight 9810 N/m^3.
- H_b the gross head in m.
- h_{hydr} the hydraulic circuit losses for the used flow Q_{i_u} in m.
- Q_{i_u} the used flow in m^3/s, which is the flow that is forwarded to the turbine.
- h_{flood} the flood losses for the incoming flow Q_i in m.
- η_t the turbine efficiency for the used flow Q_{i_u}.
- η_g the generator efficiency, which is assumed to be constant.
- $\eta_{transfo}$ the transformer efficiency, which is assumed to be constant.
- p_{other} other losses, e.g., in the auxiliary services of the power station, which are assumed to be constant.

A residual flow must be left in the river so that the riverbed is not dry. We, therefore, define the available flow, Q_i', as

$$Q_i' = max(Q_i - Q_r; 0) \quad (5.2)$$

where Q_r is the residual flow. The residual flow must be known, normally it is about 3–5% of the rated flow.

In this model, it is assumed that the maximum turbined flow is equal to the rated flow. We recall that no α_1 and α_2 apply in the detailed model. Thus, the used flow, i.e., the flow that is forwarded to the turbine, is

$$Q_{i\_u} = min(Q_i'; Q_N) \quad (5.3)$$

The hydraulic circuit losses are losses inside the pipes (penstock) that carry the water from the higher elevation to the powerhouse. The hydraulic circuit head losses are proportional to the square of the flow; in the same way, the Joule losses are proportional to the square of the current, in electrical circuits. Accordingly, we can write

$$h_{hydr} = H_b p_{hydr}^{max}\left(\frac{Q_{i\_u}}{Q_N}\right)^2 \quad (5.4)$$

where p_{hydr}^{max} is the maximum value of the hydraulic circuit losses as a percentage of the gross head, H_b. Its value must be known, a common value being about 3–5%. Of course, the hydraulic circuit losses depend upon the used flow, as this is the turbined flow, the one that travels inside the penstock. These losses are maximum for the rated flow, which is the maximum turbined flow.

The flood flow losses are not related to the used flow, but instead to the incoming flow. In fact, a flood flow causes a reduction in the available head; this reduction is accounted for in this parameter. The flood losses are maximum for the maximum incoming flow and are null for the rated flow. It makes no sense to compute the flood losses for flows lower than the rated flow because these flows do not cause a flood. Hence, the flood flow losses can be written as

$$h_{flood} = h_{flood}^{max} \left(\frac{Q_i - Q_N}{Q_{max} - Q_N} \right)^2 \qquad (5.5)$$

where h_{flood}^{max} is the maximum value of the flood losses in m, which must be known or estimated somehow. Q_{max} is the maximum incoming flow (the maximum of the FDC).

The efficiency of the turbine is specific to each turbine. Analysing several efficiency curves for the different types of turbines, as the ones shown in Fig. 9, the following general-purpose model for the efficiency of any turbine was built:

$$\eta_t = \left\{ 1 - \left[\alpha \left| 1 - \beta \frac{Q_{i\_u}}{Q_N} \right|^{\chi} \right] \right\} \delta \qquad (5.6)$$

The efficiency of the turbine depends on the used flow as would be expected. The four parameters of Eq. 5.6 depend on the type of turbine as shown in Table 4.

Regarding Table 4, it can be seen that the Francis turbine efficiency depends on the useful head at rated flow, h'_u. The useful head, h_u, is defined as

$$h_u = H_b - h_{hydr}(Q_{i\_u}) - h_{flood}(Q_i) \qquad (5.7)$$

Table 4 Parameters of the efficiency model; h'_u: useful head at rated flow; n: number of nozzle jets

	Propeller	Kaplan	Francis	Pelton
α	1.25	3.5	1.25	$1.31 + 0.025n$
β	1.00	1.333	$1.1173{h'_u}^{0.025}$	$(0.662 + 0.001n)^{-1}$
χ	1.13	6.0	$3.94 - 11.7{h'_u}^{-0.5}$	$5.6 + 0.4n$
δ	0.905	0.905	0.919	0.864

Table 5 Minimum turbined flow

Turbine type	$\frac{Q_{mT}}{Q_N}$ (%)
Propeller	65
Kaplan	15
Francis	30
Pelton	10

The useful head at rated flow is

$$h'_u = h_u(Q_N) = H_b - h_{hydr}(Q_N) - h_{flood}(Q_N)$$

$$h'(u) = h_u(Q_N) = H_b - p_{hydr}^{max}H_b - 0 = H_b\left(1 - p_{hydr}^{max}\right) \qquad (5.8)$$

When the flow reduces, the efficiency of the turbine is also reduced. As far as the detailed model is concerned, a minimum turbined flow, Q_{mT}, is defined. For used flows lower than this minimum, the turbine is disconnected, because the efficiency is too low. Table 5 shows the minimum turbined flow as a percentage of the rated flow.

5.2 Installed Capacity

To compute the installed, or rated capacity, Eq. 5.1 is used as follows:

$$P_N = P(Q_N) = \gamma Q_N\left[H_b - h_{hydr}(Q_N)\right]\eta_t(Q_N)\eta_g\eta_{transfo}(1 - p_{other}) \qquad (5.9)$$

5.3 The Power Curve and Electrical Energy Yield

We recall that we assumed we had an FDC discretized in several points. For each point (t_i, Q_i), the output power, for each incoming flow Q_i, is computed through Eq. 5.1. In the FDC, to each incoming flow, Q_i, corresponds to a time in which it is exceeded. Therefore, a power curve can be obtained, which plots the evolution of the output power as a function of time.

It is well-known that the area below the power curve is the electricity produced by the power plant. Assuming an FDC discretization in equal time intervals of ΔT (in %), in n points, using the trapezoidal integration method, we can write

$$E_a = \sum_{k=1}^{n}\left(\frac{P_{k-1}(Q) + P_k(Q)}{2}\right)\Delta T(1 - p_{unav})8760 \qquad (5.10)$$

where p_{unav} are the losses due to the unavailability of either the power plants (e.g., maintenance) or the grid (e.g., failures or maintenance).

Let us look at an example that illustrates how the model works.

Example 5—1

Consider an SHP with a gross head of 6.35 m where the FDC is described by $Q(t) = 25exp\left(-\frac{t}{100}\right)$ *(t in days and Q in m^3/s). The maximum hydraulic circuit head losses are 4% and the maximum flood losses are 6.1 m. The efficiencies of the generator and transformer are 95% and 99%, respectively, and other losses can be fixed in 2%. The unavailability losses are estimated at 4%. The SHP is equipped with a double regulation Kaplan turbine. The rated flow is 11.25 m^3/s and the residual ecologic flow is 1 m^3/s. Compute the annual electricity yield of the SHP.*

Solution:

First, let us handle the flows: incoming flow (Q_i)*, available flow* (Q_i')*, and used flow* (Q_{i_u})*. The FDC is discretized in time intervals of 5% of the number of days in one year. The results are shown in* Fig. 14.

Then, let us handle the hydraulic losses—the hydraulic circuit head losses that depend upon the used flow, and the losses due to flood flow that depend on the incoming flow. We recall that it is $Q_N = 11.25\,\mathrm{m^3/s}$ *and* $Q_{max} = 25\,\mathrm{m^3/s}$ *(the maximum incoming flow). The results are presented in* Figs. 15 and 16.

We now proceed to the computation of the efficiencies of the SHP components—the Kaplan turbine, which depends on the used flow (the parameters are $\alpha = 3.5; \beta = 1.333; \chi = 6.0; \delta = 0.905;$ *see* Table 4*) and the generator,*

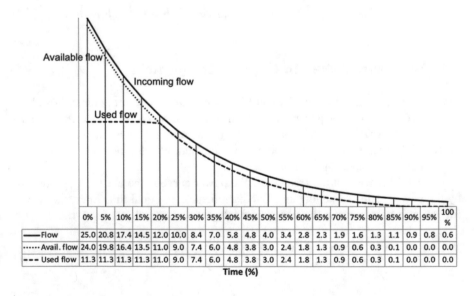

	0%	5%	10%	15%	20%	25%	30%	35%	40%	45%	50%	55%	60%	65%	70%	75%	80%	85%	90%	95%	100%
——Flow	25.0	20.8	17.4	14.5	12.0	10.0	8.4	7.0	5.8	4.8	4.0	3.4	2.8	2.3	1.9	1.6	1.3	1.1	0.9	0.8	0.6
······Avail. flow	24.0	19.8	16.4	13.5	11.0	9.0	7.4	6.0	4.8	3.8	3.0	2.4	1.8	1.3	0.9	0.6	0.3	0.1	0.0	0.0	0.0
- - - Used flow	11.3	11.3	11.3	11.3	11.0	9.0	7.4	6.0	4.8	3.8	3.0	2.4	1.8	1.3	0.9	0.6	0.3	0.1	0.0	0.0	0.0

Time (%)

Fig. 14 Incoming, available, and used flows in a discretized FDC (Example 5—1)

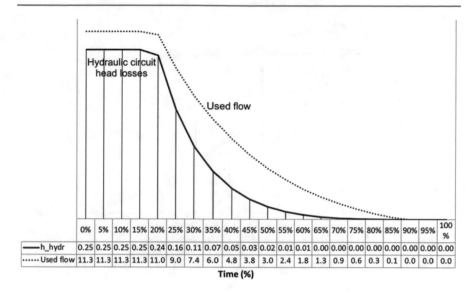

	0%	5%	10%	15%	20%	25%	30%	35%	40%	45%	50%	55%	60%	65%	70%	75%	80%	85%	90%	95%	100%
—h_hydr	0.25	0.25	0.25	0.25	0.24	0.16	0.11	0.07	0.05	0.03	0.02	0.01	0.01	0.00	0.00	0.00	0.00	0.00	0.00	0.00	0.00
······Used flow	11.3	11.3	11.3	11.3	11.0	9.0	7.4	6.0	4.8	3.8	3.0	2.4	1.8	1.3	0.9	0.6	0.3	0.1	0.0	0.0	0.0

Time (%)

Fig. 15 Hydraulic circuit head losses (Example 5—1)

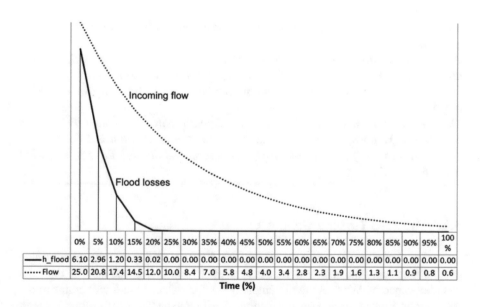

	0%	5%	10%	15%	20%	25%	30%	35%	40%	45%	50%	55%	60%	65%	70%	75%	80%	85%	90%	95%	100%
—h_flood	6.10	2.96	1.20	0.33	0.02	0.00	0.00	0.00	0.00	0.00	0.00	0.00	0.00	0.00	0.00	0.00	0.00	0.00	0.00	0.00	0.00
······Flow	25.0	20.8	17.4	14.5	12.0	10.0	8.4	7.0	5.8	4.8	4.0	3.4	2.8	2.3	1.9	1.6	1.3	1.1	0.9	0.8	0.6

Time (%)

Fig. 16 Flood losses (Example 5—1)

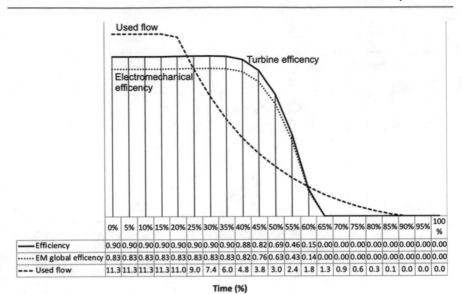

Fig. 17 Efficiencies of the components (Example 5—1)

transformer, and other losses, which are constant. Figure 17 *displays the obtained results.*

We note that for used flows lower than 15% of the rated flow (Table 5), *the Kaplan turbine is disconnected because the minimum turbined flow is attained.*

Now, for each point, we can build the power curve using Eq. 5.1, as depicted in Fig. 18.

The annual electrical energy yield is now straightforwardly computed through the area below the power curve, using Eq. 5.10. The obtained result is 1502 MWh.

The rated (installed) capacity of the SHP is determined through Eq. 5.9, the result being 558.5 kW.

5.4   The Two Turbine Generator Case

Let us now analyse the case in which the SHP is equipped with two turbine generator groups, the installed capacity of each one being one-half of the total installed capacity. We will take Example 5—1 for reference and will compare the case of one Kaplan turbine (as in Example 5—1) versus two Kaplan turbines with the half-rated flow. Let us focus, for example, at the point in which the used flow is $Q_{50} = 3.03\,\mathrm{m^3/s}$. If one Kaplan turbine is installed, the efficiency would be

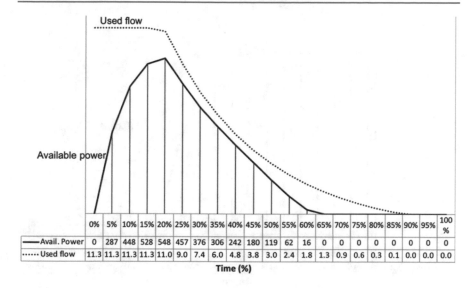

	0%	5%	10%	15%	20%	25%	30%	35%	40%	45%	50%	55%	60%	65%	70%	75%	80%	85%	90%	95%	100%
—Avail. Power	0	287	448	528	548	457	376	306	242	180	119	62	16	0	0	0	0	0	0	0	0
⋯⋯ Used flow	11.3	11.3	11.3	11.3	11.0	9.0	7.4	6.0	4.8	3.8	3.0	2.4	1.8	1.3	0.9	0.6	0.3	0.1	0.0	0.0	0.0

Time (%)

Fig. 18 Output power curve (Example 5—1)

$$\eta_{t1} = \left\{ 1 - \left[3.5 \left| 1 - 1.333 \frac{3.03}{11.25} \right|^6 \right] \right\} 0.905 = 0.686 \qquad (5.11)$$

If two half-sized Kaplan turbines are installed, the rated flow of each one is $Q_{N1} = Q_{N2} = 5.625 \, \text{m}^3/\text{s}$. For flows lower than this, only one turbine is operating, its efficiency being

$$\eta_{t2} = \left\{ 1 - \left[3.5 \left| 1 - 1.333 \frac{3.03}{5.625} \right|^6 \right] \right\} 0.905 = 0.903 \qquad (5.12)$$

As one can see, the efficiency is much higher in the second case than in the first. If we repeat the very same process for the other flows, we will obtain the graphic depicted in Fig. 19.

For flows larger than half the rated flow, the efficiency is equal, whether we have one or two turbines. For lower flows, only one of the two turbines is operating and with much higher efficiency than in the case where only a single turbine is installed.

Furthermore, more water is used to produce energy because one of the two turbines that is operating is disconnected for lower flows (its rated flow is lower). Of course, this will increase the annual energy yield both because of this and because the global efficiency is higher. In the case we have been studying, the annual produced electricity increases from 1502 to 1602 MWh, the total installed capacity remaining the same. This increase in the annual energy yield must be confronted with the increase in investment and O&M costs: two half-sized turbines are more expensive than a single full-rated turbine.

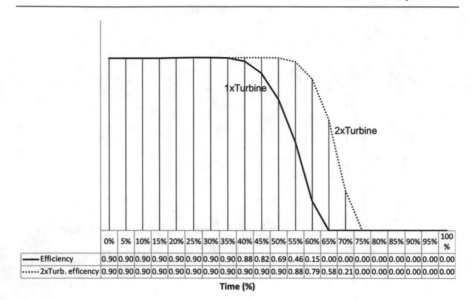

	0%	5%	10%	15%	20%	25%	30%	35%	40%	45%	50%	55%	60%	65%	70%	75%	80%	85%	90%	95%	100%
——Efficiency	0.90	0.90	0.90	0.90	0.90	0.90	0.90	0.90	0.88	0.82	0.69	0.46	0.15	0.00	0.00	0.00	0.00	0.00	0.00	0.00	0.00
······2xTurb. efficency	0.90	0.90	0.90	0.90	0.90	0.90	0.90	0.90	0.90	0.90	0.90	0.90	0.88	0.79	0.58	0.21	0.00	0.00	0.00	0.00	0.00

Time (%)

Fig. 19 Efficiency of one turbine versus two turbines (Example 5—1)

6   Conclusions

The Small Hydro Plants (SHP), defined as hydroelectric power plants with an installed capacity below 10 MW, were addressed in this chapter. Two models—a simplified one and a more detailed one—were presented to allow the computation of the electrical energy that can be produced in a given year. Both models are based on the knowledge of the flow duration curve, in which the incoming flows are ordered by decreasing magnitude as a function of time, on a yearly basis.

In the simplified model, the SHP was assumed to be described by a global efficiency—of the hydraulic circuit, turbine, and generator—which was assumed to be constant. Considering the standard characteristics of each type of turbine that equips the plant, an operating area was marked on the flow duration curve, and it was shown that the energy produced is proportional to that area.

Regarding the detailed full model, specific sub-models were presented for assessing the individual performance of each component of the plant. Given its crucial importance, particular emphasis was placed on the hydro turbine model, which was detailed for each type of turbine—Pelton, Francis, Kaplan, and Propeller. Furthermore, appropriate models were introduced for the calculation of the losses in the hydraulic circuit and for the losses due to flood flows.

The two models—simplified and detailed—were extended to the case where the SHP is equipped with two equal turbine generator groups. It was shown that, in this case, an increase in the produced energy is always obtained. However, the surplus energy benefit must be evaluated in relation to the respective over cost.

Table 6 Flow duration curve for Problem SHP1

t (%)	0%	20%	40%	60%	80%	100%
Q (m³/s)	20	4	1	0.5	0.1	0.01

7 Proposed Exercises

Problem SHP1

The flow duration curve from a river is given in Table 6.

In this river, in a place where the gross head is 100 m, it is intended to study the installation of an SHP equipped with a 3 MW Francis turbine and rated flow equal to 4 m³/s. The SHP is going to be operated in the following way. For the flows that are exceeded in 20% of the year, the turbine is operated at constant rated power; for the flows lower than the ones that are exceeded in 40% of the year, the power plant is disconnected, the incoming flows being used for other purposes. Use the simplified model to compute

(1) The annual energy produced.
(2) The annual energy not produced due to the adopted operation mode.

Solution

(1)

$$\eta_C = \frac{P_N}{9.81 Q_N H_b} = 76.45\%$$

$$E_a = 9.81 \eta_C H_b \left(20\% Q_N + \frac{40\% - 20\%}{2} (Q_N + Q_{40\%}) \right) \times 365 \times 24 = 8541 \text{ MWh}$$

(2)

$$A_1 = \frac{20\% - 0\%}{2} (Q_{0\%} + Q_N) - 20\% Q_N = 1.60$$

$$A_2 = \frac{60\% - 40\%}{2} (Q_{60\%} + Q_{40\%}) + \frac{80\% - 60\%}{2} (Q_{80\%} + Q_{60\%}) + \frac{100\% - 80\%}{2} (Q_{100\%} + Q_{80\%}) = 0.22$$

$$E_{np} = 9.81 \eta_C H_b (A_1 + A_2) \times 365 \times 24 = 11,964 \text{ MWh}$$

Problem SHP2

The flow duration curve of a river can be approximated by the following expression, where t is in day and Q in m^3/s:

$$\begin{cases} Q(t) = -1.5\ln t + 8; \, 1 \le t \le 207 \\ \qquad Q(t) = 0; \, 208 \le t \le 365 \end{cases}$$

It is foreseen to install a Small Hydro Plant (SHP) in a river place where the gross head is 350 m. Assume that (i) the nominal flow is equal to the average flow; (ii) the SHP will stop for 15 days due to flood flow; the SHP is to be equipped with one Pelton turbine. Remember that $\int \ln x \, dx = x \ln x - x$. Use the simplified model to compute

(1) The SHP nominal capacity.
(2) The electrical energy produced in an average year.

Solution

(1)

$$Q_{avg} = Q_N = \frac{1}{365} \int_1^{207} -1.5\ln t + 8 \, dt = \frac{1}{365} [-1.5(t\ln t - t) + 8t]_1^{207}$$
$$= 0.8252 \, m^3/s$$

$$P_N = 7 Q_N H_b = 2021.69 \, \text{kW}$$

The closest rated power in the datasheet is

$$P_N = 2000 \text{kW}$$

We should now correct the rated flow.

(2)

$$Q_N = \frac{P_N}{7 H_b} = 0.8163 \, m^3/s$$

$$t_0 = 15 \, \text{day}$$

For a Pelton turbine:

$$\alpha_1 Q_N = 0.1224 \, m^3/s \xrightarrow{yields} t_2 = 190.89 \, \text{day}$$

$$\alpha_2 Q_N = 0.9388 \, m^3/s \xrightarrow{yields} t_1 = 110.77 \, \text{day}$$

$$E_a = 7 H_b \left(\int_{t_1}^{t_2} Q(t) dt + (t_1 - t_0)\alpha_2 Q_N \right) 24 = 7613 \, \text{MWh}$$

Problem SHP3

Consider a river where the maximum registered flow equals 50 m³/s. The flow duration curve may be represented by a straight line. The river is dry for 115 days per year. A small hydro plant is to be installed in this river, in a location where the gross head is 40 m. The capacity of this power plant equals 6 MW, and the nominal flow is 20 m³/s. The turbines to be installed are two identical double regulated Kaplan turbines with a nominal capacity of 3 MW each. Assume that the overall efficiency of the power plant equals 0.75. The flood volumetric flow equals 40 m³/s. Using the simplified model, compute the expected annual energy produced.

Solution

Given: $Q_{max}; t_{dry}; H_b; P_N; Q_N; 2 \times KAPLAN(DR); \alpha_1; \alpha_2; \eta_G; Q_c$

The FDC is a straight line:

$$m = \frac{Q_{max} - 0}{0 - (365 - t_{dry})} = -0.2$$

$$b = Q_{max} = 50$$

$$Q(t) = mt + b$$

$$Q_{MT} = \alpha_2 Q_N = 25 \, \text{m}^3\text{s}^{-1}$$

$$Q_{mT} = \alpha_1 \frac{Q_N}{2} = 2.5 \, \text{m}^3\text{s}^{-1}$$

$$Q_{MT} = Q(t_1) = mt_1 + b \leftrightarrow t_1 = 125 \, \text{day}$$

$$Q_{mT} = Q(t_3) = mt_3 + b \leftrightarrow t_3 = 237.5 \, \text{day}$$

$$Q_c = Q(t_0) = mt_0 + b \leftrightarrow t_0 = 50 \, \text{day}$$

$$A_1 = (t_1 - t_0)Q_{MT} = 1875 \, \text{m}^3\text{s}^{-1}\text{day}$$

$$A_2 = (Q_{MT} + Q_{mT})\frac{(t_3 - t_1)}{2} = 1546.88 \, \text{m}^3\text{s}^{-1}\text{day}$$

$$E_a = 9.81\eta_G H_b(A_1 + A_2) \times 24 \times 10^{-3} = 24,169 \, \text{MWh}$$

$$h_a = \frac{E_a}{P_N} = 4028 \, \text{h}$$

Problem SHP4

Take again Problem SHP but now use the detailed model. Consider a single Kaplan turbine and take note of the additional data:

- Maximum hydraulic circuit head loss = 5%.
- Maximum head loss under flood conditions = 10 m.
- Reserved flow = 0.
- Efficiencies of the generator and transformer = 95% and 99%.
- Other electrical losses = 2%.
- Forced outage rate = 4%.

Compute the annual energy produced using only the following points: maximum flow (50 m³/s—Point 1), rated flow (20 m³/s—Point 2), and minimum flow (15% of the rated flow—Point 3).

Solution

We will solve for the rated flow $Q_N = 20 \, \text{m}^3/\text{s} \overset{yields}{\longrightarrow} t_N = 150 \, \text{day}$.

$$h_{hidr2} = H_b p_{hidr}^{max} \left(\frac{Q_N}{Q_N} \right)^2 = 2 \, \text{m}$$

$$h_{flood2} = h_{flood}^{max} \left(\frac{Q_N - Q_N}{Q_{max} - Q_N} \right)^2 = 0 \, \text{m}$$

$$h_{useful2} = H_b - h_{hidr2} - h_{flood2} = 38 \, \text{m}$$

$$\eta_{t2} = \left\{ 1 - \left[\alpha \left| 1 - \beta \frac{Q_N}{Q_N} \right|^{\chi} \right] \right\} \delta = 0.901$$

Kaplan turbine: $\alpha = 3.500; \beta = 1.333; \chi = 6.000; \delta = 0.905$

$$\eta_2 = \eta_{t1} \eta_{gen} \eta_{transfo} (1 - p_{other}) = 0.830$$

$$P_2 = P(Q_N) = P_N = \gamma Q_N h_{useful2} \eta_2 = 6189.10 \, \text{kW}$$

$$P_1 = P(Q_{Max}) = \gamma Q_N h_{useful1} \eta_1 = 4560.39 \, \text{kW}$$

$$P_3 = 80.88 \, \text{kW}$$

$$E_1 = \frac{P_1 + P_2}{2} \times (150 - 0) \times 24 \times (1 - p_{unav}) \times 10^{-3} = 18,575.105 \, \text{MWh}$$

$$E_2 = 6,139.561 \, \text{MWh}$$

$$E_a = E_1 + E_2 = 24,715 \, \text{MWh}$$

Table 7 Flow duration curve for Problem SHP5	t (%)	0	25	50	75	100
	Q (m³/s)	33	1.76	1.64	1.59	1.48

Problem SHP5

A Small Hydropower Plant (SHP) is to be installed in a location where the gross head equals 146 m. The flow duration curve is shown in Table 7.

The maximum head loss due to the hydraulic circuit equals 5% of the gross head under nominal flow conditions. The maximum head loss under flood conditions equals 5 m. The reserved flow due to environmental conditions equals 0.25 m³/s. The efficiencies of the generator and transformer are 95% and 99%, respectively. The auxiliary services of the SHP absorb 2% of the net electrical power. The forced outage rate of the SHP is 4%. A Francis turbine is to be installed in this SHP. An assessment of the economic feasibility of this project is requested. Two options are under assessment:

Option A:

- 1xFrancis turbine with a nominal flow equal to 4 m³/s.
- Investment cost: 1400 €/kW, totally accomplished in t = 0.
- Maintenance cost: annually, 3% of total investment.

Option B:

- 1xFrancis turbine with a nominal flow equal to 2 m³/s.
- Investment cost: 2000 €/kW, totally accomplished in t = 0.
- Maintenance cost: annually, 4% of the total investment.

The average value of the Feed-In Tariff (FIT) equals 70 €/MWh. Compute

(1) The annual energy produced, the installed capacity, and the utilization of the installed capacity for options A and B.
(2) The net present value considering a 15-year lifetime of the plant and a discount rate of 7%, for options A and B.

Solution

Given: $H_b; FDC; p_{hidr}^{max}; h_{flood}^{max}; Q_r; \eta_{gen}; \eta_{transfo}; P_{other}; P_{unav}; 1 \times FRANCIS; Q_N; I_{01}; c_{O\&M}; FiT; n; a$

We will exemplify the first two points of the FDC in Option B: $Q_N = 2\,\text{m}^3\text{s}^{-1}$.

$$t_1 = 0\%; Q(t_1) = Q_1 = 33\,\text{m}^3\text{s}^{-1}$$

$$Q_{avail} = Q_1' = Q_1 - Q_r = 32.75\,\text{m}^3\text{s}^{-1}$$

$$Q_{used} = Q_{1\_u} = Q_N = 2\,\text{m}^3\text{s}^{-1}$$

$$h_{hidr1} = H_b p_{hidr}^{max}\left(\frac{Q_{1\_u}}{Q_N}\right)^2 = 7.30\,\text{m}$$

$$h_{flood1} = h_{flood}^{max}\left(\frac{Q_1 - Q_N}{Q_{max} - Q_N}\right)^2 = 5\,\text{m}$$

$$h_{useful1} = h_{u1} = H_b - h_{hidr1} - h_{flood1} = 133.7\,\text{m}$$

$$\eta_{t1} = \left\{1 - \left[\alpha\left|1 - \beta\frac{Q_{1\_u}}{Q_N}\right|^{\chi}\right]\right\}\delta = 0.896$$

Francis turbine: $h_u' = h_u(Q_N) = H_b - h_{hidr}(Q_N) - 0 = 138.7\,\text{m}; \alpha = 1.25;$
$\beta = 1.1173 h_u'^{0.025} = 1.264; \chi = 3.94 - 11.7 h_u'^{-0.5} = 2.947; \delta = 0.919$

$$\eta_1 = \eta_{t1}\eta_{gen}\eta_{transfo}(1 - p_{other}) = 0.83$$

$$P_1 = P(Q_1) = \gamma Q_{1\_u} h_{u1} \eta_1 = 2167.10\,\text{kW}$$

$$P_2 = P(Q_2) = \gamma Q_{2\_u} h_{u2} \eta_2 = 1779.43\,\text{kW}$$

$$E_1 = \frac{P_1 + P_2}{2} \times 25\% \times 8760(1 - p_{unav}) \times 10^{-3} = 4148.59\,\text{MWh}$$

$$E_{aB} = \sum_{k=1}^{4} E_k = 14,310.48\,\text{MWh}$$

$$P_{NB} = P(Q_N) = \gamma Q_N h_u(Q_N)\eta(Q_N) \times 10^{-3} = 2.248\,\text{MW}$$

$$h_{aB} = \frac{E_{aB}}{P_{NB}} = 6365\,\text{h}$$

Results for option A

$$E_{aA} = 14,058.70\,\text{MWh}$$

$$P_{NA} = 4.496\,\text{MW}$$

$$h_{aA} = 3127\,\text{h}$$

Economic analysis

$$NPV_B = (FiT \times E_{aB} - c_{O\&MB}I_{01B}P_{NB})k_a - I_{01B}P_{NB} = 2,989,340\,\text{€}$$

$$NPV_A = (FiT \times E_{aA} - c_{O\&MA}I_{01A}P_{NA})k_a - I_{01A}P_{NA} = 948,391\,\text{€}$$

Combined Heat and Power

9

Abstract

Combined Heat and Power (CHP) is the simultaneous productions of electricity and heat from the combustion of a single fuel. CHP may be renewable if renewable fuels (biomass, biofuels,...) are used. In general, it is not renewable. A typical CHP configuration shows a natural gas fired Internal Combustion Engine (ICE). Despite, in general, not being renewable, CHP portrays a more efficient way of producing energy. CHP avoids the much less efficient double system to separately produce heat and electricity. CHP recycles the available heat not converted into electricity in a useful local heating application. In this Chapter, we introduce the main technologies commonly used in CHP installations, from the conventional (ICE, gas, and steam turbines) to the emerging ones (fuel cells, micro-turbines). Heat exchangers are important components of CHP system as they allow to obtain useful heat. These components are addressed, and a model is introduced to compute the different temperatures involved in the heat transfer process (log mean temperature difference model). Finally, a model for the technical and economic assessment of CHP systems is proposed and the main assessment indexes are outlined. The detailed computation of the Energy Utilization Factor and the fuel variable cost is offered. In the final part of the chapter, a list of application problems is proposed, together with the guidelines of the solution.

1 Introduction

Nowadays, the necessity of producing and consuming energy in an efficient way is becoming more and more apparent. Energy efficiency measures can be identified in several sectors. For instance, in the domestic sector at the households level (isolation, climate, lighting, appliances, …), construction sector (location, orientation, building quality, windows, natural ventilation, shadows, painting, renewables' use, …), recovery of waste, transportation sector, industrial and services sector, etc. Talking about the industrial and services sector, the use of Combined Heat and Power (CHP), also called cogeneration, is a very relevant energy efficiency measure.

The first question to be asked is "What is CHP?". CHP is the simultaneous generation of multiple forms of useful energy, most commonly, electricity and heat, from a single source of primary energy. The primary source can be either a fossil fuel, usually natural gas, or diesel, or a renewable fuel, for instance, biomass or biogas.

This means that, although it can be, in general CHP is not a renewable energy. However, even when it is not a Renewable Energy Source (RES), it is always a way of producing energy in a more efficient way, and therefore, is often coming alongside with RES. This justifies the inclusion of a chapter about CHP in this textbook.

The 2nd Law of Thermodynamics states that it is impossible to convert all the heat stored in a thermal source into electricity. Therefore, when producing electricity from a thermal source, a part of the heat is converted into electricity and the remaining, excluding the heat losses, is available to be used. We say we have a CHP plant when it is possible to find a useful application for the non-converted heat. We highlight that the heat is freely available to be used.

The Combined Cycle Gas Turbine (CCGT) power plants are an example of a CHP process. In this scheme, gas turbines are used together with steam turbines. The exhausted gases of the gas turbines are used to produce steam that runs the steam turbines, therefore, producing additional electricity (see Fig. 1). The overall efficiency of CCGT systems is about 55%, much higher than the average efficiency of conventional steam-only power plants (coal-fired, for instance) that is about 35%.

CHP can also go renewable if renewable fuels are used.[1] There are three categories of renewable fuels:

- Solid biomass
- Liquid biofuels
- Gaseous biofuels

These fuels are considered carbon neutral, and therefore, renewable in the sense that they only emit the CO_2 they absorbed while growing. It is considered that the CO_2 emissions associated with cultivation, processing, transportation, etc., are considerably lower than those from fossil fuels.

Solid biomass fuel refers to non-fossilized carbon-based solid materials derived from plant or animal matter. Examples of solid biomass fuels include: commercial-grade wood fuels (such as clean woodchips, logs, and wood pellets),

---

[1]For more information about renewable CHP consult, for instance, a report from the Department for Business, Energy & Industrial Strategy of the UK government, available at https://assets. publishing.service.gov.uk/government/uploads/system/uploads/attachment_data/file/345189/Part_ 2_CHP_Technology.pdf.

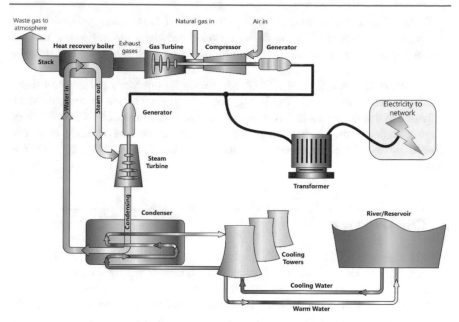

Fig. 1 CCGT scheme

agricultural residues, waste wood, straw, milling residues, pruning, and sewage treatment residues. Solid biomass fuels can be utilized for direct combustion or converted into a secondary gaseous or liquid renewable fuel via a fuel conversion process.

Manufactured liquid biofuels such as bioethanol and biodiesel are produced primarily for use as transport fuel by processing solid biomass via industrial conversion processes. The production of manufactured liquid biofuels is highly energy intensive and is not favoured for use as fuel in CHP.

The main gaseous biofuels are the biogas, which is produced by the decomposition of organic matter by anaerobic bacteria in anaerobic digesters and landfills, and synthesis gas, which is produced from solid biomass via advanced thermal conversion processes known as gasification and pyrolysis.

The main difference between renewable and conventional CHP is the combustion process allowing the heat to be obtained. In this textbook, we are interested in other CHP aspects rather than in the combustion process, which is outside the scope of this textbook. We will introduce the main technologies currently used in CHP installations, namely the Internal Combustion Engines (ICE), encompassing the spark-ignition type and the compression-ignition type, the gas turbines, and the steam turbines. Also, some emerging technologies, the micro-turbines and the fuel cells (with special emphasis on Alkaline and Proton Exchange Membrane technologies) are going to be presented.

To turn the available exhaust heat into a useful local heat application, heat exchangers are required. The main heat exchanger types are reviewed, and the known log mean temperature difference model is used to study the heat transfer process.

The technical assessment of a CHP installation is commonly performed using a global efficiency index called the Energy Utilization Factor, whose computation is addressed in this chapter. Moreover, the economic assessment of a CHP installation is introduced, offering a detailed model to compute the fuel variable cost, distributed by the three output components of the installation: heat, cooling, and electricity costs.

2 CHP Overview

As said before, in a CHP power plant, electricity and heat are produced. The use of the electricity product is apparent: it is used in a bunch of applications whose listing is needless. As for the use of the heat, it is less obvious, since contrariwise to the electricity, its transmission is difficult. Therefore, the heat is to be used locally, for instance, in industrial factories with process heating needs (chemical, ceramic, paper, …), or in buildings (hospitals, hotels, and shopping centres) requiring heat for heating and cooling purposes.

This last sentence leads us to the concept of Combined Heat, Cooling and Power (CHCP), also called trigeneration. The heat recovered in a CHCP power plant, or a part of it, can be used to produce cooling, using absorption chillers. The study of the thermodynamic cycles related to the absorption chillers is outside the scope of this textbook. For now, what matters is that an absorption chiller is a thermal machine able to transform heating into cooling. Figure 2 shows a CHCP diagram, where we

Fig. 2 CHCP diagram

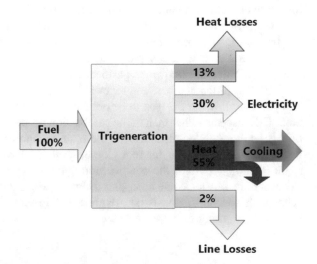

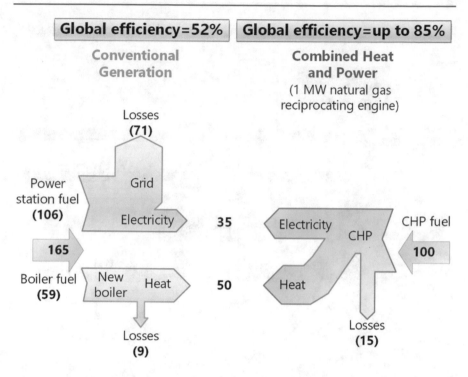

Fig. 3 Efficiencies of a CHP system and a separate generator + boiler system

can see that the output products are electricity, heat and cooling, cooling being produced from heating.

For application purposes, the alternative to a CHP system is a generator, to produce electricity, and a boiler, to produce heat. The CHP system can be much more efficient than the alternative system, depending on how much of the available heat is used in useful applications. For a CHP system with full use of the available heat, which is not usual to find, the efficiencies are shown in Fig. 3.

To produce the same amount of electricity and heat, the CHP system requires only 100 units of fuel, whereas the separate system requires 165 units. We stress that the CHP figures relate to a system with full local use of the available heat, which is difficult to find because heat cannot be transmitted over long distances. The mentioned efficiency of 85% for the CHP system must be seen as a theoretical maximum.

In CHP systems, the most used technology for the thermal machine is the Internal Combustion Engine (ICE), which can be spark-ignition type, running an Otto cycle, or a compression-ignition type, under a Diesel cycle. As an alternative, steam turbines (Rankine cycle) or gas turbines (Brayton cycle) are also used. Micro-turbines and fuel cells are emerging technologies that are likely to be used in

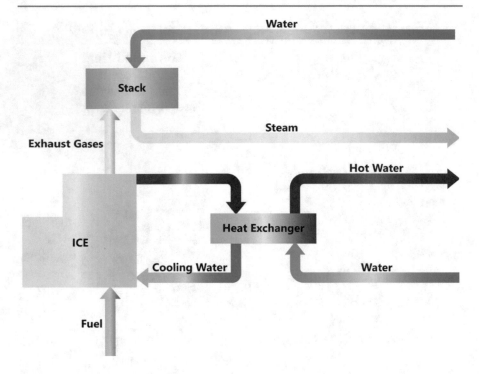

Fig. 4 Use of heat in a CHP system

the future. We will come to CHP thermal machines technologies later, with a special focus on ICE.

Whether ICE or turbines are used as thermal machines, there is always an electrical generator, normally the traditional alternator, coupled in the same shaft to produce electricity. In what concerns the heat, Fig. 4 depicts the principle of using heat in a CHP system with ICE. The heat available in the cooling circuits of the ICE and in the exhausted gases flowing in the stack is recovered through heat exchangers and used in a useful application, for instance, to produce hot water or steam that is required.

The heat sources in an ICE are mainly two: the cooling system and the exhausted gases in the stack. In normal operation, the ICE needs to be refrigerated. For this purpose, High-Temperature (HT) water, Low-Temperature (LT) water, and lubricating oil is used. These fluids are input in the ICE at a lower temperature and output at a higher temperature. Using an outside heat exchanger per cooling circuit, it is possible to lower again the temperature and recover heat. In what concerns the exhausted gases in the stack, they still present a high thermal content that can be recovered using another heat exchanger.

3   CHP Technologies

The technological solutions featuring CHP plants can be divided into:

- Conventional technologies
  - Internal Combustion Engines
- Spark-ignition
- Compression-ignition
  - Gas turbines
  - Back-pressure steam turbines
- Emerging technologies
- Micro-turbines
- Fuel cells

In the next sections, we briefly describe each type of technology. The appropriate models to study these systems are based on advanced concepts of Thermodynamics, which are outside the scope of this textbook. Therefore, a qualitative analysis is offered, based on simplified models.

The conversion of energy in conventional thermal power plants, coal-fired, for instance, is performed according to the Rankine cycle, in such a way the working fluid, the water, changes phase along the thermodynamic cycle. In other technologies, like ICE and gas turbines, a gas is the working fluid. However, the gas changes its structure along the cycle: first, the air is the working fluid; then, during the combustion process, fuel is added, finally transforming the working fluid into a mixture of air and fuel, known as combustion products. It is said that the latter is an internal combustion equipment, while the former is called external combustion because the heat is transferred from the combustion products to the working fluid, which is always the same.

In the internal combustion thermal machines, the working fluid does not cover a closed cycle, operating in open cycle instead. However, from the perspective of studying the phenomena using thermodynamic cycles, it is convenient to work with closed cycles that approximate the real open cycles. One approximation that is commonly used is the standard-air approximation based on the following simplifying hypothesis:

(1) The working fluid is always the air, considered as an ideal gas, which continuously circulates in a closed loop (cycle).
(2) The combustion process is modelled by a heat-addition process from an external source.
(3) The exhaust process is modelled by a heat-rejection process that restores the working fluid (air) at its initial state.
(4) Air specific heat is constant.

3.1   Internal Combustion Engines

There are two types of ICE, also called reciprocating engines: spark-ignition, which uses normally natural gas as fuel, and compression-ignition, which operates with diesel.

The spark-ignition engine was invented by the German engineer Gottlieb Daimler, back in 1885, and its main components are depicted in Fig. 5. The combustion chamber contains a cylinder, two valves (one for intake and one for exhaust), and a spark plug. The piston that moves inside the cylinder is coupled to the connecting rod, which articulates with the crankshaft. The crankshaft turns the reciprocating movement into a rotating movement.

Figure 6 shows the operating principle of a spark-ignition engine.

One may identify four strokes: intake, compression, power, and exhaust.

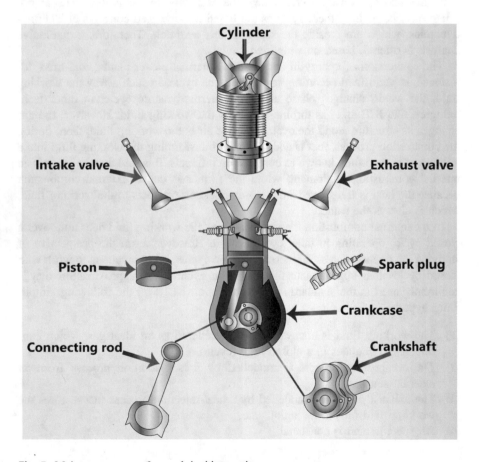

Fig. 5 Main components of a spark-ignition engine

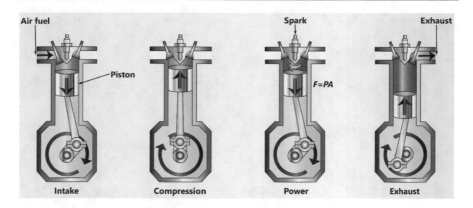

Fig. 6 Operating principle of a spark-ignition engine

- The intake stroke begins as the piston starts its downward travel. When this happens, the intake valve opens, and the fuel/air mixture is drawn into the cylinder.
- The compression stroke begins when the intake valve closes, and the piston starts moving back to the top of the cylinder, compressing the mixture.
- The power stroke begins when the fuel/air mixture is ignited by the spark plug. This causes a tremendous pressure increase in the cylinder, and forces the piston downward, creating the power that turns the crankshaft.
- The exhaust stroke is used to purge the cylinder of burned gases. It begins when the exhaust valve opens, and the piston starts to move towards the cylinder head once again.

In 1894, another German engineer, Rudolf Diesel, invented an engine which eliminated the necessity of an electrical circuit (spark plug) to initiate the combustion. This way, the Diesel engine was born, in which, highly compressed hot air is used to ignite the fuel rather than using a spark plug (compression-ignition rather than spark-ignition). Furthermore, a high compression ratio increases the engine's efficiency. Figure 7 depicts the four strokes of a compression-ignition engine.

Only air is initially introduced into the combustion chamber. The air is compressed with a compression ratio between 15:1 and 22:1 compared to 7:1–10:1 in the spark engine. This high compression heats the air to more than 500 °C. High pressure fuel is injected directly into the compressed air by a fuel injector. The air is so hot that the fuel instantly ignites and explodes. This controlled explosion makes the piston push back out of the cylinder, producing power. When the piston goes back into the cylinder, the exhaust gases are pushed out through an exhaust valve.

Increasing the compression ratio in a spark-ignition engine where fuel and air are mixed before entry to the cylinder is limited by the need to prevent damaging pre-ignition. Since only air is compressed in a diesel engine, premature detonation is not an issue and compression ratios are much higher.

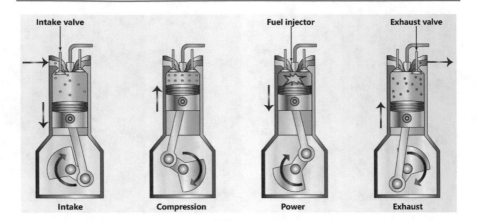

Fig. 7 Operating principle of a compression-ignition engine

The Otto and Diesel cycles are normally used to describe the behaviour of the spark-ignition and the compression-ignition engines, respectively. The (P, v) pressure–volume and (T, s) temperature-entropy diagrams are shown in Figs. 8 and 9, respectively, for the Otto and Diesel cycles.

Regarding, for instance, the Otto cycle (Fig. 8) four processes can be identified:

- 1–2: Isentropic compression (piston goes up)
- 2–3: Constant-volume heat addition (air + fuel ignition)
- 3–4: Isentropic expansion (piston goes down)
- 4–1: Constant-volume heat rejection (exhaust valve opens)

The difference between the Otto and Diesel cycles is in the 2–3 process. In the Otto cycle, the heat addition is isochoric (constant volume), whereas in the Diesel cycle, it is isobaric (constant pressure).

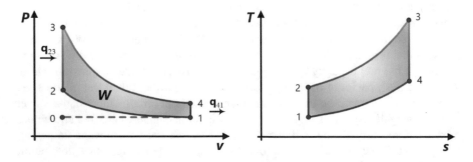

Fig. 8 Otto cycle of a spark-ignition engine

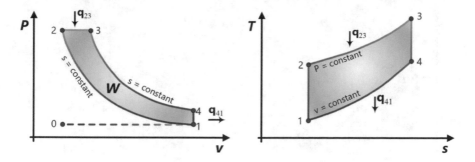

Fig. 9 Diesel cycle of a compression-ignition engine

3.2 Gas Turbines

Gas turbines operate in open cycle in real world. The operating principle of a gas turbine is as follows. Atmospheric air flows through the compressor where the temperature and pressure are high. In the combustion chamber air is mixed with fuel (typically natural gas) that is burning at constant pressure. The resulting gases, at high temperature, enter the turbine where they are expanded to produce shaft work. The useful work is the difference between the work delivered by the turbine and the work delivered to the compressor. The exhausted gases are rejected, and we can recover still useful heat. The temperature of the exhausted gas is relatively high—from 400 °C in small turbines to 600 °C in larger turbines, therefore, it is possible to obtain steam or hot water by installing a heat exchanger.

In Fig. 10, it is shown a scheme of a real gas turbine power plant (open cycle) alongside with a scheme of the same system used for modelling purposes (closed cycle).

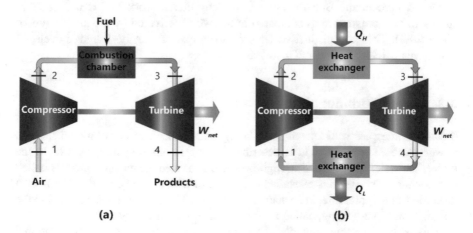

Fig. 10 Gas turbine scheme; **a** real system; **b** system model

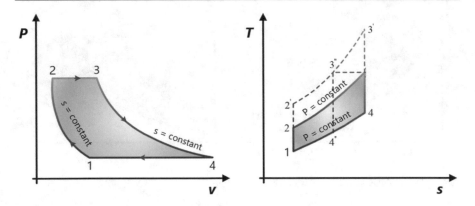

Fig. 11 Diagrams (pressure, volume) and (temperature, entropy) of the Brayton cycle

The combustion process is replaced by a constant pressure heat addition (Q_H) from the outside. The exhaust process is replaced by a constant pressure heat rejection (Q_L) to the outside.

These simplifications (standard-air approximation) allow the gas turbines to be studied using the Brayton cycle, with the following four steps:

- 1–2: isentropic compression at the compressor.
- 2–3: constant pressure heat addition.
- 3–4: isentropic expansion in the turbine.
- 4–1: constant pressure heat recover.

Figure 11 shows the (P, v)—pressure–volume and (T, s)—temperature-entropy diagrams of the Brayton cycle.

One feature of the Brayton cycle is the significant work that is necessary to supply to the compressor, as compared to the work delivered by the turbine (about 40 to 80%). This is a distinctive feature in comparison to the Rankine cycle, in which this percentage is about 1 to 2%.

3.3   Steam Turbines

Steam turbines are well-known from being widely used in conventional thermal power plants. Their operation is described by the Rankine cycle (Fig. 12). In the boiler, the water is converted to high-pressure saturated steam (in a process called overheating). The steam is expanded in a turbine, with at least one reheat; it is then discarded, at low pressure, in a vacuum condenser. The condensate is pumped to the boiler, with a pre-heating, called regeneration, to restart the cycle.

This is the operating principle of the condensing turbines used in conventional thermal power plants, which are designed to optimize the efficiency of the electrical

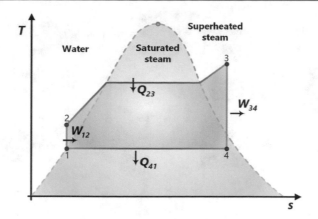

Fig. 12 Diagram (temperature, entropy) of the Rankine cycle

conversion. In CHP power plants, another type of turbine is used, the back-pressure or non-condensing steam turbines. The main difference is that the steam, after expansion in the turbine, is directly used in an industrial process. The pressure of steam leaving the turbine is around atmospheric pressure, therefore, higher than the condenser vacuum. This explains why these turbines are called back-pressure turbines. The efficiency of the Rankine cycle drops because the heat is rejected at a higher temperature as compared to a condensing turbine. Nevertheless, the characteristics of the steam (pressure and temperature) are more suited to the ends it is required to be used.

3.4 Micro-turbines

The operating principle of micro-turbines is similar to gas turbines, both being described by the Brayton cycle. The size is what distinguishes the two technologies. Micro-turbines range from 30–300 kW, whereas gas turbines may be found with up to 250 MW installed capacity. Figure 13 shows a scheme of a micro-turbine.

The speed in the shaft is very high, usually in the range 50,000–60,000 rpm. To allow power to be injected in 50 Hz grid, the use of a high-frequency generator coupled to a rectifier and inverter is required.

3.5 Fuel Cells

Fuel cells portray a completely different way of producing electricity. Fuel cells are somehow similar to batteries in the sense that both generate DC power through an electrochemical process without combustion and intermediate processing into mechanical energy.

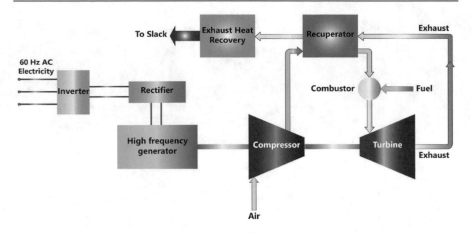

Fig. 13 Micro-turbine scheme

However, there are significant differences. Batteries convert a finite and limited quantity of stored chemical energy into electrical energy. Fuel cells do not store chemical energy, instead they require a constant renewal of the reactants and removal of products. In theory, fuel cells can operate indefinitely provided they are continuously supplied by a fuel source—hydrogen. Hydrogen can be obtained on-site from a hydrocarbon, for instance, natural gas, and oxygen is removed from the ambient air.

The simplest fuel cell is called Alkaline Fuel Cell (AFC), a scheme being shown in Fig. 14. In this fuel cell, two electrodes are immersed in an alkaline electrolyte. Hydrogen is introduced in the anode and oxygen is introduced in the cathode.

At the anode, hydrogen is ionized in contact with hydroxyl ion OH^- (electrolyte mobile ions) releasing electrons and energy and producing water

$$H_2 + 2(OH)^- \rightarrow 2H_2O + 2e^-$$

At the cathode, oxygen reacts with electrons removed from the electrode and with water contained in the electrolyte to form new ions OH^-

$$1/2O_2 + H_2O + 2e^- \rightarrow 2(OH)^-$$

In order that these reactions may continuously occur, the ions should flow through the electrolyte and an external electrical circuit must carry the electrons from the anode to the cathode, therefore, producing an electrical DC current.

Another type of fuel cell is the Proton Exchange Membrane Fuel Cell (PEMFC), also known as Polymer Electrolyte Membrane. Pressurized hydrogen gas enters the fuel cell on the anode side. When an H_2 molecule meets the platinum on the catalyst, it splits into two H^+ ions (protons) and two electrons e^-. The polymer electrolyte membrane only allows the H^+ ions to flow from the anode to the

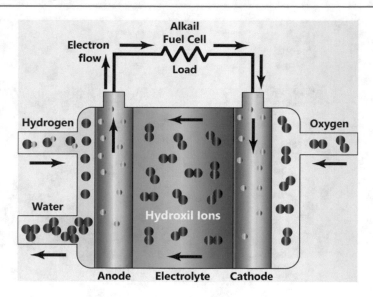

Fig. 14 Alkaline fuel-cell scheme

cathode, not the electrons. The electrons are forced to travel along an external circuit to the cathode, therefore, producing an electrical DC current. Meanwhile, on the cathode, electrons and protons combine with oxygen to produce water. Figure 15 depicts a scheme of a PEMFC.

The two mentioned types of fuel cells are just examples of technologies the market is offering. There are some more technologies available in the fuel cells market. In Table 1, we show a non-exhaustive list of fuel cells technologies together with the typical operating temperatures and electrical efficiencies. The acronyms are: AFC—Alkaline Fuel Cell; MCFC—Molten Carbonate Fuel Cell; PAFC— Phosphoric Acid Fuel Cell; PEMFC—Proton Exchange Membrane Fuel Cell; SOFC—Solid Oxide Fuel Cell.

MCFC and SOFC are the most used fuel-cell types in CHP applications. Due to the high operating temperatures, at a relatively high pressure, they allow to obtain medium pressure steam. Figure 16 shows a schematic of a CHP fuel cell-based system.

The available heat sources in a fuel cell are the reaction products—water and heat, and the cooling circuits of the fuel cell (responsible for about 25 to 45% of the available heat). Finally, it goes without saying that in contrast to reciprocating engines and gas turbines, fuel cells generate electricity and heat without combusting the fuel.

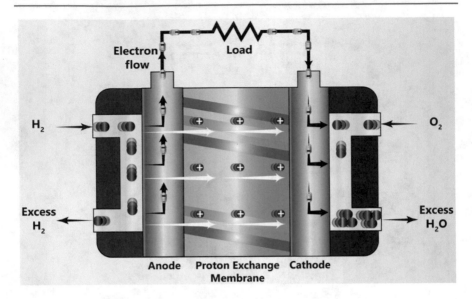

Fig. 15 Proton Exchange Membrane fuel-cell scheme

Table 1 Fuel cell technologies, operating temperatures, and electrical efficiencies

Acronym	Temperature (°C)	Electrical Efficiency (%)
PEMFC	65–85	35–45
PAFC	190–210	35–45
AFC	90–260	55–60
MCFC	650–700	45–55
SOFC	750–1000	45–55

4   Heat Exchangers

An equipment whose purpose is to exchange heat between two fluids is called a heat exchanger. It is usual to separate heat exchangers in: regenerators, where the hot and cold fluid flow alternately in the same space; open type exchangers, where the hot and cold fluids mix, resulting in a fluid with an intermediate temperature, and close type exchangers or recuperators.

Recuperators are used in CHP plants, and therefore, will be the subject of our study. In recuperators, there is no mixing of the hot and cold fluids. Both hot and cold streams flow continuously and are separated by a tube wall or surface. Energy exchange occurs through convection, from one fluid to the wall, conduction, through the wall, and finally by convection again, from the wall to the other fluid. Figure 17 shows an image of a typical recuperator.

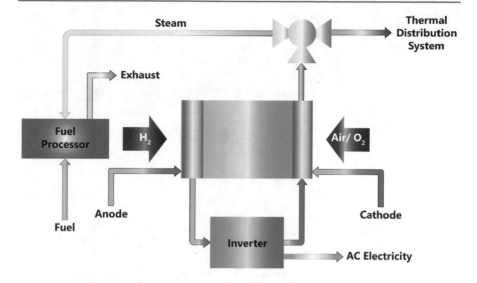

Fig. 16 CHP fuel cell-based system

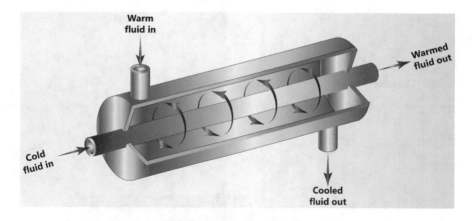

Fig. 17 Recuperator

4.1   Types of Recuperators

Recuperators can be classified as single pass, if each fluid flows only once in the recuperator, or multiple pass, if it flows more than once. In what concerns, the direction of the fluid, recuperators may be of co-current type (Fig. 18), if the two fluids flow in the same direction, counter-current (Fig. 19), if they flow in opposite directions, and crossflow (Fig. 20), if the streams flow at right angles to each other.

Fig. 18 Co-current
recuperator

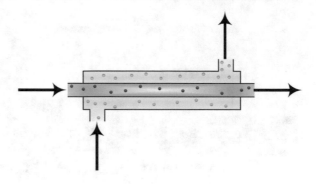

Fig. 19 Counter-current
recuperator

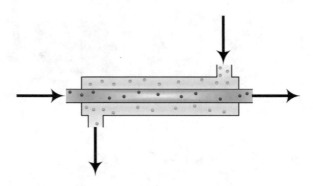

Fig. 20 Crossflow
recuperator

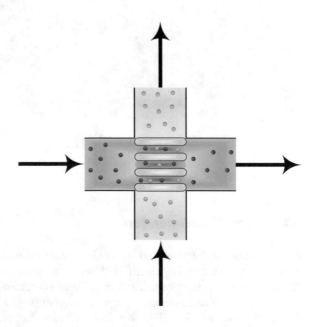

We will focus only on single pass recuperators, of co-current and counter-current types.

In Fig. 21, we show the temperature profiles along a co-current recuperator and a counter-current recuperator. We define the following temperatures: $T_{h,in}$—inlet temperature of the hot fluid; $T_{c,in}$—inlet temperature of the cold fluid; $T_{h,out}$—outlet temperature of the hot fluid; $T_{c,out}$—outlet temperature of the cold fluid. It is apparent that the two temperature profiles are quite different. In the co-current recuperator, the exit temperatures of the two fluids are closer; in the counter-current, the exit temperature of the hot fluid may even be lower than the exit temperature of the cold fluid, which is not possible in the co-current recuperator.

We can, therefore, conclude that the heat exchange per unit area of the recuperator is bigger in the counter-current type than in the co-current. This explains why the former are used more often than the latter. Co-current recuperators are used whenever we want to obtain a uniform temperature of the two fluids, not when the maximization of the heat transfer is the target.

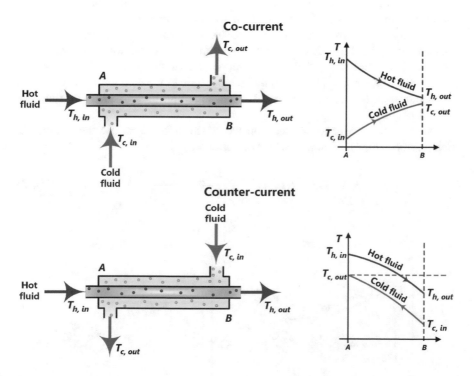

Fig. 21 Temperature profiles along co-current and counter-current recuperators

4.2 Log Mean Temperature Difference

The objective of this section is to learn how to design a recuperator. The main step involved in the design of a recuperator consists of determining its area, A. The general equation of heat exchangers is used for this purpose

$$P_t = UA(T_h - T_c) = UA\Delta T \tag{4.1}$$

where P_t is the thermal power to be transferred in W, U is the overall heat transfer coefficient in W/(m^{2}K), A is the heat transfer area in m^2 and ΔT is the temperature difference in K. The overall heat transfer coefficient depends on the heat transmission coefficients of the two fluids and on the thermal conductivity of the heat exchanger wall.

One obvious question that can be posed is what temperatures to use in the temperature difference? We recall that there are four temperatures to consider, as can be seen in Fig. 22.

In Fig. 22, we defined arbitrarily two points 1 and 2. Depending on the type of recuperator, the temperatures can have different meanings, with respect to the inlet and outlet temperatures of the two fluids. h stands for hot and c stands for cold. For instance, for a co-current recuperator and for the location of the two arbitrarily points indicated in Fig. 22, the four temperatures are as follows: T_{h1}—inlet temperature of the hot fluid; T_{c1}—inlet temperature of the cold fluid; T_{h2}—outlet temperature of the hot fluid; T_{c2}—outlet temperature of the cold fluid. This changes for a counter-current recuperator and for a different position of the two arbitrarily defined points.

One obvious answer to the question we posed before is to use the Arithmetic Mean Temperature Difference (AMTD), defined as

$$AMTD = \Delta T_{avg} = \frac{T_{h2} + T_{h1}}{2} - \frac{T_{c2} + T_{c1}}{2} \tag{4.2}$$

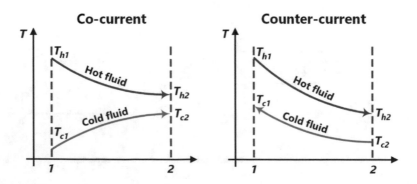

Fig. 22 Temperatures in co-current and counter-current heat exchangers

However, it can be demonstrated that better results can be obtained if the Log Mean Temperature Difference (LMTD) is used instead.

$$LMTD = \Delta T_{lm} = \frac{\Delta T_2 - \Delta T_1}{\ln\left(\frac{\Delta T_2}{\Delta T_1}\right)} = \frac{(T_{h2} - T_{c2}) - (T_{h1} - T_{c1})}{\ln\left(\frac{T_{h2} - T_{c2}}{T_{h1} - T_{c1}}\right)} \tag{4.3}$$

A brief and simplified demonstration of Eq. (4.3) follows.
The infinitesimal thermal power transferred from the cold fluid is

$$dP_t = d\dot{q} = \left(\dot{m}C_p\right)_c dT_c = C_c dT_c \tag{4.4}$$

and received in the hot fluid

$$dP_t = d\dot{q} = \left(\dot{m}C_p\right)_h dT_h = C_h dT_h \tag{4.5}$$

where C_p is the specific heat capacity of the respective fluid, q is the heat transferred, $\dot{q} = dq/dt$, and $\dot{m} = dm/dt$ is the mass flow rate. It is assumed that the mass flow and the specific heats of the fluids remain constant.
We can write

$$\frac{C_h}{C_c} = \frac{dT_c}{dT_h} \approx \frac{\Delta T_c}{\Delta T_h} = \frac{T_{c2} - T_{c1}}{T_{h2} - T_{h1}} \tag{4.6}$$

On the other side, in infinitesimal units (see Eq. (4.1))

$$dP_t = d\dot{q} = UdA(T_h - T_c) = UdA\Delta T \tag{4.7}$$

The temperature differential is (see Eqs. 4.4 and 4.5)

$$d\Delta T = dT_h - dT_c = d\dot{q}\left(\frac{1}{C_h} - \frac{1}{C_c}\right) = \frac{d\dot{q}}{C_h}\left(1 - \frac{C_h}{C_c}\right) \tag{4.8}$$

From Eq. (4.6), we can write

$$d\Delta T = \frac{d\dot{q}}{C_h}\left(\frac{(T_{h2} - T_{h1}) - (T_{c2} - T_{c1})}{(T_{h2} - T_{h1})}\right) \tag{4.9}$$

Rearranging Eq. (4.9)

$$d\Delta T = \frac{d\dot{q}}{C_h}\left(\frac{(T_{h2} - T_{c2}) - (T_{h1} - T_{c1})}{(T_{h2} - T_{h1})}\right) = \frac{d\dot{q}}{C_h}\left(\frac{\Delta T_2 - \Delta T_1}{\Delta T_h}\right) \tag{4.10}$$

Solving Eq. (4.10) for $d\dot{q}$ and considering Eq. (4.5)

$$d\dot{q} = \frac{C_h \Delta T_h d(\Delta T)}{\Delta T_2 - \Delta T_1} = \frac{\dot{q}d(\Delta T)}{\Delta T_2 - \Delta T_1} \tag{4.11}$$

Integrating and considering Eq. (4.7)

$$\int_{\Delta T_1}^{\Delta T_2} \frac{d(\Delta T)}{\Delta T} = \frac{U}{\dot{q}} (\Delta T_2 - \Delta T_1) \int_0^A dA \tag{4.12}$$

The result of the integration is

$$P_t = \dot{q} = UA \frac{\Delta T_2 - \Delta T_1}{\ln\left(\frac{\Delta T_2}{\Delta T_1}\right)} = UA \Delta T_{lm} \tag{4.13}$$

where

$$\Delta T_{lm} = \frac{\Delta T_2 - \Delta T_1}{\ln\left(\frac{\Delta T_2}{\Delta T_1}\right)} = \frac{(T_{h2} - T_{c2}) - (T_{h1} - T_{c1})}{\ln\left(\frac{T_{h2} - T_{c2}}{T_{h1} - T_{c1}}\right)} \tag{14}$$

is the LMTD.

The practical use of Eq. (4.13) is to compute the area of the heat exchanger

$$A = \frac{P_t}{U \Delta T_{lm}} \tag{4.15}$$

In general, we have $AMTD > LMTD$. This means that when the AMTD is used instead of the LMTD, the heat exchanger becomes undersized. The LMTD for a counter-flow recuperator is always higher than for a co-current type. This is why, for the same thermal power to be transferred, the area of a counter-flow recuperator is smaller than the area of a co-current heat exchanger.

Let us look at an example that illustrates how the model works.

Example 4—1

The lubricating oil of an ICE exits the engine at 102°C and must enter at 77°C. To use the released heat, a heat exchanger is installed with the purpose of heating water which is at an initial temperature of 7°C. The mass flows of oil and water are 0.5 and 0.201 kgs⁻¹, respectively. The specific heat capacities of oil and water are 2090 and 4177 Jkg⁻¹ K⁻¹, respectively. The heat transfer coefficient of the heat exchanger is 250 Wm⁻² K⁻¹. Design the heat exchanger considering that it is: (a) counter-current; (b) co-current.

Solution:

(a) A schematic of the problem is shown in Fig. 23. The reader is highlighted for the directions of the fluids in the heat exchanger. As it is a counter-current recuperator, the fluids flow in opposite directions. Also, note that the two points 1 and 2 have been arbitrarily chosen.

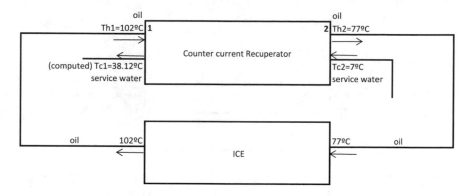

Fig. 23 Schematic of the heat recovery using a counter-current heat exchanger (Example 4—1)

The first step is to compute the outlet temperature of the water, the cold fluid. The cold fluid is the fluid that is initially at a lower temperature. We use the following equation to compute the transferred heat power from the hot fluid to the cold fluid

$$P_t = (\dot{m}C_p)_h (T_{h2} - T_{h1}) = -26.13 \text{kW}$$

The minus sign indicates the hot fluid (oil) releases heat, by convention. This heat is totally transferred to the cold fluid (we admit no heat losses). The heat power received by the cold fluid is now positive by convention. The same equation, but now applied to the cold fluid, is used to compute the outlet temperature of the cold fluid, T_{c1}.

$$-P_t = (\dot{m}C_p)_c (T_{c1} - T_{c2}) \xrightarrow{yields} T_{c1} = 38.12 \text{°}C$$

Now, we know the four temperatures, and therefore, can compute the LMTD

$$\Delta T_{lm} = \frac{\Delta T_2 - \Delta T_1}{\ln\left(\frac{\Delta T_2}{\Delta T_1}\right)} = \frac{(T_{h2} - T_{c2}) - (T_{h1} - T_{c1})}{\ln\left(\frac{T_{h2} - T_{c2}}{T_{h1} - T_{c1}}\right)} = 66.89 \text{°}C$$

As a side note: AMTD = 66.94 °C.

Finally, the area of the counter-current recuperator is

$$A = \frac{|P_t|}{U \Delta T_{lm}} = 1.56 \text{m}^2$$

(b) If a co-current heat exchanger was used instead, the schematic is the one presented in Fig. 24. We note the directions of the fluids. In a co-current recuperator, the two fluids flow in the same direction.

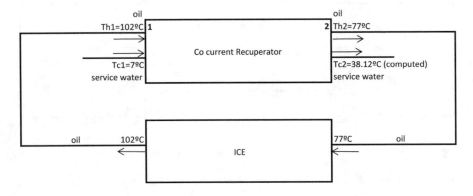

Fig. 24 Schematic of the heat recovery using a co-current heat exchanger (Example 4—1)

The heat power transferred between the two fluids is the same as before, as well as the outlet temperature of the cold fluid (water). The difference is that now $T_{c1} = 7\,^{\circ}C$ and $T_{c2} = 38.12\,^{\circ}C$. Because of this the LMTD is different.

$$\Delta T_{lm} = \frac{\Delta T_2 - \Delta T_1}{\ln\left(\frac{\Delta T_2}{\Delta T_1}\right)} = \frac{(T_{h2} - T_{c2}) - (T_{h1} - T_{c1})}{\ln\left(\frac{T_{h2}-T_{c2}}{T_{h1}-T_{c1}}\right)} = 62.82\,^{\circ}C$$

As a side note: AMTD = 66.94 °C.

For the same transferred heat power, the area of a co-current recuperator is bigger, as expected

$$A = \frac{|P_t|}{U \Delta T_{lm}} = 1.66 \text{m}^2$$

5 Technical and Economic Assessment

5.1 Technical Assessment

A common metrics to evaluate the efficiency of a CHP plant is the *EUF*, Energy Utilization Factor, defined as

$$EUF = \frac{E+Q}{F} \tag{5.1}$$

where E is the electrical energy produced, Q is the useful heat consumed, and F is the fuel energy input measured from the Lower Heating Value (LHV) of the used fuels. All quantities are referred to the same time period and should be in the same units. The usual time period is one year.

The useful heat consumed includes the parcel used for heating purposes (Q_{h2h}) and the parcel used for cooling purposes, with absorption chillers, as mentioned before, (Q_{h2c}). Therefore

$$Q = Q_{h2h} + Q_{h2c} \tag{5.2}$$

The utilization factor of the different uses of the CHP installation may be different. Another way to write Eq. (5.1) is

$$EUF = \frac{P_e h + P_{h2h} h_{h2h} + P_{h2c} h_{h2c}}{P_F h} \tag{5.3}$$

where: P_e—electrical power; h—utilization factor of the heat engine; P_{h2h}—thermal power used for heating purposes; h_{h2h}—utilization factor of the heating-to-heating use; P_{h2c}—thermal power used for cooling purposes; h_{h2c}—utilization factor of the heating to cooling use, P_F—thermal power of the fuel input.

The electric efficiency is defined as

$$EE = \frac{E}{F} \tag{5.4}$$

The reciprocal of the electrical efficiency is the heat rate defined as

$$HR = \frac{F}{E} \tag{5.5}$$

The heat rate represents the number of input energy units that are needed to produce one unit of electricity.

Similarly, the thermal efficiency is

$$TE = \frac{Q}{F} \qquad (5.6)$$

We stress that Q is the useful heat actually consumed on-site by the installation, not the heat available to be consumed. For instance, let us assume that the electric efficiency of an ICE is 40%. Disregarding losses, it means that 60% of the fuel energy input is available to be consumed as heat, but it does not mean the thermal efficiency is 60%. From all this available heat, only a part of it will be really consumed in a useful application. This part is what we call Q. The meaning of E is more obvious: it is the electricity produced that is going to be consumed by the installation or injected in the grid.

Of course, the heat losses cannot be disregarded. This is linked with the heat to power ratio parameter, which is defined as

$$HPR = \frac{Q_{avail}}{E} \qquad (5.7)$$

where Q_{avail} is the heat available to be used. It is different from Q, which is the heat actually used on-site. Of course, $Q_{avail} \geq Q$. The HPR parameter defines the number of available heat units per each unit of produced electricity.

The maximum theoretical thermal efficiency is

$$TE_{MAX} = \frac{Q_{avail}}{F} \qquad (5.8)$$

The maximum theoretical CHP installation efficiency that is found in the datasheets is

$$EUF_{MAX} = EE + TE_{MAX} \qquad (5.9)$$

Figures as high as $EUF_{MAX} = 80\%$ are displayed in the literature. However, these figures must be regarded with caution as they assume that all the heat available is used on-site in a useful application.

Typical electrical efficiencies and heat to power ratios for the CHP technologies are displayed in Table 2.

Table 2 Typical electrical efficiencies and HPR of the CHP technologies. Source: https://www.epa.gov/sites/production/files/2015-07/documents/catalog_of_chp_technologies.pdf

Technology	Electrical efficiency (%)	HPR
Reciprocating engine	27–41	0.8–2
Steam turbine	5–40	10–14
Gas turbine	24–36	0.9–1.7
Micro-turbine	22–28	1.4–2
Fuel cell	30–63	0.5–1

Oftentimes, the choice is between an engine and a turbine. The decision is made upon an economic analysis. In general, engines are preferred if the application calls for low-temperature heat recovery and turbines are the correct choice for heat recovery applications as high-pressure steam.

Example 5—1

A gas natural (LHV = 38 MJ/Nm^3) fired 1860 kW$_e$ ICE is going to be installed in an industry. The annual consumption of heat is 1.560 GWh$_t$ and the annual consumption of cooling is 0.8 GWh$_t$. The thermal needs of the industry are supplied through the cooling and exhaust circuits of the ICE, which are able to supply a maximum thermal power of 1953 kW$_t$. The efficiency of the heating production equipment is 100%. The efficiency of the absorption chiller that is producing cooling is 64%. The heat rate of the ICE is 9080 kJ/kWh$_e$ and the annual utilization factor is 7500 h. Compute: (1) the annual consumption of natural gas; (2) the electrical efficiency; (3) the thermal power used in useful applications; (4) the Energy Utilization Factor (EUF) and the maximum theoretical EUF.

Solution

kWh$_e$ stands for electrical kWh and kWh$_t$ stands for thermal kWh, which can be from heating or cooling.

(1)

First, let us put all quantities in the appropriate units.

$$1\,\text{kWh} = 1000\,\text{W} \times 3600\,\text{s} = 3.6\ \text{MJ}$$

$$HR = \frac{F}{E} = 9080\,\frac{\text{kJ}}{\text{kWh}_e} = 2.52\,\frac{\text{kWh}_t}{\text{kWh}_e}$$

$$F = HR \times P_e \times h_a = 35.19\text{GWh}_t$$

To obtain the annual consumption of natural gas, we must divide the fuel input energy by the natural gas LHV, in the appropriate units

$$LHV = 38\,\frac{\text{MJ}}{\text{Nm}^3} = 10.56\,\frac{\text{kWh}_t}{\text{Nm}^3}$$

$$C_{NG} = \frac{F}{LHV} = 3.33 \times 10^6\text{Nm}^3$$

(2)

$$EE = \frac{E}{F} = \frac{1}{HR} = 0.4\,\frac{\text{kWh}_e}{\text{kWh}_t}$$

(3)

The heat is used to produce heating, in a 100% efficiency equipment, and to produce cooling, in a 64% efficiency equipment

$$Q = 1.56 + \frac{0.8}{0.64} = 2.81 \text{GWh}_t$$

The used thermal power is

$$P_t = \frac{Q}{h_a} = 374.67 \text{kW}_t$$

(4)

The thermal efficiency is

$$TE = \frac{Q}{F} = \frac{2.81}{35.19} = 8\%$$

And the EUF becomes

$$EUF = EE + TE = 0.4 + 0.08 = 48\%$$

The maximum theoretical EUF is

$$EUF_{MAX} = EE + TE_{MAX} = 0.4 + \frac{1953 \times 10^{-6} \times 7500}{35.19} = 0.4 + 0.42 = 82\%$$

This example highlights the significant difference that might exist between the real EUF and the maximum theoretical EUF.

5.2 Economic Assessment—Fuel Variable Cost

In Chap. 3 of this textbook, we approached the economic assessment of renewable energy projects. We recall that the discounted average cost, also known as Levelized Cost Of Energy (LCOE), includes fixed costs (investment and operation & maintenance) and variable costs (fuel and CO_2 emissions). In renewable energy projects the variable cost is zero. In CHP installations, it is not. In this section, we will be concerned about the fuel variable cost of the heat and cooling used and electricity consumed in CHP installations.

In general, in a CHP installation, a single fuel originates the production of electricity and the use of heat, whether it is for heating or cooling purposes. The question is how to distribute the fuel cost for these three components? What part of the fuel cost should be assigned to heating, to cooling and to electricity?

In the following, it is assumed that the quantity of fuel is measured in Nm^3. Nm^3 stands for Normal cubic meter, i.e., a cubic meter in normal conditions of pressure and temperature (temperature equal to 0 °C and pressure equal to 1.01325 bar).

Let us begin with the heating parcel of the fuel variable cost. We can write, considering the fuel cost per unit of heat consumed, c_{fh}(€/MWh$_t$):

$$c_{fh} = \frac{p_f F_h}{Q_h} = \frac{p_f \frac{Q_{h2h}}{LHV_f}}{Q_h} = \frac{p_f}{LHV_f} \frac{1}{\eta_{h2h}}$$

(5.10)

where: c_{fq}—heat unit fuel cost (€/MWh$_t$); p_f—fuel unit price (€/Nm^3); F_h—fuel consumption to produce Q_h (Nm^3); Q_h—heat consumed in a useful heating application (MWh$_t$); Q_{h2h}—heat used to produce heat Q_h (MWh$_t$); LHV_f—fuel Lower Heating Value (LHV) (MWh$_t$/Nm^3); $\eta_{h2h} = Q_h/Q_{h2h}$—efficiency of the heating production equipment.

Some notes on Eq. (5.10) follow. MWh$_t$ stands for Megawatt hour thermal, i.e., 1 MWh of heat. In rough terms, the LHV is the heating released when one burns a normal cubic meter of fuel. Heat Q_h is produced from heat Q_{h2h} in a heat production equipment, whose efficiency is η_{h2h}. A value close to 1 is normally taken for η_{h2h}. The unit fuel cost of heating is independent of the quantity of consumed heat.

We will now apply the same reasoning for the fuel cost per unit of cooling consumption, c_{fh}(€/MWh$_t$). We must bear in mind that the cooling is produced from heat. Therefore, we can write

$$c_{fc} = \frac{p_f F_{h2c}}{Q_c} = \frac{p_f \frac{Q_{h2c}}{LHV_f}}{Q_c} = \frac{p_f}{LHV_f} \frac{Q_{h2c}}{Q_c} = \frac{p_f}{LHV_f} \frac{1}{\eta_{h2c}}$$

(5.11)

where: c_{fc}—cooling unit fuel cost (€/MWh$_t$); p_f—fuel unit price (€/Nm^3); F_{h2c}—fuel consumption to produce Q_{h2c} (Nm^3); Q_c—cooling consumed in a useful application (MWh$_t$); Q_{h2c}—heat used to produce cooling (MWh$_t$); LHV_f—fuel LHV (MWh$_t$/Nm^3); $\eta_{h2c} = Q_c/Q_{h2c}$—absorption chiller efficiency.

Cooling Q_c is produced from heat Q_{h2c} in an absorption chiller, whose efficiency is η_{h2c}. The cooling fuel cost is normally higher than the heating fuel cost because the efficiency of the absorption chiller is $\eta_{h2c} < \eta_{h2h}$. We recall that the total heat consumed in useful applications is $Q = Q_{h2h} + Q_{h2c}$.

Finally, for the electricity, the following equation is applied to compute the fuel cost per unit of electricity consumption c_{fe}

$$c_{fe} = \frac{p_f F_e}{E} = \frac{p_f \frac{F-Q}{LHV_f}}{E} = \frac{p_f}{LHV_f} \frac{F - Q}{E}$$

(5.12)

where: c_{fe}—unit electricity fuel cost (€/MWh$_e$); p_F—fuel unit price (€/Nm^3); F_e—remaining fuel consumption (Nm^3); E—electricity consumed (MWh$_e$); Q—heat consumed in useful applications (MWh$_t$); F—fuel energy input (MWh$_t$); LHV_f—fuel LHV (MWh$_t$/Nm^3).

MWh$_e$ stands for electrical MWh. It is noted that the unit electricity fuel cost includes the cost of losses, $F - Q = E + Q_L$, where Q_L are the heat losses. The definition of heat losses comprises the available heat that was not used in useful applications plus other heat losses.

The total fuel cost per unit of useful energy consumed is

$$c_f = \frac{p_f F_t}{Q_h + Q_c + E} = \frac{p_f \frac{F}{LHV_f}}{Q_h + Q_c + E} = \frac{p_f}{LHV_f} \frac{F}{Q_h + Q_c + E} \tag{5.13}$$

where F_t is the total fuel consumption (Nm^3).

Example 5—2

A reciprocating engine of 1860 kW_e is fed by natural gas LHV = 38 MJ/Nm^3; p_f= 0.3 €/Nm^3. The thermal power not converted into electricity, including available heat and heat losses, is 2793 kW_t. Assume that half of this thermal power is used to produce cooling in an absorption chiller with an efficiency of 70%. The monthly utilization factor is 720 h. Compute: (1) the efficiency of the CHP system; (2) the unit fuel cost of cooling and electricity; (3) the monthly fuel cost.

Solution.

(1) As the utilization factor of all the uses of the CHP system is common (720 h for both the production of electricity and cooling), we can consider powers instead of energies. Nonetheless, we stress that, in general, we should always consider energies because the utilization factors of the different uses might be different.

$$Q = Q_{h2c} = 0.5 \times 2793 = 1396.5\,kW_t$$

$$Q_c = 0.7 \times Q_{h2c} = 977.55\,kW_t$$

$$E = 1860\,kW_e$$

$$F = Q_{nc} + E = 2793 + 1860 = 4653\,kW_t$$

$$EUF = \frac{Q+E}{F} = 70\%$$

(2)
We recall that

$$LHV = 38\,\frac{MJ}{Nm^3} = 10.56 \times 10^{-3}\,\frac{MWh_t}{Nm^3}$$

The unit fuel costs are :

$$c_{fc} = \frac{p_f}{LHV_f} \frac{1}{\eta_{h2c}} = \frac{0.3}{10.56 \times 10^{-3} \times 0.7} = 40.6 \, \text{€/MWh}_t$$

$$c_{fe} = \frac{p_f}{LHV_f} \frac{F - Q}{E} = 49.76 \, \text{€/MWh}_e$$

(3)
The monthly consumption of natural gas is

$$C_{NG} = \frac{F \times 720}{LHV} = 317{,}384 \text{Nm}^3$$

which costs $0.3 \times 317{,}384 = 95{,}215$€
The distribution of this cost is
Total cost assigned to cooling: $40.6 \times 977.55 \times 10^{-3} \times 720 = 28{,}576$ €
Total cost assigned to electricity: $49.76 \times 1860 \times 10^{-3} \times 720 = 66{,}639$ €

6 Conclusions

Combined Heat and Power (CHP) can be renewable if renewable fuels, biomass or biofuels, are used. However, in most cases, CHP is non-renewable as it uses a fossil fuel, for instance, the natural gas. So, the question that may be raised is "Why including a CHP chapter on a renewable energy textbook?" Despite, in general, CHP is not a Renewable Energy Source (RES), it portrays an efficient way of producing two, or, in certain cases, three forms of energy: heat, cooling, and electricity. The alternative would be to have three separated systems to produce each of the said forms of energy, what would be by far much less efficient. This explains the reason behind including this chapter in this textbook.

The global efficiency of the CHP process depends on the amount of heat for which a useful application can be found. The heat cannot be transported, so these local-only applications may be difficult to find. In optimal conditions, the global efficiency of a CHP process may reach 85%, which is remarkable. We highlight that for this figure to be possible, most of the freely available heat must be used, what, as already mentioned, is very difficult to achieve.

The difference between renewable CHP and non-renewable CHP lies in how the heat to drive the thermal machine is produced. The study of these processes is a subject of Thermodynamics, which is outside the scope of this textbook. Instead, we are interested in the technical and economic aspects of CHP plant.

In this chapter, we have reviewed the main technologies used in CHP systems. The working principles of the Internal Combustion Engines (ICE)—spark-ignition and compression-ignition—,gas turbines, steam turbines, micro-turbines, and fuel cells have been reviewed, with a brief view to the thermodynamic cycles that govern each one.

Heat exchangers are one of the most important components of a CHP power plant, as they allow for the available exhaust heat to be transferred to a useful local heating application. The computation of the four temperatures involved (input/output hot/cold fluids temperatures) was offered by introducing the log mean difference temperature model. It was found that the counter-current recuperators are more efficient than the co-current recuperators, which explain the use of the former in CHP applications.

The global efficiency of a CHP power plant is measured using the Energy Utilization Factor (EUF). When computing this index, one must take into account that only the heat effectively used in a useful heating application is to be considered. The correct computation of the EUF refutes some extremely high EUF values that are completely unrealistic.

Besides electricity and heat, cooling can also be produced in CHP plants using absorption chillers that transform heat into cooling. In this way, three products can be produced. When the economic profitability of a CHP installation is to be assessed, it is important to distribute the fuel cost by each one of the obtained products. In this chapter, we proposed a model to assign this distribution by separating the fuel variable cost into three components: heat, cooling, and electricity.

7 Proposed Exercises

Problem CHP1 We wish to heat water from 30 to 75 °C from 540 kW of heat transferred from the waste heat of an internal combustion engine (ICE). The heat exchanger is of the counter-flow type, and has a heat transfer coefficient, U, equal to 0.25 $kWm^{-2} K^{-1}$. The heat exchanger is to be placed either in the lubricating oil cooling system, the high temperature (HT) water cooling system or the low temperature (LT) water cooling system. The characteristics of these systems are presented in Table 3.

(1) Which source of waste heat from the ICE is appropriate?
(2) Find the heat exchanger transfer area and the heated water mass flow rate.

Table 3 Characteristics of the systems from problem CHP1

	$T_{in}(°C)$	$T_{out}(°C)$	$\dot{m}(kg/s)$	$C_p(kJ/kgK)$
Lubricat Oil	75	85.6	11.74	2.090
HT water	91.1	100	15.62	4.177
LT water	40	43.3	17.20	4.177

Solution:

Given: characteristics of the ICE refrigerating circuits; $P_t; T_{c1}; T_{c2}; U$
(1)

$$P_t^{HT} = \dot{m}_{HT} C_p^{water} \Delta T_{HT} = 580.68 kW$$

$$P_t^{LT} = \dot{m}_{LT} C_p^{water} \Delta T_{LT} = 237.09 kW$$

$$P_t^{oil} = \dot{m}_{oil} C_p^{oil} \Delta T_{oil} = 260.09 kW$$

The HT refrigerating circuit is the appropriate choice.
(2)

$$P_t = \dot{m}_{water} C_p^{water} \Delta T_{water}$$

$$\dot{m}_{water} = 2.87 kgs^{-1}$$

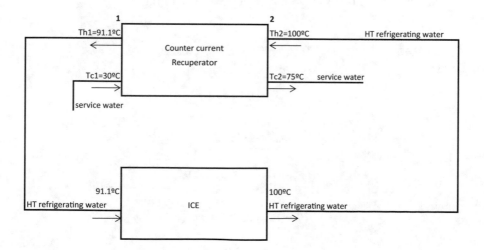

Fig. 25 Diagram of system from problem CHP1

$$A = \frac{P_t}{U\Delta T_{lm}}$$

$$\Delta T_{lm} = \frac{\Delta T_2 - \Delta T_1}{\ln\left(\frac{\Delta T_2}{\Delta T_1}\right)} = \frac{(T_{h2} - T_{c2}) - (T_{h1} - T_{c1})}{\ln\left(\frac{T_{h2} - T_{c2}}{T_{h1} - T_{c1}}\right)} = 40.40°C$$

$$A = 53.47\text{m}^2$$

Problem CHP2

To obtain cooling thermal power, we use an absorption chiller which provides 420 kW of cooling power with an efficiency of 67%. In order that the chiller may deliver this power, we must deliver water at 100 °C. At the output of the chiller this water has a temperature of 90 °C. The heating of the water provided to the chiller is carried with a counter-flow heat exchanger placed in the stack. The heat transfer coefficient, U, of the heat exchanger equals 0.25 kW/(m²K). At the input to the stack, the temperature of air and combustion products equals 357.2 °C. A decrease in the temperature of the exhaust gases to 121.1 °C results in the transfer of a maximum of 875.4 kW of waste heat.

Determine:

(1) The mass flow rates of the combustion products and of the water fed to the chiller.
(2) The heat transfer area of the heat exchanger.

Data: $c_p(\text{air}) = 1.004$ kJ/(kgK); $c_p(\text{water}) = 4.177$ kJ/(kgK).

Solution.

Given: $P_c; \eta_{h2c}; T_{in}^{chiller}; T_{out}^{chiller}; U; T_{in}^{stack}; \min T_{out}^{stack}; \max P_t$

(1)

$$P_t^{max} = \dot{m}_{air} C_p^{air} \Delta T_{air}$$

$$\dot{m}_{air} = 3.69\text{kgs}^{-1}$$

$$P_t = \frac{P_c}{\eta_{h2c}} = 626.87\text{kW}$$

$$P_t = \dot{m}_{water} C_p^{water} \Delta T_{water}$$

$$\dot{m}_{water} = 15.01\text{kgs}^{-1}$$

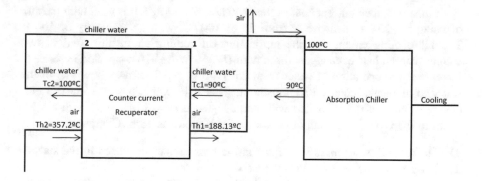

Fig. 26 Diagram of the system from problem CHP2

(2)

$$P_t = \dot{m}_{air} C_p^{air} \left(T_{in}^{stack} - T_{out}^{stack} \right)_{air}$$

$$T_{out}^{stack} = 188.13^\circ C$$

$$A = \frac{P_t}{U \Delta T_{lm}}$$

$$\Delta T_{lm} = \frac{\Delta T_2 - \Delta T_1}{\ln \left(\frac{\Delta T_2}{\Delta T_1} \right)} = \frac{(T_{h2} - T_{c2}) - (T_{h1} - T_{c1})}{\ln \left(\frac{T_{h2} - T_{c2}}{T_{h1} - T_{c1}} \right)} = 165.09^\circ C$$

$$A = 15.19 m^2$$

Problem CHP3 An 1860 kW internal combustion engine (ICE) is installed in an industry. The thermal power available, temperatures, and constant pressure specific heats are listed in Table 4.

Table 4 Characteristics of the systems from Problem CHP3

Circuit	P_{th}(kW)	T_{in}(°C)	T_{out} (°C)	c_p(kJ/(kg. K))
Stack	875.4	357.2	121.1	1.000
HT Cooling	580.7	91.1	100	4.177
LT Cooling	237.1	40	43.3	4.177
Oil	260.1	75	85.6	2.09

In the stack, a co-current heat exchanger (U=250 W/(m²K)) is used with the aim of using 500 kW to heat water from 25 to 100 °C, with an efficiency of 100%. The LT cooling circuit and the lubricating oil cooling circuit are used in two counter-current heat exchangers (both with U=250 W/(m²K)) connected in series to increase the temperature of water from 25 °C to T_f °C. The mass flow rate of the water to be heated equals 17.1 tons/hour. This ICE, which has an electrical efficiency equal to 40%, is fuelled with natural gas (LHV=38 MJ/Nm³) with a price equal to 0.3 €/Nm³. The daily operation of this motor is 18 h. Compute:

(1) The air and water mass flow rates in the heat exchanger placed in the stack.
(2) The area of the heat exchanger placed in the stack.
(3) The output temperature of the water heated by the two heat exchangers connected in series.
(4) The total daily variable cost of the: (i) production of useful heat; (ii) production of electricity.

Solution

Given: characteristics of the reciprocating engine refrigerating circuits; $P_{t1}; T_{i1}; T_{f1}; T_{i2}; U_1; \dot{m}_2; U_2; P_N; EE; LHV_{NG}; p_{NG}; h_d$

(1)

$$P_t^{max-stack} = \dot{m}_{air} C_p^{air} \left(T_{in}^{stack} - T_{out\_min}^{stack} \right)_{air}^{max}$$

$$\dot{m}_{air} = 3.69 \text{kgs}^{-1}$$

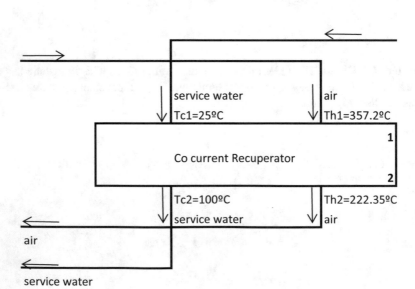

Fig. 27 Diagram of the system from problem CHP3-2

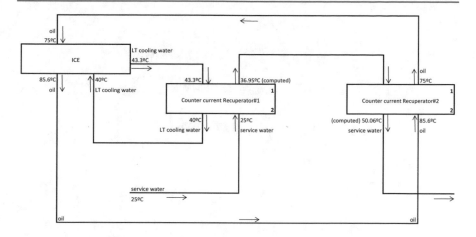

Fig. 28 Diagram of the system from problem CHP3-3

$$P_{t1}^{stack} = \dot{m}_{water} C_p^{water} \left(T_{f1}^{stack} - T_{i1}^{stack} \right)_{water}$$

$$\dot{m}_{water} = 1.60 \text{kgs}^{-1}$$

(2)

$$P_{t1}^{stack} = \dot{m}_{air} C_p^{air} \left(T_{in}^{stack} - T_{out}^{stack} \right)_{air}$$

$$T_{out}^{stack} = 222.35\underline{^\circ}C$$

$$A = \frac{P_t}{U \Delta T_{lm}}$$

$$\Delta T_{lm} = \frac{\Delta T_2 - \Delta T_1}{\ln\left(\frac{\Delta T_2}{\Delta T_1}\right)} = \frac{(T_{h2} - T_{c2}) - (T_{h1} - T_{c1})}{\ln\left(\frac{T_{h2} - T_{c2}}{T_{h1} - T_{c1}}\right)} = 210.09\underline{^\circ}C$$

$$A = 9.52 \text{m}^2$$

(3)

$$P_{t2}^{LT + oil} = \dot{m}_2^{water} C_p^{water} \left(T_{f2}^{water} - T_{i2}^{water} \right)$$

$$T_{f2}^{water} = 50.06\underline{^\circ}C$$

(4)

$$Q = \left(P_{t1} + P_{t2}^{LT+oil}\right)h_d = 17.95 \, \text{MWh}$$

$$E = P_N h_d = 33.48 \text{MWh}$$

$$F = \frac{E}{EE} = 83.70 \text{MWh}$$

$C_{fh} = \frac{p_f}{LHV_f} \frac{1}{\eta_{h2h}} Q = 510.13 \, \text{€}$

$C_{fe} = \frac{p_f}{LHV_f} \frac{F-Q}{E} E = 1868.71 \text{€}$